CW01635799

HICL

HICL PROCEEDINGS

Editors: Kersten, W., Blecker, T., Ringle, C.M. and Jahn, C. 2019
Artificial Intelligence and Digital Transformation in Supply Chain Management.
Digital Transformation in Maritime and City Logistics.

Editors: Kersten, W., Blecker, T., Ringle, C.M. and Jahn, C. 2018
The Road to a Digitalized Supply Chain Management.
ISBN: 978-3-746765-35-8
Logistics 4.0 and Sustainable Supply Chain Management.
ISBN: 978-3-746765-36-5

Editors: Kersten, W., Blecker, T., Ringle, C.M. and Jahn, C. 2017
Digitalization in Supply Chain Management and Logistics.
ISBN 978-3-7450-4328-0
Digitalization in Maritime Logistics and Sustainable Supply Chain Management.
ISBN 978-3-7450-4332-7

Editors: Kersten, W., Blecker, T. and Ringle, C.M. 2015
Innovations and Strategies for Logistics and Supply Chains.
ISBN 978-3-7375-6206-5
Sustainability in Logistics and Supply Chain Management.
ISBN 978-3-7375-4057-5
Operational Excellence in Logistics and Supply Chains.
ISBN 978-3-7375-4056-8

Editors: Kersten, W., Blecker, T. and Ringle, C.M. 2014
Innovative Methods in Logistics and Supply Chain Management.
ISBN 978-3-7375-0341-9
Next Generation Supply Chains. *ISBN 978-3-7375-0339-6*
...find more proceedings on hicl.org/publications

Proceedings of the Hamburg International Conference of Logistics 27

Artificial Intelligence and Digital Transformation in Supply Chain Management

Innovative Approaches for Supply Chains

Prof. Dr. Dr. h. c. Wolfgang Kersten
Prof. Dr. Thorsten Blecker
Prof. Dr. Christian M. Ringle
(Editors)

Edition	1st edition, September 2019
Publisher	epubli GmbH, Berlin, www.epubli.de
Editors	Wolfgang Kersten, Thorsten Blecker and Christian M. Ringle

Cover design	Martin Brylowski
Cover photo	Photo by Julius Drost on Unsplash
Layout	Michelle Dietrich, Ashwin Kudva, Ayman Nagi and Hamza Bin Sohail

ISBN	978-3-750249-47-9
ISSN (print)	2635-4430
ISSN (online)	2365-5070

Preface

Digitalization trends continue to shape the industrial world opening up new opportunities across a wide range of sectors. Artificial intelligence (AI) is considered a key driver of digital transformation that has the potential to introduce new sources of growth. Besides AI, the recent advances in machine learning and automation have created a whole new business ecosystem.

This year's edition of the HICL proceedings complements the last year's volume: The Road to a Digitalized Supply Chain Management. All entities along the supply chain are challenged to adapt new business models, techniques and processes to enable a smooth transition into a digitalized supply chain management.

This book focuses on core topics of artificial intelligence and digitalization in the supply chain. It contains manuscripts by international authors providing comprehensive insights into topics such as digital logistics, robot-human learning, risk management or gamification and provide future research opportunities in the field of supply chain management.

We would like to thank the authors for their excellent contributions, which advance the logistics research process. Without their support and hard work, the creation of this volume would not have been possible.

Hamburg, September 2019

Prof. Dr. Dr. h. c. Wolfgang Kersten

Prof. Dr. Thorsten Blecker

Prof. Dr. Christian M. Ringle

Table of Contents

I.
Advanced Manufacturing and Industry 4.0

Digital Twin for Real-Time Data Processing in Logistics

Hendrik Haße[1], Bin Li[1], Norbet Weißenberg[1], Jan Cirullies[1] and Boris Otto[1]

1 – Fraunhofer Institute for Software and Systems Engineering ISST

Purpose: Key performance indicators (KPIs) are an essential management tool. Real-time KPIs for production and logistics form the basis for flexible and adaptive production systems. These indicators unfold their full potential if they are seamlessly integrated into the "Digital Twin" of a company for data analytics.

Methodology: We apply the Design Science Research Methodology for Information Systems Research for deriving a digital twin architecture.

Findings: Research in the field of digital twins is at an early state, where the main objective is to find new applications for this technology. The majority of digital twin applications relate to the fields of manufacturing. Finally, it became apparent that existing architectures are too generic for usage in logistics.

Originality: The approach presented is an affordable solution for stakeholders to start with a digital transformation, based on standards and therefore highly technology-independent. The combined use of a lambda architecture with a semantic layer for flexible KPI definition is a special case.

Keywords: Digital Twin, Real-time, KPI, IoT

First received: 19.May.2019 **Revised:** 28.May.2019 **Accepted:** 11.June.2019

1 Introduction

Every day, logistics generates a vast amount of data, which is mainly generated by controlling and monitoring enormous flows of goods (Jeske, Grüner and Weiß 2014, p. 9). The data generated in this way holds considerable potential for optimization. A central challenge is the intelligent use of data (Spangenberg, et al., 2017, p. 44). The value of data is not measured by the amount of data collected, but by the applications made possible by the data. For this purpose, the collected data must be prepared in such a way that it can form the basis for optimization measures.

Making use of such data requires a substantial and valid data basis. Data collection, for example, is no longer a particular challenge due to increasingly improved and cheaper sensor technology. What is essential, however, is how this data is evaluated and how this evaluated data contributes to improving the specific process. Another important aspect is the processing time required to evaluate the collected data. The processing of large amounts of data, such as that generated by IoT (Internet of Things) applications, requires a particular framework in order to evaluate these enormous amounts of data (Mishra, Lin and Chang 2014, p. 124). Enterprises have to cope with an ever-increasing amount of data, which be-comes increasingly more efficient with the use of big data frameworks (Gupta, et al. 2017, p. 9).

Hence, it is generally no coincidence that the field of big data analytics plays such an important role in logistics. Logistics, with its cross-sectional function, is a key success factor, making big data analytics increasingly a strategic tool (Spangenberg, et al., 2017, p. 44; Hazen, et al. 2016, p. 592). The determination of key performance indicators (KPIs) is an essential management tool that allows a variety of different evaluations and analyses (Chae 2009, p. 423). This form of data processing and data visualization is made possible by the digital representation of physical assets in the form of a digital twin.

Particularly in logistics, the use of real-time data is an important instrument for visualizing events immediately (Park et al. 2014, 5). However, an industrial application of digital twin frameworks can be found mainly in the context of product management, shop floor and production management (Zhuong, et al. 2018, p. 1153; Qi, et al. 2018, p. 238). In addition, Hopkins and Hawking point out that there is a lack of real life use cases in logistics for both IoT applications and big data analytics (Hopkins and Hawking 2018, p. 579).

Motivated by these aspects, the approach chosen in this paper is to develop a data processing architecture that is tailored to the needs of logistics in particular. The architecture presented in this article is essentially based on IoT applications and big data analytics and therefore enables the evaluation of large amounts of data and the generation of user-defined KPIs in real-time. The digital twin is thus an essential component of these architectures, as it enables an extensive exchange of information (Mičieta, Gašo and Krajčovič, 2014 cited in Furmann, Furmannová and Więcek, 2017, p. 208).

2 Research Design

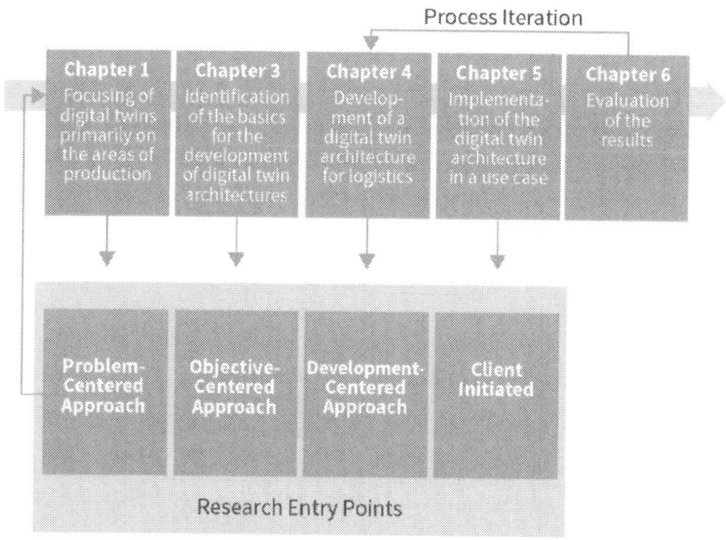

Figure 1: The DSRM for the development of a digital twin architecture in logistics based on Peffers et al. 2007

The structure of this paper and the approach to the development of a digital twin architecture are based on the Design Science Research Methodology (DSRM) for Information Systems Research according to Peffers et al. (2007). This methodology is structured into six different steps and begins with a problem-centered approach that identifies and motivates a problem. In the second step, objectives of the solution are presented, in which the necessary solution approaches are determined. Based on steps one and two, a central artifact is designed and developed in step three. The fourth step is the demonstration of the artifact in a specific context. Steps five and six evaluate and finally communicate the solution. They represent evaluation and communication, and are used for process iterations, which in turn influence the structure of the central artifact (Peffers, et al. 2007, pp. 12-14). As shown in Figure 1, the sixth step of this paper is excluded because communication has not yet taken place.

3 State of the Art of Real-Time Data Processing in Logistics

Following the DSRM by Peffers et al., the problem identification presented in chapter 1, according to which digital twins are mainly found in the area of manufacturing, is the motivation for considering this technology in the context of logistics. It is also about the question whether current architectures are suitable for an application in the field of logistics.

3.1 Internet of Things and Big Data in Logistics

Digitalization and the associated digital transformation of processes affect almost all areas of the economy and industry (Kersten, et al., 2017, p. 8). For its implementation new technological concepts are needed, which primarily relate to data management and analytics. These include comprehensive sensor technology, which serves as a data source for monitoring and improvement, as well as predictive analyses and artificial intelligence, which form the basis for the optimization of logistics processes (Kersten, et al., 2017, p. 12). The core technologies are therefore the realization of extensive sensor technologies and the development of algorithms capable of processing large amounts of data. IoT and Big Data technologies are proving to be the most promising way to process large amounts of data in real-time (Malek, et al., 2017, p. 429). The ability to extract already processed raw data in the format of KPIs from running processes and to visualize them in real-time will bring fundamental improvements in the area of data management. This is especially true in the field of logistics, where many data sets are generated (Wang, et al, 2016, p. 104).

IoT applications and Big Data analysis in combination form a considerable potential for various applications in the field of data management. In general, IoT refers to the vision of a continuous networking of objects so that these objects can communicate with each other and with their environment (Bousonville 2017, p. 25). In this context, IoT refers to a network of sensors by which data can be obtained from various processes (Hopkins and Hawking 2018, p. 576-578). IoT applications thus form the basis for comprehensive data generation. The number of data collected is substantial, depending on the area of application. However, data collection is only the first step.

At this point, it is still completely unclear what value the generated data has for the processes from which it was collected. This means that data from different sources must be merged in order to be processed further, which requires Big Data analysis (Bousonville 2017, p. 25). The term Big Data can essentially be described with the 4 Vs that stand for Volume, Velocity, Variety, Value (Dijcks 2014, pp. 3-4). Value is a particularly im-portant parameter in this context, since the analysis of large amounts of data must focus on the aspect of generating only data of relevance (Dijcks 2014, p. 4; Bousonville 2017, p. 26). The combination of IoT applications and big data analytics is done with architectures that enable end-to-end data management from data collection through data preparation up to data visualization.

Such architectures provide components for three stages (Malek, et al., 2017, p. 431): Data acquisition, data processing and data visualization.

This becomes more obvious with the consideration of an exemplary IoT architecture in Figure 2 (ISO/IEC 2016, p. 41). It shows the Inside Domain Functions of an IoT architecture, which was developed for the definition of an IoT Reference. The lowest level refers to the Physical Entity Domain, which

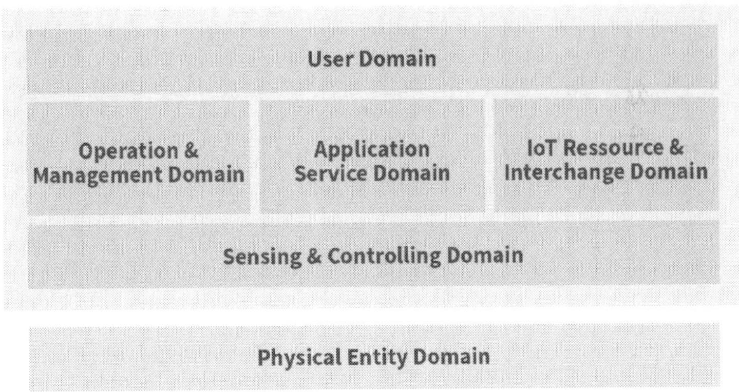

Figure 2: Exemplary IoT architecture according to (ISO/IEC 2016, p. 41)

describes the physical asset under consideration. The next level covers the Sensing and Controlling Domain, which in relation to Malek et al. represents the level of data acquisition. The data processing level is divided into the three domains Operation and Management, Application Service, IoT Resource and Interchange. This level contains the essential analysis functions for real-time data processing. The last level shows the user domain, which enables data visualization. However, a more detailed look reveals that the basics for an exact semantic description of systems is missing, as it is specifically necessary for the area of logistics.

3.2 Digital Twins in Logistics

The continuous development in the field of big data analytics and sensor technologies and the associated progress in the field of IoT technologies finally led to the advancement in digital twins (Tao, et al., 2019, p. 2405). With these platform architectures, the collected data can be processed and visualized in real-time. They provide a technical framework on which concrete industrial applications can be developed. Nevertheless, it is evident that concrete use cases for the implementation of such architectures are rarely found in logistics, although the implementation of IoT technologies in logistics offers considerable additional value (Hopkins and Hawking 2017, p. 579). If logistics objects are equipped with a comprehensive sensor system, a digital image of the respective logistics objects is created, a so-called digital twin. More precisely, a digital twin is the digital representation of a physical asset (Wohlfeld 2019, p. 65). Digital twins enable the connection between the physical and digital world, which must be based on a complete database (Tao, et al., 2019, p. 2405; Wohlfeld 2019, p. 65).

A digital twin is far more than the digital representation of a physical asset. A digital twin represents a comprehensive physical and functional depiction of an asset that provides all the information necessary to process it throughout its lifecycle. (Boschert and Rosen, 2016, p. 59). The exact definition of a digital twin depends on the integration level. A distinction must be made between a digital model, a digital shadow and a digital twin. The core of the consideration is in all cases a physical and a digital object. In a digital model, there is only a manual data flow between the physical and the digital object. A change in the physical object has no effect on the digital object and vice versa. In a digital shadow, there is an automatic data flow in at least one direction, whereby the change of the physical object leads to a change of the digital object. However, this does not apply to the reverse case. In a digital twin, the data flow between the two objects is automatic. Thus, a change to the physical object leads directly to a change to the digital object and vice versa. (Kitzinger et al. 2018, p. 1017).

The use of digital twins enables real-time communication between assets and different systems. With regard to logistics, data collection alone does not represent a major challenge. The decisive factor here is how this data must be further processed in order to offer real added value. In this context, the added value is created with the help of KPIs tailored precisely to the respective application. Depending on the sensors used, different KPIs can be determined from the same data sources in real-time, exactly as required for the respective process. This technology thus offers consider-able potential for logistics and contributes to targeted decision-making (Wang et al. 2016, p. 99).

3.3 Applicability of Lambda Architecture in Logistics Systems

The Industrial Internet of Thing (IIOT) produces massive quantities of sensor data, which arrives in a streaming fashion. The lambda architecture is an efficient big data solution for generic, scalable and fault-tolerant data processing (Gröger, C. 2018).

In the context of IoT data processing, two layers of the lambda architecture consume incoming data simultaneously. The batch layer enables time-consuming analyses on stored raw data, therefore the results are provided to the serving layer and can be consumed by the users. Using a distributed storage topology, the vast amount of sensor data is stored in the batch layer efficiently. In the meanwhile, the speed layer enables real-time analysis of the incoming data streams.

Because of the limitation of computational resources, it is often impossible to load whole datasets at once and analyzing them with classical machine learning models. On the contrary, visiting each instance of the data stream only once and analyzing it with either an adaptive online model or a robust batch model provides the chance to get fresh knowledge from data streams in time, which is of vital importance for IoT data analysis applications.

3.4 Research Gap on Digital Twin Architectures in Logistics

As already pointed out at the beginning of this article, the majority of publications on digital twins relate to the area of shop floor and production management (Zhuong, et al. 2018, p. 1153; Qi, et al. 2018, p. 238). Kitzinger et al. (2018) offer a comprehensive literature analysis on this topic. It shows that more than half of the publications on the subject of digital twin initially describe basic concepts only. Just a quarter of the publications refer to concrete use cases, but most of them are in the area of simulation, product lifecycle management and manufacturing in general (Kitzinger et al. 2018, pp. 1018-1020).

In the context of logistics, Hopkins and Hawking contribute to the application of IoT and Big Data Analytics in logistics. This contribution is based on a literature review on these topics. The result of this investigation is a lack of concrete use cases of both topics, from which the claim is derived to close the gap between theory and logistics practice. Finally, a Big Data Framework is examined using a case study approach. However, this study does not explicitly focus on digital twins, but on the influence of IoT and Big Data on various problems in transport logistics (Hopkins and Hawking 2018). By looking at existing reference architectures, it also becomes apparent that these are too generic to be used in logistics. Furthermore, a functionality to consider a semantic description of the systems to be considered is missing.

Thus, research in the field of digital twins is at an early state, where the main objective is to find new applications for this technology (Negri, et al. 2017 p. 946). Therefore, this paper makes a contribution to the use of digital twin architectures in logistics. It shows how a digital twin architecture can be set up to achieve a seamless integration of logistics systems.

4 A Digital Twin Architecture for Logistics

Considering Big Data Analytics in logistics as well as existing architectures, a digital twin architecture for logistics is now being developed in the third step of the DSRM according to Peffers et al. Therefore, this architecture forms the central artifact. Figure 3 shows a real-time IoT data processing and analyzing platform with a lambda architecture, which aims to provide a scalable and powerful infrastructure for IoT data acquisition, processing and visualization. As an IoT solution for logistics, it is flexible and industrial-application-oriented. The architecture is composed of four layers, as described in detail in the next sections. It is a digital twin architecture with an optional data acquisition layer. The digital twin architecture itself has layers for data visualization, data processing and a semantic layer for providing the overall system model and data integration. These layers are used to enrich, integrate and process the data from the sensors to values that are finally visualized in real-time.

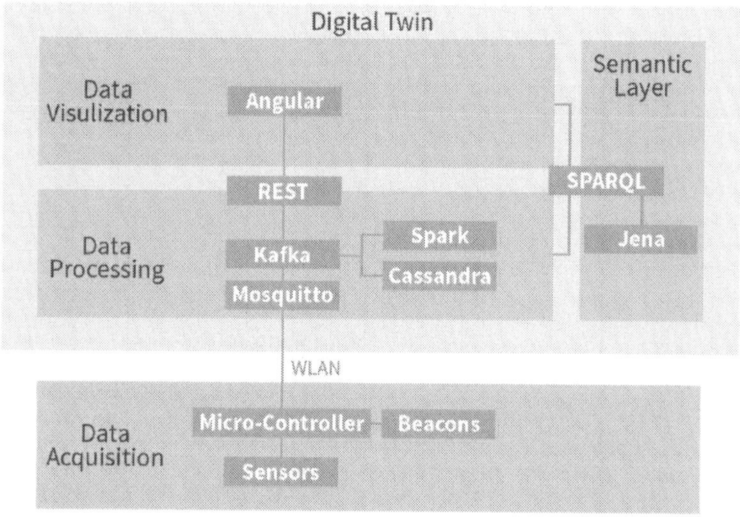

Figure 3: Lambda architecture for real-time IoT analytics in logistics

The implementation of this lambda architecture was realized with a modified SMACK (Spark, Mesos, Akka, Cassandra and Kafka) stack, which is a proven distributed big data stack for streaming data processing. The following sections describe how the individual layers of this infrastructure operate. Furthermore, it is shown which software tools were used to implement these layers and why these software components are best suited for this layer.

4.1 Data Acquisition

With regard to data acquisition, the architecture described here is sensor-independent. Therefore, the description of data collection in this article is a secondary aspect. The selection of a sensor system ultimately decides on the limitation of the possible analyses, since these can only be as good as the sensor system itself. For data acquisition it is useful to select a microcontroller, which can to hold a multitude of different sensors. A decisive aspect in the selection of these modules is the costs. It is important to emphasize that even with a simple sensor system the most diverse evaluations are possible. When such microcontrollers and sensors are used, it must always be ensured that they are also suitable for industrial applications. They must be resistant to vibrations and temperature fluctuations.

4.2 Data Processing

The architecture shown in Figure 3 uses a modified SMACK stack to perform real-time and batch KPI analysis such as shock detection, indoor localization, and usage analysis. Unlike the classic SMACK stack, Apache Akka has been replaced by Apache Nifi, which provides similar features with a more straightforward structure. In addition, several backend functions have been implemented with the Java Spring Boot Framework.

The sensor data is transferred to the infrastructure via the microcontroller. These raw data arrive in the MQTT data broker Mosquitto. There the raw data is pre-processed and distributed to different target units. Kafka transforms the data streams in the overall system and thus forms a distributed data processing platform that enables real-time data stream processing with high throughput. Streaming data is also stored in a Cassandra database, an extensive NoSQL database, for batch analysis. Spark is used as a real-time data analysis engine where the data stream is analyzed in near real-time using the native MLlib machine learning library. After backend processing, the raw data stream and analysis results are visualized on the web frontend.

Data processing is used in industrial applications to generate the KPIs required for the respective process in real-time. The definition, calculation and visualization of KPIs for a specific application is, therefore, the central analysis function of a digital twin system. The combination of lambda architecture and digital twin enables powerful and scalable KPI calculations in real-time. The KPIs generated by this kind of architecture enables companies to quickly determine the condition of their assets. Three steps are required to define and store a new KPI function for a specific scenario:

1. Implementation of the KPI function
2. Implementation of KPI visualization
3. Adding a semantic description to an ontology

The KPI functions are calculated with statistical and machine learning models in batch or real-time. A distinction must be made between whether it is really necessary to generate a KPI in real-time, or not. In general, each KPI function is visualized on the frontend, allowing the user to monitor all relevant indicators. Thus, it makes sense to build the frontend of such an architecture component-based. This means that each KPI function is organized as an isolated component, which makes it much easier to implement new KPIs into the architecture. The components communicate with the backend via a REST API. The real-time KPIs are visualized dynamically from the streaming data. After the implementation of a new KPI function and the visualization of this indicator, the relevant physical objects as well as the analysis functions are described in the ontology within the semantic layer. The concrete advantage here is the standard semantic annotation in an overall model.

4.3 Data Visualization

In addition to data processing, data visualization is another important component, since end users have access to the processed data here. KPIs and the digital description of physical objects are visualized on the frontend. Finally, the optimization options can be identified on the basis of the data visualized in the frontend. For example, Angular is used to create a component-based Web interface. This also enables flexible extensibility of the functions for the frontend. The raw data and analysis results are transferred to the frontend in the data stream and displayed dynamically so that the user can monitor the systems according to the real-time conditions. In this context, it is useful to display the key figures and graphical evaluations on a mobile device, since these are particularly suitable for monitoring running processes. A large number of different KPIs that are relevant for an application in the logistics context can be displayed on the frontend or on the user interface.

4.4 Semantic Layer for Digital Twins in Logistics

The introduction of digital twins faces difficulties due to a lack of semantic interoperability between architectures, standards and ontologies (Datta 2016, p. 1). A digital twin needs a detailed model of its physical counterpart and its relevant environment. This can be a business-oriented semantic model to provide an integrated view of all relevant units in detail, based on the use and extension of standard ontologies. This includes, for example, the relevant assets of the company for which a digital twin is defined and the microcontrollers and sensors associated with those assets. For example, the heterogeneity of the various sensors used is managed with a standard sensor ontology such as W3C SSN (Haller, et al., 2017). New sensors and new assets can be easily connected and configured by instantiating ontology concepts.

The semantic layer of a digital architecture mainly consist of software components and ontologies in a RDF format. The software components are primarily Triple Store and Reasoner. These ontologies are stored in the Triple Store and are used by semantic SPARQL queries executed by the Reasoner. To keep license costs low, open source software can be used, such as the free open source Jena Framework (Apache Software Foundation 2019), which can serve as the basis for the implementation of a se-mantic layer. The composed ontology is a structured semantic model of all relevant entities such as IoT devices, assets and their relationships. The top level of the ontology architecture describes the digital twin and its analyses. The company and its assets follow in the next ontology layer. To support the digital twin with values, IoT devices connected to assets are described in the lowest ontology layer.

4.5 Research Progress by the Presented Architecture

The concept of a digital twin architecture is a very flexible and cost-effective IoT solution. In order to become flexible, analytical modules with logic and display functionality are semantically combined in a lambda architecture. For each KPI function on the frontend, there is a corresponding semantic description in the semantic layer. In the Triple Store, not only the digital description of physical objects is stored, but also respective services of the object, for example the machine learning model used for this object and the relevant sensor types. Furthermore, the semantic model can be easily updated when changes are made to sensors or when machine learning models are updated.

Another aspect is the use of modular micro services. The analysis functions of digital twin architecture can be implemented as modular micro-services, with semantic annotation. The analysis functions and frontend services are fully customizable. For changing sensors or analysis models, the micro-services can be easily extended by adapting semantic annotations and function changes. Also a component-based user interface with Angular has been implemented for flexible customization. For installation, the entire application is packaged in docker images so that it can be used on different platforms at any time.

The large data tools used for the architecture are flexibly scalable. As a large data analysis solution for industrial applications, a digital twin architecture is suitable for various scenarios and applications. Depending on the type and number of sensors, the complexity of the analysis models and the availability of computing resources, the performance of a digital twin architecture can be adapted to the respective application area. By using a combination of these distributed large data tools, a certain fault tolerance is ensured by storing computing information redundantly over different computing nodes. This ensures that the data is not lost in the event of a system failure. Communication between the components is usually implemented either with the REST API or with specific connectors, which are also easily expandable.

The flexibility of this architecture is further demonstrated by its independence from the sensor technology used. The specific sensor types can vary from user to user. The metadata of sensors and analysis functions are stored and linked in the semantic layer. The Data Broker receives all data and its metadata in JSON format, so that further processing and analysis is planned on the basis of metadata and information from the semantic layer.

5 Application Scenario of the Digital Twin Architecture for Real-Time Data Processing Based on Artificial Intelligence

In order to demonstrate the framework presented in section 4, RIOTANA was developed in the context of logistics, which contains all described properties of the digital twin architecture presented here. According to the DSRM by Peffers et al., in chapter 5 the implementation and demonstration of the artifact, the digital twin architecture developed here, is realized. RIOTANA stands for real-time Internet of Things analytics and represents a digital twin architecture with which KPIs can be updated in real-time.

In intralogistics it is possible to transform industrial trucks into a virtual asset with the help of comparatively inexpensive sensor technology. Hence, analyses can be carried out with which related internal processes can be fundamentally optimized. With RIOTANA, a comprehensive forklift control system can be implemented without having to access the elec-tronics of the industrial trucks. Based on discussions with stakeholders and the analysis of existing forklift control systems, a comprehensive system for controlling forklift fleets was developed. The modular structure of the architecture makes changes easy to implement, as the mathematical calculations implemented in the architecture can be adapted to new application conditions and new sensors. The sensor modules collect data on position, acceleration and localization, for example. These different data types are merged by the RIOTANA architecture, processed in real-time and displayed on a web frontend.

Already with these three sensor types in combination, various analyses are possible, which offer a complete overview of the running processes. This includes the current workload, the current location as well as detected shocks and collisions. Accordingly, an entire forklift fleet can be equipped with such sensors, which in turn allows conclusions to be drawn about the overall effectiveness of this fleet.

Figure 4 shows a section of the RIOTANA user interface during a field test. This field test was particularly concerned with a test of the implemented machine learning algorithms for the detection of shocks, as this is an important KPI for a large number of industrial scenarios that must be generated in real-time. Especially this indicator is regarded as a classification problem, for which a K-Nearest Neighborhoods (KNNs) model is used. This is done by collecting time series data from motion sensors attached to the devices. For this purpose, a sliding window is transferred to the time series, whereby each window is regarded as a pattern. Shock patterns are classified on the basis of standard patterns, allowing each incoming pattern to be classified in real-time.

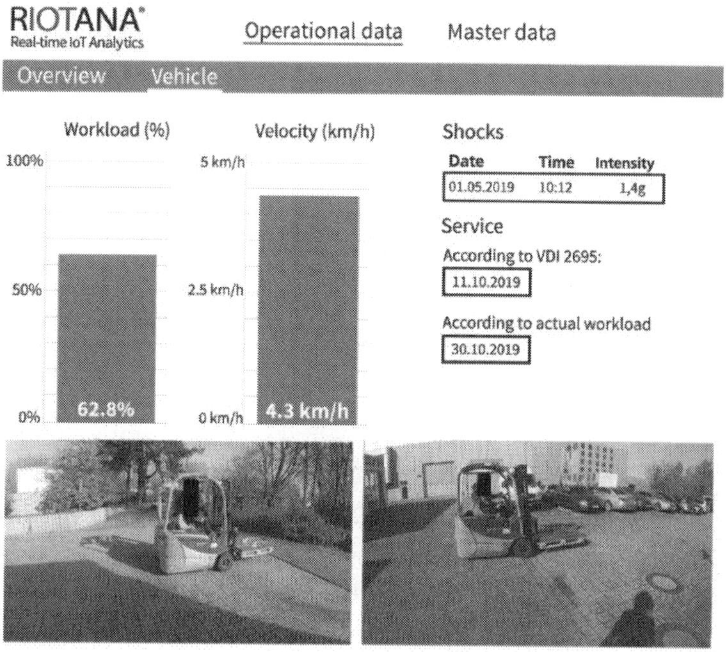

Figure 4: The RIOTANA user interface during a field test

Using the sensor modules RIOTANA can be integrated into existing processes as a "retrofitting solution". By using cheap sensor technology, previously "non-intelligent" objects can become virtual assets that are available for a variety of analyses and can be used to optimize processes. In addition, the use of the sensor modules is optional. RIOTANA can also be integrated into processes in which data are already collected but not yet evaluated. The application of such architectures is conceivable on the basis of digital platforms that operate according to the "as-a-service" principle (Otto, et al. 2019, p. 115).

6 Conclusion and Future Work

In this paper a digital twin architecture was presented, which enables the analysis and processing of large amounts of data in real-time on the basis of IoT applications and big data analytics. It was also shown that the realization of such architectures can be realized with open source software components (Holtkamp 2019, p. 10). The special feature is the description of a digital twin architecture with reference to a concrete application in logistics. It is exactly this practical relevance that presents a particular challenge in the further development of this architecture. This is expressed in an iterative process according to DSRM by Peffers et al., shown in Figure 1. There have to be further investigations on how such architectures can be used in logistics, which in turn has an influence on the structure of the architecture presented here.

The collection of data in an industrial context is always a critical topic that must be considered with special attention. This is particularly the case for personal data. In order to ensure that the processed data is only made available to those who are authorized to do so, a corresponding sensor connector must be implemented in the sensor module. In this way, access to the data can be considerably restricted.

Another important technical aspect is the further development of the machine learning functions in RIOTANA in order to achieve even more precise results with the shock detection. In addition to the further development of machine learning functions to recognize patterns and anomalies and the implementation of software components to ensure data sovereignty, there are also conceptual questions. These include questions about the criteria that determine whether an asset needs a digital representation. Furthermore, it will be necessary to clarify which processes or systems require real-time data processing at all. Beyond that, there are no descriptions of how such architectures can be implemented in processes. Finally, it becomes evident that due to the focus of digital twins on the area of manufacturing, further investigations are necessary with regard to logistics.

References

Apache Software Foundation, 2019. Apache Jena - A free and open source Java framework for building Semantic Web and Linked Data applications. Available at: <https://jena.apache.org/index.html> [Accessed 3 April 2019].

Boschert, S., Rosen, R. 2016. Digital Twin-The Simulation Aspect. In: Hehenberger, P., Bradley, D.: Mechatronic Futures - Challenges and Solutions for Mechatronic Systems and their Designers. Springer, Int. Pub. Switzerland, 2016, pp. 59-74.

Bousonville, T., 2017. Logistik 4.0. Die digitale Transformation der Wertschöp-fungs-kette. Wiesbaden: Springer Gabler.

Chae, B., 2009. Developing key performance indicators for supply chain: an industry perspective. In: Supply Chain Management: An International Journal, [e-journal] 14 (6), pp.422 - 428. https://doi.org/10.1108/13598540910995192

Datta, S.P.A., 2016. Emergence of Digital Twins. MIT Auto-ID Labs, Massachusetts Institute of Technology, Available at: <https://dspace.mit.edu/bitstream/handle/1721.1/104429/Digital%20Twins.pdf?sequence=152> [Accessed 3 April 2019].

Dijcks, J.-P., 2014. Oracle: Big Data for the Enterprise. [pdf] An Oracle White Paper from September 2014. Available at: <https://www.oracle.com/assets/wp-bigdatawithoracle-1453236.pdf> [Accessed 3 Mai 2019].

Furmann, R. Furmannová, B., Więcek, D., 2017. Interactive design of reconfigurable logistics systems. In: Procedia Engineering, [e-journal] 192, pp. 2017 - 212. https://doi.org/10.1016/j.proeng.2017.06.036

Gröger, C., 2018. Building an Industry 4.0 Analytics Platform. In: Datenbank Spektrum, [e-journal] 18 (1), pp. 5 - 14. https://doi.org/10.1007/s13222-018-0273-1.

Gupta, A, Thakur, H. K., Shrivastava, R., Nag, S., Kumar, P., 2017. A Big Data Analysis Framework Using Apache Spark and Deep Learning. In: IEEE, International Conference on Data Mining Workshops (ICDMW). New Orleans, LA, USA, 18-21 Nov 2017. https://doi.org/10.1109/ICDMW.2017.9.

Haller, A.; Janowicz, K.; Cox, S.; Le Phuoc, D.; Taylor, K., Lefrançois, M., 2017. Semantic Sensor Network Ontology. W3C Recommendation. Available at: <https://www.w3.org/TR/vocab-ssn/> [Accessed 3 April 2019].

Holtkamp, B., 2019. RIOTANA: Echtzeit-Ermittlung von Kennzahlen in Produktion und Logistik. Ein leichtgewichtiger Einstieg in Industrie 4.0 für den Mittelstand. [pdf] Fraunhofer Institute for Software and Systems Engineering: Logistics. Available at: < https://www.isst.fraunhofer.de/de/geschaeftsfelder/digitization-in-logistics/RIOTANA.html> [Accessed 3 Mai 2019].

Hopkins, J., Hawking, P., 2018. Big Data Analytics and IoT in logistics: a case study. The International Journal of Logistics Management, [e-journal] 29(2), pp.575 - 591. https://doi.org/10.1108/IJLM-05-2017-0109.

ISO/IEC, 2016. ISO/IEC CD 30141:20160910(E) [pdf]. Available at: <https://www.w3.org/WoT/IG/wiki/images/9/9a/10N0536_CD_text_of_ISO_IEC_30141.pdf> [Accessed 16 May 2019].

Jeske, M., Grüner, M., Weiß,F., 2014. White Paper: Big Data in der Logistik. [pdf] T-Systems. Available at: <https://www.t-systems.com/blob/219384/067ce0047c88fc357ee42b102204512d/DL_WP_Big_Data_in_Logistik.pdf> [Accessed 3 Mai 2019].

Hazen, B.T., Skipper, J.B., Ezell, J.D., Boone, C.A., 2016. Big data and predictive analytics for supply chain sustainability: A theory-driven research agenda. Computers & Industrial Engineering, [e-journal] 101, pp. 592 - 598. http://dx.doi.org/10.1016/j.cie.2016.06.030.

Kersten, W., Seiter, M., von See, B., Hackius, N., Maurer, T., 2017. Trends und Strategien in Logistik und Supply Chain Management. Chancen der digitalen Transformation. Bremen: DVV Media Group GmbH.

Kitzinger, W., Karner, M., Traar, G., Henjes, J., Sihn, W., 2018. Digital Twin in manufacturing: A categorical literature review and classification. In: IFAC-PapersOnLine, [e-journal] 51 (11), pp. 1016 - 1022. https://doi.org/10.1016/j.ifacol.2018.08.474.

Malek, Y.N., Kharbouch, A., El Khoukhi, H., Bakhouya, M., De Florio, V., El ouadghiri, D., Latre, S., Blondia, C., 2017. On the use of IoT and Big Data Technologies for Real-time Monitoring and Data Processing. In: Procedia Computer Science, [e-journal] 133, pp.429 - 434. https://doi.org/10.1016/j.procs.2017.08.281.

Mishra, N., Lin, C.-C., Chang, H.-T., 2014. A Cognitive Oriented Framework for IoT Big-data Management Prospective. http://dx.doi.org/ 10.1109/IC-CPS.2014.7062233.

Negri, E., Fumagalli, L., Macchi, M., 2017. A Review of the Roles of Digital Twin in CPS-based Production Systems. A Review of the Roles of Digital Twin in CPS-based Production Systems. In: Procedia Manufacturing, [e-journal] 11, pp. 939 - 948. https://doi.org/10.1016/j.promfg.2017.07.198.

Otto, B., ten Hompel, M., Wrobel, S., 2019. Industrial Data Space. Referenzarchitektur für die Digitalisierung der Wirtschaft. In: R. Neugebauer, ed. 2019. Digitalisierung. Schlüsseltechnologien für Wirtschaft & Gesellschaft. München: Springer Vieweg. pp. 113-133.

Park, H., Srinivasan, A., Bolsius. R., Hsiao, E., Toillion, J. Shukla, V., Oracle International Corporation 2014. Real-Time Monitoring Event Analysis and Monitoring. US 2014/0095537 A1 (Patent Application Publication).

Peffers, K; Tuunanen, T., Rothenberger, M. A., Chatterjee, S., 2007. A Design Science Research Methodology for Information Systems Research, [e-journal] 24 (3), pp. 45 - 77. https://doi.org/10.2753/MIS0742-1222240302.

Qi, Q., Tao, F., Zuo, Y., Zhao, D., 2018. Digital Twin Service towards Smart Manufacturing. In: Procedia CIRP, [e-journal] 72, pp. 237 - 242. https://doi.org/10.1016/j.procir.2018.03.103.

Spangenberg, N., Roth, M., Mutke, S., Franczyk, B., 2017. Big Data in der Logistik. Ein ganzheitlicher Ansatz für die datengetriebene Logistikplanung, -überwachung und -steuerung. Industrie 4.0 Management, [e-journal] 33(4), pp. 43-47.

Tao, F., Zhang, H., Liu, A., Nee, A. Y.C., 2019. Digital Twin in Industry: State-of-the-Art. IEEE Transactions on Industrial Informatics, [e-journal] 15(4), pp.2405-2415. https://doi.org/10.1109/TII.2018.2873186

Wang, G., Gunasekaran, A., Ngai, E. W. T., Papadopoulos, T., 2016. Big data analystics in logistics and supply chain management: Certain investigations for research and applications. Int. J. Production Economics, [e-journal] 176, pp.98-110. https://doi.org/10.1016/j.ijpe.2016.03.014

Wohlfeld, D., 2019. Digitaler Zwilling für die Produktion vom Übermorgen. Große Fortschritte auf dem Forschungscampus ARENA2036. Zeitschrift für wissenschaftlichen Fabrikbetrieb, [e-journal] 144 (1-2), pp.65-67. https://doi.org/10.3139/104.112008.

Zhuang, C., Liu, J., Xiong; H., 2018. Digital twin-based smart production management and control framework for the complex product assembly shop-floor. In: The International Journal of Advanced Manufacturing Technology, [e-journal] 96, pp. 1149–1163 https://doi.org/10.1007/s00170-018-1617-6.

Breaking Through the Bottlenecks Using Artificial Intelligence

Julia Feldt[1,2], Henning Kontny[2] and Axel Wagenitz[2,3]

1 – University of the West of Scotland, Paisley

2 – HAW Hamburg

3 – Fraunhofer IML Dortmund

Purpose: Performance of Supply Chain is highly dependent on weak spots, so-called bottlenecks. This research paper presents the findings from the analysis of operation processes of a mid-sized producing company and the digital solution for opening up the bottlenecks in order to achieve effectiveness by cutting down the order lead time.

Methodology: The study is employing several rounds of simulation based on processes and data from a manufacturing company.

Findings: Simulation results demonstrate that by allowing a system to take autonomous decisions for production planning based on current changes in environment such as new customer order or available capacity, the order lead time can be shortened significantly, while granting additional flexibility and robustness to the whole supply chain.

Originality: The findings of this research reveal new insights on potentials of artificial intelligence in solving of existing issues within supply chain IT systems.

Keywords: Artificial Intelligence, Assembly-to-Order, Bottlenecks, Supply Chain

First received: 19.May.2019 **Revised:** 31.May.2019 **Accepted:** 20.June.2019

1 Introduction

Recent developments such as Internet of Things (IoT), Industry 4.0, Artificial Intelligence (AI) and other digital technologies are transforming Supply Chains, allowing them to operate based on autonomous decisions analyzing collected data in real-time modus. Thus, granting access to previously inaccessible software solutions and new levels of automation (Calatayud, Mangan and Christopher, 2019; Shmeleva et al., 2018). Consequently, information that was formerly collected by humans will gradually be machine-generated, allowing more precise decisions as well as faster reactions to any disruptions, changing supply chain into a robust interconnected system (Buxmann and Schmidt, 2019; Monostori et al., 2010). Future supply chains will be able to steer themselves continuously, monitoring the environment and react to changes, autonomously learning from previous situations and simulating possible scenarios, developing advanced dimensions of flexibility and agility (Fisel et al., 2019; Tjahjono et al., 2017; Wagner and Kontny, 2017).

Despite all the promising gains, there is still no confidence in what artificial intelligence stands for. In popular cultures, such as Chanel 4's series "Humans", the focus lies on mimicking humans, which may be the long-term goal of research on machine intelligence. Still, current research should be focused on the more practical use of artificial intelligence, such as support of humans in decision-making processes in everyday operations in the form of self-learning software instead of focusing on a recreation of a workers body (Tredinnick, 2017). According to the Accenture Study (Plastino and Purdy, 2018), manufacturing is one of the three most meaningful sectors, which would benefit from AI technologies in the next years, since AI could provide tremendous support in dealing with an increasing number of product types, customization and other growing customer expectations (Lv and Lin, 2017). Given the circumstances that supply chains are confronted with disruptions daily, companies should aim to increase their flexibility by development and implementation of AI solutions customized to the company-specific operations (Scholten, Sharkey Scott and Fynes, 2019).

This paper aims to present the AI-based assembly-to-order supply chain solution for a mid-sized manufacturing company and thus, to make a contribution to the research with practical focus as well as provide support for companies, searching for the ways to improve their operations.

2 Theoretical Background

2.1 Definitions and History of Artificial Intelligence

There are many different definitions of Artificial Intelligence, referring to it as:

"a cluster of technologies and approaches to computing focused on the ability of computers to make flexible rational decisions in response to often unpredictable environmental conditions" (Tredinnick, 2017),

"a subject that studies theories, methods, and applications with respect to simulation, extension, and expansion of human intelligence for problem-solving. AI aims to understand the essence of intelligence and design intelligent machines that can act as human behavior" (Niu et al., 2016),

"AI concerned with creation of computational system that imitates the intelligent behavior of expertise" (Leo Kumar, 2017).

Other Authors emphasize that AI systems "can learn by experiencing, universalize where direct experience is absent, and map from the inputs to the outputs" (Mohammadi and Minaei, 2019; Chaturvedi, 2008). At the same time, the authors agree that machine learning should not provide the same level of complexity as human learning (Niu et al., 2016; Mohammadi and Minaei, 2019).

In order to develop a better understanding of the definitions of artificial intelligence, a summary of essential step stones in its history is provided below. Already in the 1940s at the start of computing, the idea of "machine intelligence" was discussed. In 1950s, Turing described the famous "Turing Test" for the test of machine intelligence, claiming that "by the end of the century it will be possible to programme a machine to answer questions in such a way that it will be extremely difficult to guess whether the answers are being given by a man or machine" (Tredinnick, 2017). Some years later, in 1956, John McCarthy introduced the term Artificial Intelligence, arguing that a machine could solve problems and improve itself on the same level as a human being (Leo Kumar, 2017). Some researchers (Tredinnick, 2017) suggest, that the first big step towards AI was Eliza, the chatbot from Joseph Weizenbaum, demonstrated in 1966 for psychotherapeutic conversations with people. Despite the success, it took researchers another 30 years till in 1997 the IBM's Deep Blue famously won a chess game with world champion Garry Kasparov. Later, in 2011, the quiz show Jeopardy was played by IBM's Watson, marking with its win the intelligence to analyze unstructured data to find answers to questions, asked in "natural" language. Three years later, 2014, chatbot named Eugene Gootsman could persuade 1/3 of the jurors that it is human (Tredinnick, 2017).

Presently, AI solutions have been successfully tested in areas such as autonomous unmanned vehicles, medical diagnosis, speech recognition, video games and others (Mohammadi and Minaei, 2019). Although the focus turned away from a very general simulation of the human brain toward problem-solving in a real work environment, i.e.:

- Speech recognition,
- Semantic reasoning,
- Machine learning ("the ability to improve at performing tasks on the basis of iteration"),
- Intelligent data processing (Tredinnick, 2017).

In order to support assembly processes, this article focuses on the last two application since they are most interesting for autonomous decision-making.

2.2 Related Work

Although the researchers do not provide a clear statement on how the Artificial Intelligence (AI) in manufacturing is defined in comparison to Machine Learning (ML), they agree that both concepts are valuable for the Industry 4.0 and especially for the operations, regarding to it as a Smart Factory which uses "new innovative developments in digital technology including advanced robotics and artificial intelligence" (Tjahjono et al., 2017).

Daehn and Taub (2018) introduce the concept of "Robotic Blacksmith" in order to investigate the ways of using an autonomous system based on closed-loop Machine Learning for metal forming within metamorphic manufacturing, which includes all metal forming operations. The authors provide a general framework and two case studies with a 3D simulation of a corresponding practice. One of the main benefits outlined in the study is the ability of measurement of the environment with sensors, precise control of actions and thus the reproducible results, which are especially essential in industries working with safety-critical products, such as aerospace and nuclear. Another benefit is lower energy consumption of the machine-based solution in comparison to "classic" manufacturing or additive manufacturing.

Mourtzis and Doukas (2015) provide two case studies from automotive industry with highly customized products using the concept of Artificial Intelligence, arguing that in a very complex global supply chains some decisions are nearly impossible to calculate, since the number of possible solutions even for a simple case is calculated at 12,266,496 and in more complicated situation at 48x1017. Such high complexity in decision-making processes, as well as the need for real-time information, makes the Machine Learning or Artificial Intelligence technologies indispensable for (self)-adaptive Smart Supply Chains.

Monostori (2018) indicates increased transparency as well as higher robustness of supply chains through faster identification of the probable disruptions by use of cyber-physical solutions. Verdouw (2016) illustrates such an increase of transparency on the example of food supply chains in general and fish distribution in particular. Furthermore, supply chains can achieve robustness and competitiveness by the implementation of adaptive and IoT-based solutions where decisions made by machine-intelligence will be aligned with high-level decisions taken by humans as explained in the mathematical programming model from Rezaei et al. (2017).

Other authors (Wu et al., 2016) state that the use of the above technologies transforms supply chains into Smart Supply Chains (SSC) with six unique characteristics:

- Instrumented; information is mostly obtained by machines using sensors, RFIDs etc.
- Interconnected; the entire operations and assets are connected.
- Intelligent; SSC optimizes their performance by taking decisions.
- Automated; most of the processes are automated in order to replace less efficient resources.
- Integrated; information is shared across all SC departments;
- Innovative; new solutions will be developed to solve any occurring issues.

Despite the acknowledgement of all the positive characteristics of the intelligent solutions, Jede and Teuteberg (2016) warn about the challenges of their implementation across the supply chain. They argue that researchers should pay more attention to the security aspects, explore the technical details such as interface configuration among different SC partners and definitions of connections between sub-processes in order to provide valuable support for practitioners. Others, (Merlino and Sproģe, 2017) predict that in near future smart factories supported by artificial intelligence and IoT "will make running a supply chain as easy as pushing buttons".

3 AI-Based Assembly-to-Order Supply Chain

3.1 Project Phases

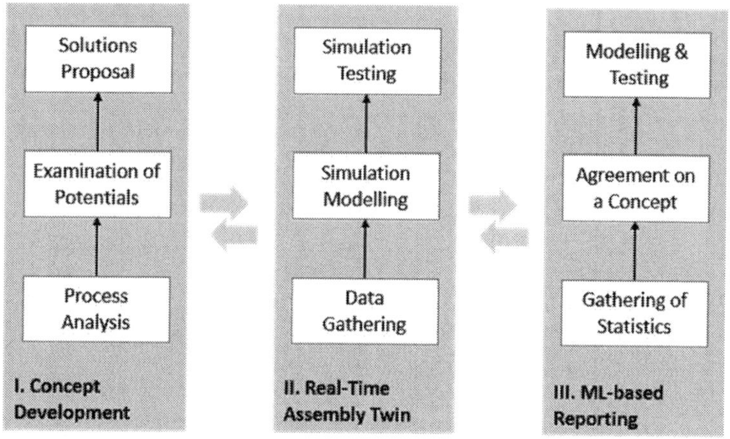

Figure 1: Project Phases

In order to make the project manageable and trace the progress of significant mile-stones, it was divided into three phases (as shown in Fig.1):

- Concept Development
- Modelling of working Real-Time Assembly Twin
- Use of Machine Learning based Reporting.

The overall process is iterative, which means that despite the clearly defined process order, some steps were performed more than once. For example, after simulation testing in phase II, the solutions proposal undergone various changes. Each phase is described in a corresponding chapter below.

3.2 Planning Processes as Main Source of Bottlenecks within the Supply Chain

The first step of concept development was to provide an analysis and description of all the Supply Chain processes of a company, which is a leading company in filter fans production. Although the company has other products such as alarm lights and electronic devices, the focus has lied on filter fans. High-level Supply Chain of the company is quite similar to many manufacturing companies; it consists of different independent departments:

- Purchasing (Raw Materials, Spare Parts and Packaging)
- Production of Components (Molding of Plastic Components as Mass Production)
- Assembly (in Assembly Cells using workforce)
- Warehouse (Stock Management, Transportation, In- and Outbound Logistics).

After a deep-dive into processes, the following conclusions were made:

- Customer orders contained no seasonality, and demand is quite stable from month-to-month (max. deviation 18%, mostly based on delivery of big orders in containers in overseas).
- Order-Lead-Time in most of the cases was around three weeks.
- Minimal production time per batch (several hundred pieces) is one hour (plus 1-2 hours to change molding components).
- Average assembly time 6 minutes per product.
- Stock levels for finished goods are unnecessary high (approx. 5 weeks)

All planning processes (Production Planning, Assembly Planning, Human Resource Planning) are performed in different departments in different Excel sheets (Wagner and Kontny, 2017). The High-level intercompany supply chain is shown in Fig.2, where the order-lead-time (here as the time from customer order in ERP system to the point, when finished goods are shipped) was used to define main Bottlenecks. The left side (As Is Supply Chain) shows the order lead time for the planning processes with the support of Excel sheets. Once per week assembly planners (each is responsible for different products) decide on volumes for the assembly planning for the next four weeks. Mainly basis for the decisions is information on available stocks (should products be delivered from available stocks or manufactured) as well as available workforce. Then, two days later, similar process takes place within a production department, which produce the spare parts for the assembly. At the end, purchasing planners will decide if they need to order raw materials for production or spare parts for the assembly. Despite the fact that the logic of such a decision is always the same (with given priorities), there was no automated solution implemented, which lead to unnecessarily extended order-lead-time of approx. three weeks. IT-based solution with the capacity to take decisions in (near) real-time modus would allow synchronizing of all planning processes along the supply chain at the same time significantly shortening the order-lead-time by at least half.

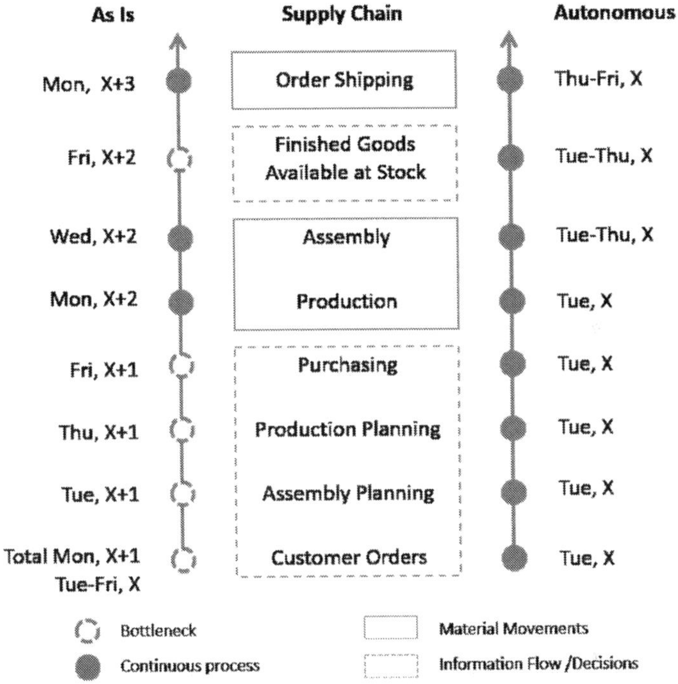

Figure 2: Bottlenecks based on Order-Lead-Time

Since the production as a process step is more time-consuming and less flexible than assembly due to time- consuming retooling of machines as well as a high volume of production batches, the assembly process was chosen as a central process, which provides others (purchasing, warehouse, production) with information relevant for their process steering. Such an approach is not new and is widely known as Assembly-to-Order (Mourtzis and Doukas, 2015) or Build-to-Order Supply Chain (Gunasekaran and Ngai, 2005). The main difference between the "classical approach" and the presented solution is the focus on automated information flow and shortening of order-lead-time instead of material movements and stocks.

The last bottleneck, which is based on a delay of goods availability according to the ERP system and goods produced, can be quickly resolved by scanning of the goods directly at assembly cell and creating additional virtual warehouse in ERP. Thus, the goods can be shown as available, before they will be shipped to the central warehouse, which will spare at least two additional days.

3.3 Solutions Proposal

Based on the fact, that assembly processes could have the most flexibility in capacity if needed (increasing from usual 100% to 400%) being at the same time the slowest process (planning processes take in average over one week), it was decided to improve the assembly planning.

Another critical argument for the automation of assembly planning is provided by Knoll, Prueglmeier and Reinhart (2016), who states that a planner uses only 20% of his time to perform planning, whereas 50% are used for data gathering and preparation. They provide three reasons for such an unfortunate time split:

- Lack of software support
- Inconsistent information
- Unavailable historical data.

In the presented case, the lack of software support leads to extensive use of Excel-Sheets by planners in each department, with data matching at jour-fixe once per week. Since the data in operations changing continuously, such work methods are very inefficient and inevitably lead to high stock level and or to the high level of delayed deliveries. Thereby the lack of software support drive towards different information status in each department and thus to inconsistent information, i.e. stock level in Excel-sheet do not display real stock level in the warehouse, open orders only consider customer orders from previous day. Moreover, ERP only shows the actual status of data and do not provide tools for the analysis of historical data. Such analysis was done only from time to time in Excel, and results were not always shared between departments, leading to a different level of professional competence in different departments by different employees.

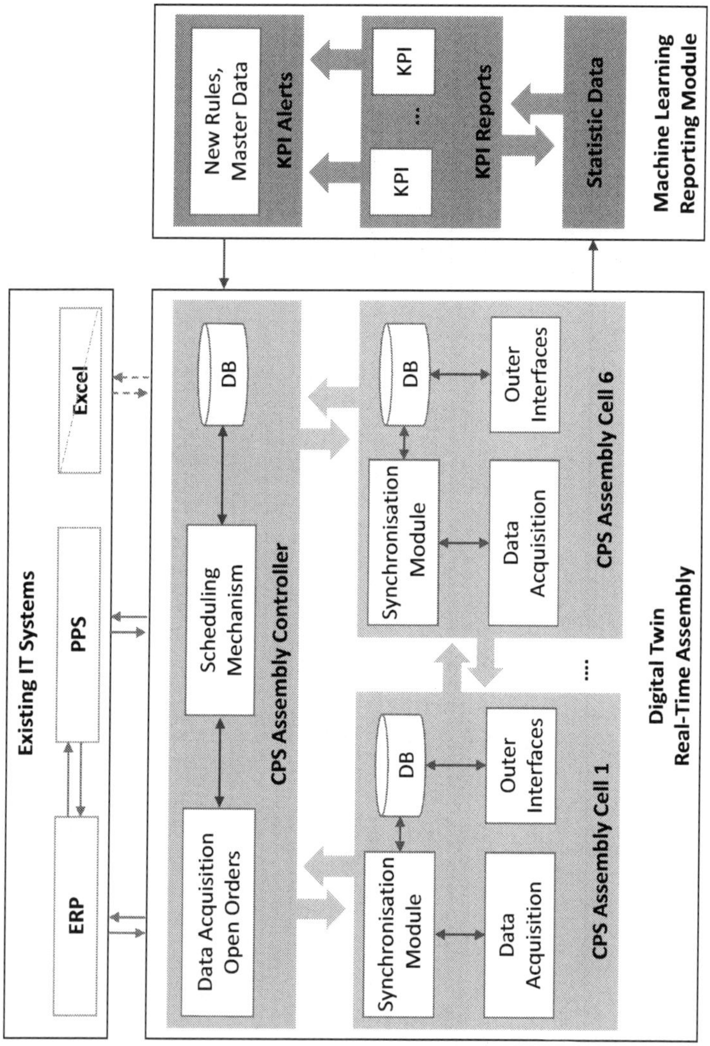

Figure 3: AI-Based Solution for the Bottlenecks

In order to enable planning processes in real-time (or near-real-time) modus with same information status at all process levels as well as the same quality of information, two solutions were created (as shown in Fig.3). Lack of software support, as well as inconsistent information, can be solved by the Real-Time Assembly Twin (Project Phase II), whereas unavailable historical data require advanced reporting module (Project Phase III).

The main distinction between Real-Time Assembly Twin and Machine Learning Reporting Module is time horizon. First concentrates on a continuous simulation of discrete events without data gathering directly in the tool (although some data is forwarded towards ERP). Second, on the contrary, should gather historical data and even overwrite the initial logic/rules for the Assembly Twin shaping it into self-learning and thus AI-based system.

3.4 Real-Time Assembly Twin

3.4.1 Functionality

Real-Time Assembly Twin was created and tested in order to support planners and workers in the assembly area. Additionally, it provides information on the current status of orders to other departments, such as warehousing and in-house transportation.

The system consists of 7 modules, one for each available assembly cell and one with controlling function (Assembly Controller, as shown in Fig.3). Assembly Controller takes the data on open orders from ERP and "translate" them via Scheduling Mechanism into Assembly Orders for each Assembly Cell. All data, needed for the scheduling, such as which cell should assembly which products, production capacity and other relevant data are stored up in the Data Bank of Assembly Controller. Each Cell shows the worker in the assembly cell the orders at the monitor, allowing the worker to update the status of each order, by pushing the button "finished" on the screen or by logging out (thus saying, there is no available worker in the cell). Since in each cell can work up to two employees in two shifts and each worker is able to assemble any product, which means he can freely move from one cell to another, the flexibility of the process is incredibly high. Unfortunately, by planning all the assembly orders weeks upfront, as it was done previous, this flexibility was seldom used. However, with the planning system in (near) real-time modus, it is possible to calculate available resources against open orders, creating a robust and flexible process.

At the end of the period (day or shift), the Assembly Controller gather the data from each cell and communicate it to the ERP, saving open orders for the next day.

3.4.2 Data Gathering

According to Uhlemann (2017), data gathering represents one of the most critical stages for modelling of a Digital Twin. Non-volatile data such as warehouse layouts, process descriptions, historical data and assembly specifications were collected during face-to-face interviews and workshops, with confirmations on the correctness of the results over the phone or via email. Volatile data, i.e. data on material movements, order volume, available stocks, were collected directly from the ERP system. Real-time modus of the digital twin of the assembly could be achieved only by using the data directly from ERP; otherwise, the general data from reporting system is already too old, since it shows the data from the day before. In summary, it can be said that data gathering is a very time-consuming process which is essential for the proper functioning of a new system.

3.4.3 Simulation Modelling

As described in Chapter 3.3, the Real-Time Assembly Twin is basically a simulation which is used to replace the non-existing operational system and for which the existing research method of simulation modelling was chosen as most appropriate.

According to Wojtusjak (2012), simulation techniques can be applied for modelling of complex systems since they can recreate the true-life system's performance. The discrete event simulation (DES) offers an opportunity to trace the alterations of a model with logical multiplex configurations by data-gathering after a defined "event" took. It contains three modifications: activity-oriented, process-oriented and event-oriented simulation. DES can be applied for systems which represent a "set of interrelated entities which only change their state at discrete points of time as a result of their behavior or the behavior of other entities" (Ullrich and Lückerath, 2017). This simulation method is broadly used for the simulation of production processes (Gong et al., 2017), which makes it suitable for our case with assembly-planning.

In the presented simulation of the assembly, the "trigger" events are either start/end of the day/shift or new customer order or change/deviation to the planned capacity in each assembly cell. At the start of the day X the tool translates open orders from the ERP system into assembly orders, creating a list of assembly orders different for each assembly cell. New customer order(s) lead to amendments in open assembly order(s), depending on the bill of materials from ERP and master data of an assembly cell. Previously unexpected changes in available capacity, i.e. if a worker did not show up in the morning due to illness or if a person new to a job works below the usual productivity level, the Real-Time Assembly Twin will consider it in a calculation and make a request to other assembly cells for the additional workforce. On the other hand, during the assembly process each assembly order will be reported to the ERP system as finished, after the worker pushed the finished button on his screen, allowing the system to update the data in near real-time modus and if needed to forward the information to other SC processes, i.e. for the warehouse or goods departure area.

In order to duplicate the sequence of all steps within assembly planning as well as analyze the data from ERP and other systems, the existing OTD-NET simulation software was significantly changed. OTD-NET is the award-winning software, which was built at the Fraunhofer Institute of Material Flow and Logistics for the simulation of production network of big automotive companies such as Daimler AG and Volkswagen (Motta et al., n.d.; Li and Fang, 2012; Liebler et al., 2013). The original software was customized to the logic and supply chain of the manufacturing company.

3.4.4 Verification and Validation

	Assembly Cell 1	**Assembly Cell 2**
Start of Day X Translate ERP „Open Orders" to Assembly Orders	Worker 1 Log In Assembly Order 1.1 Assembly Order 1.2 Assembly Order 1.3 Assembly Order 1.4 Assembly Order 1.5	Worker 2 Log In Assembly Order 2.1 Assembly Order 2.2 Assembly Order 2.3 Assembly Order 2.4
New Customer Order(s)	Assembly Order 1.6 Assembly Order 1.7 Assembly Order 1.8	Job accomplished Please go to Assembly Cell 1
Deficit in Capacity at Cell 1	Worker 1 Log Out ... Worker 2 Log In	Worker 2 Log Out

Figure 4: Assembly Interface and Simulation "Triggers

Verification can be defined as "ensuring that the computer program of the computerized model and its implementation are correct", whereas validation referred to as "substantiation that a computerized model within its domain of applicability possesses a satisfactory range of accuracy consistent with the intended application of the model" (Sargent, 2005). In other words, validity is measured concerning the purpose of the created model; if its accuracy lies within an acceptable range, the model can be concerned as valid.

In order to create new functioning systems, three steps were performed:

- Conceptual model validation
- Computerized model verification
- Operational validation

Conceptual model validation was carried out by respective assembly planning personnel, confirming that the described processes and assumptions are correct and can be executed as planned. Computerized model verification was performed on both sides: by the IT department of the company as well as the research team. It was divided into two steps; first, the "static testing" verified the correct structure of built Digital Twin, and then, by running simulations on company's historical data the "dynamic testing" verified its functionality. Operational validation was performed on a data from ERP in order to measure the time, needed for re-planning of assembly orders (which lied under 1 Minute), to confirm that the "translation" is correct, and no data was lost in the process and to prove the functionality of user interface (for an assembly worker).

3.5 Machine Learning-Based Reporting

Machine Learning (ML) can provide a basis for the assembly planning since there is high volume on "high repetition of recurring planning tasks for each material number caused by frequently changing information" (Knoll, Prüglmeier and Reinhart, 2016). This way, ML provides the best use of historical data with minimal effort.

As shown in Figure 2, the planned ML-based Reporting Module should work inde-pendently from the Real-Time Assembly. The reason for such separation is that in order to function correctly, the ML Reporting Module should first create a proper Database with statistic data. On the one hand, it will take at least a year, in order to have enough data, for reasonable planning. On the other hand, it will allow to create alerts for unusual situations (which are covered today by so-called security stock of goods) as well as rewrite current Master Data and Rules in Assembly Controller. For example, gathering information on current assembly time for product Y and comparing it with historical data on assembly time, the assembly time in the individual cell could be overwritten, providing more precise information on available capacity in this cell. Since assembly time can differentiate even by the same person from one day to another, such adaption would allow very precise planning. Additionally, the Reporting Module will allow to analyze the work of Real-Time Assembly and measure which impact it provides for the Assembly Planning as well as for the stock levels and another necessary logistic KPIs.

4 Discussion of Results

4.1 Implications for AI and SCM Theory

Previous research implied that there is a direct correlation between the rise of complexity and decline of the possible level of agility of the supply chain (Giannakis and Louis, 2016; Monostori, 2018). Application of an AI solution within the supply chain can be very efficient under such circumstances, allowing companies to react very fast to unexpected events and restart a planning process as often as needed. This way, even big companies with complex operations can stay agile and competitive without the unnecessary explosion of bureaucracy.AI and IoT technologies most frequently referred to when discussing the development of autonomous supply chain systems. Capability to analyze high volumes of data and solve complex problems make AI indispensable for future Supply Chains and thus for the research in this area (Calatayud, Mangan and Christopher, 2019).

The presented paper describes a process of finding the appropriate application area for an AI solution as opposed to the approach of the application of standardized IT tools which seldom reflect the specific requirements of an individual customer. Furthermore, we propose to split the AI solution into two parts, one responsible for the quick decisions and the other for historical data as well as steering rules. Although the simulation results could not be presented to broader public since they are based on the company data which are prohibited from publishing, the provided framework can be used by other researchers in the area.

4.2 Practical Implications

Presented solution for the assembly planning has implications on the whole supply chain:

- Order-lead-time shortens from three to one week.
- Stock levels go down automatically since the tool decides if the goods should be assembled or delivered from warehouse and due to shorter order-lead-time the finished goods, as well as spare parts will not be stored in-between, waiting for the next process step.
- Customer order can be translated into an assembly order within seconds/ minutes instead of once per week.
- Workers in the assembly have a clear overview of their orders and can report comple-tion of an order directly to the system (instead of previous paperwork).

Other processes such as logistics, purchasing and production can be provided with information on current status of assembly /stock level instead of outdated plans.Despite the high-level of presented information, it can be used by practitioners as a blueprint for the first steps in similar projects.

5 Summary

Although the terms of Artificial Intelligence and Machine Learning are discussed for almost eighty years, the development of real-life applications is still very young. Only after changing the perspective from a more human-like robot towards an attempt to duplicate merely brain part and learning capability, the researcher could present robust solutions. Presented research has no attempt to replace the workers on a shop-floor; on the contrary, a solution for the support of them in daily operations is presented. Of course, the planning process should be transformed from highly repetitive routine calculations in Excel-sheets towards autonomous Real-Time Digital Twin of Assembly, which inevitably leads to a restructuring of the planning department. Still, gained value over the whole Supply Chain leave no doubt in the meaningfulness of such change.

In summary, this research provides insights on how the AI-based digital solutions within the supply chain can allow it to become more adaptive and thus, robust to the changes, which can be used by both researchers and company executives.

Acknowledgements

The research and development of this project are funded by the German Federal Ministry for Economic Affairs and Energy (BMWi) within the "Business Innovation Lab" concept.

References

Buxmann, P. and Schmidt, H. eds., 2019. Künstliche Intelligenz: mit Algorithmen zum wirtschaftlichen Erfolg. Berlin, Germany: Springer Gabler.

Calatayud, A., Mangan, J. and Christopher, M., 2019. The self-thinking supply chain. Supply Chain Management: An International Journal, 24(1), pp.22–38.

Chaturvedi, D.K., 2008. Soft Computing [electronic Resource]: Techniques and Its Applications in Electrical Engineering. Springer.

Daehn, G.S. and Taub, A., 2018. Metamorphic manufacturing: The third wave in digital manufacturing. Manufacturing Letters. [online] Available at: <http://linkinghub.elsevier.com/retrieve/pii/S2213846318300300> [Accessed 24 Apr. 2018].

Fisel, J., Duffie, N., Moser, E. and Lanza, G., 2019. Changeability - a frequency perspective. Procedia CIRP, 79, pp.186–191.

Giannakis, M. and Louis, M., 2016. A multi-agent based system with big data processing for enhanced supply chain agility. Journal of Enterprise Information Management, 29(5), pp.706–727.

Gong, X., Landahl, J., Johannesson, H. and Jiao, R., 2017. Simulation-driven manufacturing planning for product-production variety coordination. In: 2017 IEEE International Conference on Industrial Engineering and Engineering Management (IEEM). [online] 2017 IEEE International Conference on Industrial Engineering and Engineering Management (IEEM). Singapore: IEEE.pp.2039–2043. Available at: <http://ieeexplore.ieee.org/document/8290250/> [Accessed 10 May 2019].

Gunasekaran, A. and Ngai, E., 2005. Build-to-order supply chain management: a literature review and framework for development. Journal of Operations Management, 23(5), pp.423–451.

Jede, A. and Teuteberg, F., 2016. Towards cloud-based supply chain processes: Designing a reference model and elements of a research agenda. The International Journal of Logistics Management, 27(2), pp.438–462.

Knoll, D., Prüglmeier, M. and Reinhart, G., 2016. Predicting Future Inbound Logistics Processes Using Machine Learning. Procedia CIRP, 52, pp.145–150.

Leo Kumar, S.P., 2017. State of The Art-Intense Review on Artificial Intelligence Systems Application in Process Planning and Manufacturing. Engineering Applications of Artificial Intelligence, 65, pp.294–329.

Li, C. and Fang, D., 2012. OTD-NET based simulation evaluation of (R, Q) inventory policies for complex distribution systems. In: 2012 International Conference on Information Management, Innovation Management and Industrial Engineering. [online] 2012 International Conference on Information Management, Innovation Management and Industrial Engineering (ICIII). Sanya, China: IEEE.pp.470–474. Available at: <http://ieeexplore.ieee.org/document/6339880/> [Accessed 25 Apr. 2019].

Liebler, K., Beissert, U., Motta, M. and Wagenitz, A., 2013. Introduction OTD-NET and LAS: Order-to-delivery network simulation and decision support systems in complex production and logistics networks. In: 2013 Winter Simulations Conference (WSC). [online] 2013 Winter Simulation Conference - (WSC 2013). Washington, DC, USA: IEEE.pp.439–451. Available at: <http://ieeexplore.ieee.org/document/6721440/> [Accessed 25 Apr. 2019].

Lv, Y. and Lin, D., 2017. Design an intelligent real-time operation planning system in distributed manufacturing network. Industrial Management & Data Systems, 117(4), pp.742–753.

Merlino, M. and Sproģe, I., 2017. The Augmented Supply Chain. Procedia Engineering, 178, pp.308–318.

Mohammadi, V. and Minaei, S., 2019. Artificial Intelligence in the Production Process. In: Engineering Tools in the Beverage Industry. [online] Elsevier.pp.27–63. Available at: <https://linkinghub.elsevier.com/retrieve/pii/B9780128152584000020> [Accessed 17 Apr. 2019].

Monostori, J., 2018. Supply chains robustness: Challenges and opportunities. Procedia CIRP, 67, pp.110–115.

Monostori, L., Csáji, B.Cs., Kádár, B., Pfeiffer, A., Ilie-Zudor, E., Kemény, Zs. and Szathmári, M., 2010. Towards adaptive and digital manufacturing. Annual Reviews in Control, 34(1), pp.118–128.

Motta, M., Wagenitz, A., Hellingrath, B. and Weller, R., n.d. Gestaltung logistischer Netzwerke– ein Praxisbericht. p.10.

Mourtzis, D. and Doukas, M., 2015. On the configuration of supply chains for assemble-to-order products: Case studies from the automotive and the CNC machine building sectors. Robotics and Computer-Integrated Manufacturing, 36, pp.13–24.

Niu, J., Tang, W., Xu, F., Zhou, X. and Song, Y., 2016. Global Research on Artificial Intelligence from 1990–2014: Spatially-Explicit Bibliometric Analysis. ISPRS International Journal of Geo-Information, 5(5), p.66.

Plastino, E. and Purdy, M., 2018. Game changing value from Artificial Intelligence: eight strategies. Strategy & Leadership, 46(1), pp.16–22.

Rezaei, M., Akbarpour Shirazi, M. and Karimi, B., 2017. IoT-based framework for performance measurement: A real-time supply chain decision alignment. Industrial Management & Data Systems, 117(4), pp.688–712.

Sargent, R.G., 2005. Verification and validation of simulation models. In: Proceedings of the 37th conference on Winter simulation. winter simulation conference.pp.130–143.

Scholten, K., Sharkey Scott, P. and Fynes, B., 2019. Building routines for non-routine events: supply chain resilience learning mechanisms and their antecedents. Supply Chain Management: An International Journal. [online] Available at: <https://www.emeraldinsight.com/doi/10.1108/SCM-05-2018-0186> [Accessed 17 Apr. 2019].

Shmeleva, A.G., Ladynin, A.I., Talanova, Y.V. and Galemina, E.A., 2018. Manufacturing planning information system development. In: 2018 IEEE Conference of Russian Young Researchers in Electrical and Electronic Engineering (EIConRus). [online] 2018 IEEE Conference of Russian Young Researchers in Electrical and Electronic Engineering (EIConRus). Moscow: IEEE.pp.366–369. Available at: <http://ieeexplore.ieee.org/document/8317108/> [Accessed 10 May 2019].

Tjahjono, B., Esplugues, C., Ares, E. and Pelaez, G., 2017. What does Industry 4.0 mean to Supply Chain? Procedia Manufacturing, 13, pp.1175–1182.

Tredinnick, L., 2017. Artificial intelligence and professional roles. Business Information Review, 34(1), pp.37–41.

Uhlemann, T.H.-J., Schock, C., Lehmann, C., Freiberger, S. and Steinhilper, R., 2017. The Digital Twin: Demonstrating the Potential of Real Time Data Acquisition in Production Systems. Procedia Manufacturing, 9, pp.113–120.

Ullrich, O. and Lückerath, D., 2017. An Introduction to Discrete-Event Modeling and Simulation. SNE Simulation Notes Europe, 27(1), pp.9–16.

Verdouw, C.N., Wolfert, J., Beulens, A.J.M. and Rialland, A., 2016. Virtualization of food supply chains with the internet of things. Journal of Food Engineering, 176, pp.128–136.

Wagner, J. and Kontny, H., 2017. Use case of self-organizing adaptive supply chain. epubli. [online] Available at: <https://tubdok.tub.tuhh.de/handle/11420/1474> [Accessed 30 Apr. 2019].

Wojtusiak, J., Warden, T. and Herzog, O., 2012. Machine learning in agent-based stochastic simulation: Inferential theory and evaluation in transportation logistics. Computers & Mathematics with Applications, 64(12), pp.3658–3665.

Wu, L., Yue, X., Jin, A. and Yen, D.C., 2016. Smart supply chain management: a review and implications for future research. The International Journal of Logistics Management, 27(2), pp.395–417.

Sharing Information Across Company Borders in Industry 4.0

Kai-Ingo Voigt[1], *Julian M. Müller*[2], *Johannes W. Veile*[1], *Marie-Christin Schmidt*[1]

1 – Friedrich Alexander University Erlangen-Nürnberg

2 – Salzburg University of Applied Sciences

Purpose: Industry 4.0's potentials can only unfold exhaustively if information is shared across company boarders and along supply chains. Exchanging information, however, implies several challenges for companies. This holds especially true if it is transmitted automatically and digitally, as promoted by the concept of Industry 4.0. In response to these challenges, this paper analyzes how information sharing changes in Industry 4.0 contexts.

Methodology: Expert interviews with 17 representatives from supply chain management departments of German industrial enterprises provide the study's database. Hereby, the renowned SCOR-model serves as a theoretical foundation to classify the insights gained during the interviews.

Findings: The types of information shared in supply chains, changes in the way information is shared, and initiatives to intensify information exchange in Industry 4.0 contexts are identified. These findings are analyzed against the background of the SCOR-dimensions plan, source, make, deliver and return.

Originality: Information sharing, a vital basis for intended horizontal and vertical integration within the concept of Industry 4.0 has scarcely been investigated in extant literature.

Keywords: Industry 4.0, Industrial Internet of Things, Supply Chain Management, Information Sharing

First received: 19.May.2019 **Revised:** 19.June.2019 **Accepted:** 26.June.2019

1 Introduction

Industry 4.0 refers to the horizontal and vertical digitization and interconnection of industrial value creation. It provokes numerous potentials, which constitute the base for future competitiveness of manufacturing companies (Kagermann et al., 2013; Lasi et al., 2014). A fruitful horizontal and vertical interconnection requires adequate information sharing across company borders and along supply chains. This causes several concerns and thus reluctant behavior of all relevant supply chain actors, especially small and medium-sized enterprises (SMEs) (Kagermann et al., 2013; Müller et al., 2017; Voigt et al., 2019).

So far, academia and corporate practice have regarded and analyzed Industry 4.0 mainly within a company's boundaries. Despite Industry 4.0's supply chain spanning and interconnecting nature, until recently research studies predominantly have focused on company-internal and technical aspects (Birkel et al., 2019; Müller et al., 2018a). Information sharing in supply chains within an Industry 4.0 context and resulting potentials are still not sufficiently permeated and properly understood. However, predicted potentials of intelligent and interconnected value-adding processes can only be exploited entirely if supply chain partners share information virtually, interconnected by digital means (Kagermann et al., 2013; Lasi et al., 2014). Further, sharing information is fundamental to cross-company cooperation in the context of Industry 4.0. For these reasons, analyzing the topic is of utmost importance.

Motivated by the research gap along with its practical implications, this paper's research question is: How does information sharing between companies and along supply chains change in an Industry 4.0 context?

It contributes to the current state of research by analyzing the types of information shared along supply chains, changes in the way information is shared, and initiatives to intensify information exchange in Industry 4.0

The study is of qualitative-empirical nature and bases on 17 expert interviews. These were conducted with supply chain representatives from a heterogeneous set of German industrial companies. The SCOR-model is applied to found the study theoretically and to classify the insights obtained from the empirical data.

The remainder of the paper is structured as follows. First, the theoretical fundament is outlined, which depicts Industry 4.0, supplier integration, and the Supply Chain Operations Reference (SCOR) Model. Hereafter, the methodology, dealing with research design, data sample, and analysis is presented. Subsequently, the results are illustrated, interpreted and discussed using the SCOR-Model as a theoretical framework. The paper concludes with theoretical and managerial implications and indicates proposals for future research.

2 Theoretical background

2.1 Industry 4.0

Industry 4.0 implies a transformation of future industrial value creation based on the ongoing digitization and wide-ranging interconnection. This development is characterized by a horizontal and vertical digital integration of value creation actors (Kagermann et al., 2013; Lasi et al., 2014).

It is widely agreed upon that, the global economy is on the edge of a fourth industrial revolution, heralding a paradigm shift of industrial value creation. Main technological drivers for this development are cyber-physical systems and the Internet of Things. This is why the term "Industrial Internet of Things" is frequently used synonymously to Industry 4.0 (Kagermann et al., 2013; Müller & Voigt, 2018). Given future intelligent, digitally interconnected systems in industrial value creation, products, people, machines, and facilities are enabled to communicate and cooperate in real-time. Following its extensive implications for industrial value creation, Industry 4.0 paves the way for companies to create new business models (Müller et al., 2018b; Voigt et al., 2017).

According to Platform Industry 4.0 (2017), Industry 4.0 is "a new level of organization and control of the entire value chain across the life cycle of products". As it aims at horizontally and vertically interconnecting value creation, digitalization and interconnection efforts do not only cover single corporate functions, but also individual companies, their internal value creation processes, just like global value chains (Kagermann et al., 2013).

2.2 Supplier integration

Representing a form of vertical cooperation, the concept of "supplier integration", as part of supply chain management, refers to strategic and operational cooperation efforts along the supply chain. These relate to both suppliers and customers and aim at generating and securing competitive advantages (Helmold & Terry, 2016; Schoenherr & Swink, 2012; Wiengarten et al., 2016). Main prerequisite for a successful supplier integration is systematic and strategic collaboration within a company and between different companies, interlinking internal processes with external suppliers (Fazli & Afshar, 2014; Zhao et al., 2011).

The overall strategic goal of supplier integration is contributing to a company's value creation, by managing the supplier network in an efficient manner, securing demand, and decreasing supply chain costs (Helmold & Terry, 2016). In order to fulfill this objective, supplier integration intends to adequately and properly set up value creation processes, cross-company handling practices, cooperation efforts, and integration strategies, hereby sharing both chances and risks along the value creation chain (So & Sun, 2010; Zhao et al., 2015).

Supplier integration requires financial resources, people, systems, and facilities to be coordinated and harmonized on operational, tactical, and strategical levels (Stevens, 1989). The concept can draw on numerous tools, among others, IT applications related to planning and e-business systems (So & Sun, 2010).

2.3 Related work

In the era of Industry 4.0, industrial value creation faces major obstacles, such as shortening product development and product life cycles. These in turn require fast, flexible and efficient value creation processes, which emphasizes the importance of supplier integration in the context of Industry 4.0 (Hofbauer et al., 2016).

Integrating suppliers and cooperating across company boarders encompasses major challenges. Especially in the digital era, it requires an indispensable willingness to cooperate and a great candor both from buyers and suppliers, among others (Birkel et al., 2019; Kiel et al., 2017). These characteristics, however, contradict traditional approaches to conduct interactions, exchanges, and processes within present supply chains, like aggressive price negotiations. In addition, supplier integration in the digital era asks for comprehensive technological equipment and facilities, as well as, for adequately reshaping interfaces (Kiel et al., 2017; Müller et al., 2018a). Given that many suppliers are among small and medium-sized companies, particular problems arise, e.g., as for raising resources for the required technologies (Birkel et al., 2019). However, the challenges of how to establish new ways of interaction and cooperation along the supply chain and how to strengthen suppliers' positions remain unsolved (National Research Council, 2000).

Several changes are accompanied by further integrating suppliers. On a strategic level, cooperating with suppliers, conducting common activities, and sharing capabilities paves the way for gaining and securing competitive advantages and for creating new business models (Rink & Wagner, 2007; Voigt et al., 2018). On an operational level, a further supplier integration helps to reduce costs, inventory levels, and lead times, amongst others (Giménez & Ventura, 2003). An increased cost competitiveness, a greater product individualization, a shortening of product development and product life cycles and associated fast, flexible, and efficient production processes within Industry 4.0 increase the importance of supplier integration (Kagermann et al., 2013).

Despite the importance of horizontal and vertical integration within the concept of Industry 4.0, research investigating Industry 4.0 from a supply chain management perspective remains scarce in extant literature (Birkel et al., 2019; Müller et al., 2017). The variety of data types to be exchanged, different standards used across supply chains, and different interfaces that are used represent further challenges for information sharing in Industry 4.0 (Kiel et al., 2017). Adding to technical challenges, aspects such as a lack of trust hinder cross-company information sharing and collaboration. This is especially true of SMEs, who fear that they will not benefit from information sharing, but have to make large efforts to make it possible (Müller et al., 2018a).

Although there is literature on information sharing regardless of Industry 4.0 in extant literature, as shown in the previous section, this literature is only partially applicable to Industry 4.0. This is since information sharing in Industry 4.0 shall be achieved horizontally and vertically and in real-time, which will influence information sharing, innovation, and communication processes (Kagermann et al., 2013; Lasi et al., 2014).

2.4 The Supply Chain Operations Reference (SCOR) Model

The process approach for supply chain integration manifests itself in the Supply Chain Operations Reference Model (SCOR) by the Supply Chain Council, suggesting that businesses should be managed on the basis of key processes (Stewart, 1997).

The Supply Chain Operations Reference Model (SCOR) was developed by two American management consultancies to support companies in the area of supply chain management. It was published on the Supply Chain Council in 1996, which bases on a merger of the aforementioned two consultancies in cooperation with further international companies. The SCOR model represents an approach to describe an organization's supply chain and aims at optimizing business practices (Bolstorff et al., 2007; Stewart, 1997). This normative model enables an independent, effective process assessment and a performance comparison with other companies, both within an industrial sector and across sectors. The dimensions of this model can be applied along the entire supply chain from the supplier's supplier, all the way up to the customer's customer.

The SCOR model maps process elements, performance metrics, best business practices, and characteristics of carrying out supply chain activities at four levels (Stewart, 1997):

- Level one ("Process Types") concretizes a company's supply chain objectives and maps definitions of the process types "plan", "source", "make", "deliver", and "return".
- Level two ("Process Categories") defines core process categories as possible components of a supply chain.
- Level three ("Process Elements") formulates detailed process element information for categories of Level two, enabling successful planning and definition of objectives.
- Level four ("Activities") aims at company-specific supply chain improvements, which, however, are not subject of the industrial standard model.

This study will focus on the six main strategic processes of Level one, which are briefly outlined below (Bolstorff et al., 2007; Supply Chain Council, 2012):

- Plan: Demand planning concerning available resources and capacities; Inventory planning serving sales, production, and material requirements
- Source: Procurement of goods, goods receipt and testing, storage and payment instructions for raw materials, goods, finished products, and systems
- Make: Material requirement request and material receipt, production and testing of manufactured products, intermediate product storage and release for delivery
- Deliver: Execution of order processing from quotation to delivery of goods, and coordination of the information flow (e.g., data maintenance, control and monitoring of logistical processes)
- Return: Processing of returned products with defects under warranty, handling of over delivery, including receipt, inspection, disposition, and release
- Enable: Management of the supply chain according to business rules, performance, data, operating resources, equipment, contracts, supply chain network, risk management, compliance management, and legal requirements

Figure 1 summarizes the SCOR model.

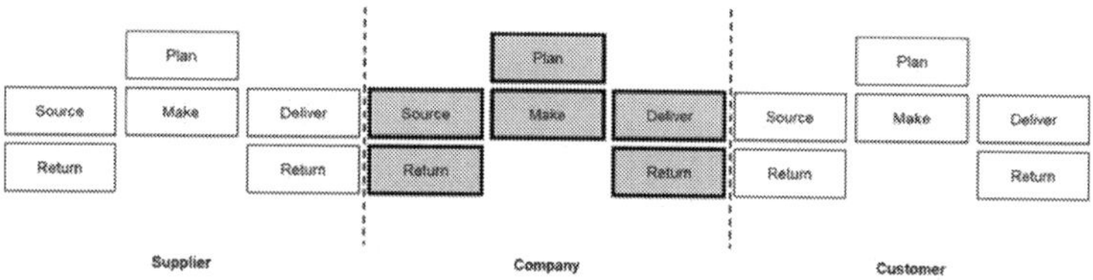

Figure 1: Strategical processes according to SCOR (Supply Chain Council, 2012

3 Methodology

The objective of this study is to analyze how information sharing between companies and along supply chains changes in an Industry 4.0 context, applying a qualitative-empirical research design (Gläser & Laudel, 2010).

The empirical database consists of interviews with 17 experts of German industrial companies. Forming a heterogeneous sample, the interviewed experts are from the areas of supply chain management, procurement, and logistics, holding lower, medium or upper management positions and possessing several years of business experience. The sample comprises German companies from various industry sectors, such as mechanical and plant engineering, automotive, electronics and electrical engineering, information and communication technology, such as chemical and pharmaceutical industry. The sizes of the companies vary from approximately 100 employees to approximately 250,000 employees respectively (data from fiscal year 2018).

The expert interviews took place between November 2018 and February 2019, lasted between 21 to 68 minutes, and were conducted in German via telephone. An interview guideline ensured a partial standardization that allowed to compare and evaluate interviews following common research standards. However, as intended by the study's nature, the interviewer could adopt questions if necessary and experts were allowed to respond openly going beyond the interview guideline, thus all interviews varied slightly in contextual length and depth (Mayring, 2015; Gläser & Laudel, 2010). With the experts' permission, all interviews were audio-recorded. Later on, the audio-files were transcribed yielding about 92 pages of text material that hereafter was analyzed as explained below.

The interviews and the respective transcripts were analyzed applying a qualitative content analysis based on Mayring (2015). Firstly, literal expert statements were paraphrased, secondly the paraphrases were reduced to their core statements, and thirdly, core statements similar in content, were analyzed and consolidated into categories. Hereafter, these categories were analyzed according to their frequency of nominations which represents an indicator for their relevance (Gläser & Laudel, 2010; Mayring, 2015).

4 Results

The results are divided into two parts. The first part consolidates findings about the current exchange of information between buyers and suppliers. The findings on the present situation include 1) ways of exchanging information, 2) types of information exchanged, and 3) frequency of exchange. The second part deals with expected changes in an Industry 4.0 context and focus on 1) the ways of exchanging information in the future, 2) changes to intensify future information exchanges, and 3) automation efforts. All findings are matched to and differentiated according to the main strategic processes of the SCOR model "plan", "source", "make", "deliver", and "return". Hereby the numbers in brackets represent the absolute number of experts indicating these aspects out of the 17 experts within the study's sample.

Table 1 gives an overview of ways and media used for exchanging information between buyers and suppliers in the present as stated by the experts in the different dimensions of the SCOR model.

Table 1: Ways of currently exchanging information

	Plan	Source	Make	Deliver	Return	Not assigned
E-Mail (17)	1	8	2	2	1	3
EDI Interface (14)	3	5	2	1	1	2
Phone (9)	3	4				2
Personally (9)	4	2				3
Networks and platforms (5)		2	1			2
Visit-on-site of suppliers (5)	2		1	1		1

	Plan	Source	Make	Deliver	Return	Not assigned
Supplier evaluation (3)	2					1
Direct contact/ Real-time exchange (1)						1
Fax and printed documents (2)		1				1
Mixed sources of information (1)						1

Table 2 depicts the types of information exchanged between buyers and suppliers at different stages of the supply chain.

Table 2: Types of information currently exchanged

	Plan	Source	Make	Deliver	Return	Not assigned
Orders (6)	1	4	1			
Technical documents/ Specifications (4)	3		1			

	Plan	Source	Make	Deliver	Return	Not assigned
Information on product, price, conditions, etc. (4)	2	1	1			
Certificates and qualifications (3)	3					
Forecasts (3)	2	1				
Supplier performance (2)	2					
Legal documents and reclamations (2)					2	
Offers (1)		1				
Information about upstream Supply Chain stages (1)			1			
Production data (1)			1			

Table 3 outlines, how often the interviewees get in contact with their suppliers for exchanging information.

Table 3: Current frequency of exchange

	Plan	Source	Make	Deliver	Return	Not assigned
Depending on the situation (4)	1	1	1			1
Depending on the supplier (3)	2		1			
Several times a day (2)	1	1				
Daily (5)	1	1	2			1
Several times a week (1)		1				
Several times a month (1)		1				
Monthly (1)						1

The second part of the results reveals changes as for buyer-supplier information sharing in an Industry 4.0 context.

Table 4 shows the ways information is exchanged in a short to medium termed Industry 4.0 context, categorized in accordance to the SCOR model.

Table 4: Ways of exchanging information in an Industry 4.0 context

	Plan	**Source**	**Make**	**Deliver**	**Return**	**Not assigned**
Networks and platforms (13)	1	2	4	1		5
EDI Interface (3)	1	1				1
Real-time exchange (3)		1	1			1
Big Data (2)	1					1

Table 5 describes the changes in information exchange, which will occur in the next 5 to 10 years that all aim at creating a more intense way of exchanging information.

Table 5: Changes aiming at a more intense information exchange

	Plan	**Source**	**Make**	**Deliver**	**Return**	**Not assigned**
Simplified sharing of information (4)	2	1				1

	Plan	Source	Make	Deliver	Return	Not assigned
Further supplier integration (2)			2			
Matching production data (2)		1	1			
Feedback processes (1)		1				
Forecasts (1)	1					

5 Interpretation and discussion using the SCOR-model

5.1 Plan

From the sample experts' point of view, quality represents a decisive procurement goal and must take precedence over other objectives. Beyond that, further goals include efficiency gains and cost reductions.

Both customers and suppliers can be involved in the product development process. Cooperating with customers helps to integrate customer-specific requirements at an early stage and to test products in collaborative innovation centers during the development. In the same manner, integrating suppliers improves the product development process, e.g., as supplier requirements and capacities can be considered from the very beginning. Innovation platforms ease interactions between various actors and therefore improve value creation, especially against the background of shorter production cycles.

A more precise planning of requirements basing on extensive data use can smooth out demand fluctuations. Pilot projects in corporate practice already apply first approaches that take advantage of comprehensive data to reduce inefficiencies caused by demand fluctuations. Collecting and evaluating different forms of data, such as historical, macroeconomic, and point of sale data, enable a targeted identification of development indications. This paves the way for precise demand forecasts, which counteracts operational inefficiencies.

A data release and transparent data exchange between suppliers and customers simplifies negotiations of delivery conditions on a data-based level, for example, regarding future suppliers' and buyers' expectations. In turn, this allows a potential supplier to analyze whether the requested demand developments can be met in the future. Company representatives can imagine sharing and releasing data with close suppliers within cross-company cooperation. This simplifies planning and ordering processes for customers. However, the results underline that a very good relationship of trust is required and data security is to be guaranteed as a prerequisite.

Price negotiations are an essential part of buyer-supplier discussions and pricing entails further optimization potentials in buyer-supplier relationships in the context of Industry 4.0. By analyzing both historical and present customer data, it is possible to determine which product attributes add the most value for customers. Knowing individual customers' willingness to pay, customer-specific and value-based pricing can be carried out, that allows to optimize margins.

Industry 4.0 can optimize supplier selection and cooperation as well. Apart from this, further potential of Industry 4.0 bases on a automation, traceability, and inter-connecting functional processes.

5.2 Source

Basing on a data- and fact-based planning, the development towards Industry 4.0 offers great benefits for procurement activities. Applying technological solutions, such as cloud systems, big data operations, sensor systems, chips and associated system integration leads to a comprehensive interconnection alongside a high degree of process automation and increased efficiency.

New solutions are required for inventory management in operative purchasing to be able to meet future requirements for flexible delivery of complex products. In the future, real-time data, a further automation of warehouse logistics, and augmented reality could potentially help in this manner. By equipping incoming and outgoing goods with sensors, movements and locations can be mapped in real time, for example using big data and cloud systems. With regard to the analysis of raw materials' usability, the experts see great potential in processing a large amount of data using Big

Data analyses. In addition, goods could autonomously be transported to their destination by means of drones and robots. Using augmented reality, warehouse employees could manage a warehouse both more efficiently and effectively. These solutions may reduce lead times, which ultimately leads to increased customer satisfaction.

Information transparency and data release play an important role for future cross-company value creation. This enables suppliers to carry out early, precise, and efficient production planning, which could have a positive effect on the customers' willingness to pay. Automation of basic and simple procurement processes is particularly desirable. Complex goods, that are of high value, individualized special products, products ordered in large quantities, and transport specifications, should still be negotiated by experienced human buyers. Anyways, the process can be streamlined and simplified through data sharing, while customers and suppliers must agree on the nature and extent of data release and exchange.

5.3 Make

By equipping components in the production process with sensors, their actual states can be digitally mapped in real time. In connection with historical and diagnostic plant data, failure prediction models could be generated, which enable both scheduling optimal maintenance intervals and minimizing downtimes.

There are concepts for generating more accurate forecasts and analyses via a comprehensive data collection in the manufacturing processes. The Virtual Plant concept promises to combine physical process data from production and data from upstream and downstream value creation stages in a digital real-time model to be able to forecast changes in production. Besides, the Virtual Plant concept offers possibilities for simulating changes concerning the processes or workplace design, and thus allows analyzing optimization potentials and ultimately reducing production costs. Using a digital twin, deviations can be simulated, analyzed, and predicted in the running production process.

Interconnecting systems in the production process enable real-time monitoring of the supplier by its customer. This leads to a close and automated coordination of requirements between suppliers and manufacturers and therefore contributes to optimize value creation.

Given a data disclosure, manufacturing processes will be subject to security risks in the course of the development towards Industry 4.0. System failures must be prevented, especially if data and systems are interconnected. In order to guarantee this, external competence is desired.

5.4 Deliver

Regarding means of transportation, delivery times, and transport conditions can be made transparent to all partners in the value chain, using sensors and digitally mapping products. This in turn reduces waiting and processing times and has the potential to increase customer satisfaction. Besides, tracing products provides the advantage to guarantee and check their quality. A greater transparency towards customers and improved quality controls enable interconnected logistics.

Instead of offering a standalone product, system solutions can be provided. Digital services, e.g., basing on the Internet of Things and cloud solutions, play a decisive role in this context. The combination of traditional products with innovative services may lead to smart product solutions.

5.5 Return

As for the return section, currently non-automated, analogue, and inefficient processes result in insufficient or wrong information concerning recycling and disassembly. This leads to resource wastage and possibly recyclable raw materials being thrown.

Technological developments, data exchange, and new forms of analyses in the context of Industry 4.0 pose the potential to reshape return processes. In addition, transaction costs can be optimized in return processes via automated orders, invoicing, and documentation processes. In this regard, returns are optimized by sharing recycling and disassembly information across the supply chain, such as manuals and certificates.

6 Conclusion

6.1 Theoretical and Managerial Implications

With regard to the "plan" process, ensuring data security and trust are key prerequisites. Although the experts see great potential in interconnected demand and procurement planning, there still seems to be certain skepticism about disclosing data, for example, to upstream value-adding stages.

Following the research results at the "source" process level, skepticism as for information transparency further increases in future buyer-supplier information sharing. Among others, risks are located in either becoming dependent on a particular supplier or in facing disadvantages due to excessive data transparency. The results reveal that it remains essential to continue negotiating procurement efforts about complex and new products as well as specific material traditionally. In the case of standardized goods, however, automated purchasing processing seems conceivable, with delivery conditions and the price range being determined in advance. Suppliers and customers have to individually agree on the extent of information transparency depending on the relationship of trust.

The study reveals hurdles in the "make" process stage. Comprehensive data security turns out to be very important, basing on two aspects. On the one hand, data security is about protecting data against third parties outside the supply chain. On the other hand, data security deals with stability in the generation and transmission of data.

Corporate practice and managers alike can make use of the insights provided by the paper and include its managerial implications into their decisions and actions. First, the basis of successful information sharing in the future is knowledge as for technologies, processes, organizational aspects, and strategy. Managers are asked to allocate sufficient resources and provide a budget to build up knowledge either internally or with the help of external partners.

Second, it becomes crucial to transform corporate culture and corporate strategy including a holistic consideration of buyers and suppliers. Thinking and acting not as a stand-alone company but as a value creation chain and ecosystem is decisive against the backdrop of further digitization and interconnection.

Third, a company is ought to build up necessary infrastructure, to shape interfaces properly, and to adapt to standards in order to prepare future information sharing. Fourth, creating mutual trust between buyers and suppliers and dispel concerns, e.g., as for data security, are further requirements. Fifth, streamlining company internal processes and conducting operational refinements, is a further step. Last but not least, sharing information is a competitive factor that can serve to differentiate from competitor. Hence, constantly checking, revising, and adapting information sharing efforts and further observing technological developments is a key to successfully manage companies in the future.

6.2 Limitations and future research

Using qualitative-empirical data, this study analyzes information sharing between buyers and suppliers based on the SCOR model. Despite its contribution to research and implications for corporate practice, the study maintains some weaknesses that are discussed in the following. Furthermore, opportunities for further research are indicated.

First, the qualitative-empirical nature of the study serves to answer the research question addressing the research gap. However, it is limited in scope and content, e.g., it does neither analyze effect sizes nor interrelations. Second, the sample exclusively comprises German companies that has to be kept in mind when transferring the study's results to different context, e.g., in terms of culture and infrastructure. Third, it must be admitted that the sample size is restricted which in turn may further limit its generalizability. Future studies could address these limits and further validate the findings for example extending the data sample and using quantitative methods.

The course of the analysis uncovered a great variety of research gaps in the context of information sharing leaving space for future research. Among others, research should shed light on future requirements as for information sharing. For instance, studies may focus on technical prerequisites and may uncover what is an adequate infrastructure.

Given numerous systems and actors in cross-company information sharing, the question of how to properly shape interfaces becomes vital but research has not sufficiently addressed this topic yet. Analyzing the perspectives of cooperating partners within a supply chain and including further stakeholders, e.g. logistics providers, complements the state of research and may help to better understand interrelations. Differentiating between various company characteristics, e.g., product manufacturer, service providers, company sizes, and industry sectors, would shed light on influence factors and drivers and help to unveil further mechanisms. Digital platforms, e.g., virtual marketplaces, interconnect various supply chain partners and hold manifold potentials for value creation in the digital era. They have a great impact on information sharing and, among others, influence the way to exchange information and the intensity wherefore research studies are ought to further analyze their implications.

A further limitation of the study can be assigned to the fact that, relating to the sparse investigation of information sharing in Industry 4.0, the study cannot build on existing literature and therefore is only partially grounded in literature. Although there is literature on information sharing regardless of Industry 4.0 in extant literature, this literature is only partially applicable to Industry 4.0. The extension of this literature base, therefore, represents a further recommendation for future research.

In a similar regard, many of the concepts and forms of information sharing mentioned do not necessarily relate to the technologies and forms of information sharing that Industry 4.0 intends. Although Industry 4.0 implementation is still at an initial stage, it seems that the sample consists of many companies that have not evolved proactively in this direction to a large extent yet. Therefore, the inclusion of more advanced companies in this regard is recommended for future research.

Going beyond platforms, ecosystems include direct actors in the value creation process and indirect actors, e.g., educational institutions. Digital and automated information sharing lead to significant changes in entire ecosystems and therefore should be subject of further research studies as well.

References

Bolstorff, P. A., Rosenbaum, R. G., and Poluha, R. G. 2007. Spitzenleistungen im Supply Chain Management. Berlin, Heidelberg: Springer.

Birkel, H. S., Veile, J. W., Müller, J. M., Hartmann, E., & Voigt, K. I., 2019. Development of a risk framework for Industry 4.0 in the context of sustainability for established manufacturers. Sustainability, 11(2), pp. 384-411.

Fazli, S., and Afshar, Z. A., 2014. The Role of Drivers on Supply Chain Integration. Asian Journal of Research in Business Economics and Management, 4(12), pp. 347–359.

Giménez, C., and Ventura, E., 2003. Supply Chain Management as a Competitive Advantage in the Spanish Grocery Sector. The International Journal of Logistics Management, 14 (1), pp.77–88.

Gläser, J. and Laudel, G., 2010. Experteninterviews und qualitative Inhaltsanalyse: Als Instrumente rekonstruierender Untersuchungen (4th Edition). Wiesbaden: Springer.

Helmold, M. and Terry, B., 2016. Lieferantenmanagement 2030: Wertschöpfung und Sicherung der Wettbewerbsfähigkeit in digitalen und globalen Märkten. Wiesbaden: Springer Gabler.

Hofbauer, G., Mashhour, T., and Fischer, M., 2016. Lieferantenmanagement: Die wert¬orientierte Gestaltung der Lieferbeziehung (3rd Edition). Berlin, Boston: De Gruyter Oldenbourg.

Kagermann, H., Wahlster, W., and Helbig, J., 2013. Umsatzempfehlungen für das Zukunfts¬projekt Industrie 4.0 – Deutschlands Zukunft als Produktions-standort sichern. [online] Available at: <https://www.bmbf.de/files/Umsetzungsempfehlungen_Industrie4_0.pdf> [Accessed 8th May 2019].

Kiel, D., Müller, J.M., Arnold, C., and Voigt, K.-I., 2017. Sustainable Industrial Value Creation: Benefits and Challenges of Industry 4.0. International Journal of Innovation Management, 21(8), p. 1740015.

Lasi, H., Fettke, P., Kemper, H.G., Feld, T., and Hoffmann, M., 2014. Industry 4.0. Business & Information Systems Engineering, 6(4), pp. 239–242.

Mayring, P., 2015. Qualitative Inhaltsanalyse: Grundlagen und Techniken (12th Edition). Weinheim: Beltz.

Müller J.M., Kiel, D., and Voigt K.-I., 2018a. What Drives the Implementation of Industry 4.0? The Role of Opportunities and Challenges in the Context of Sustainability. Sustainability, 10(1), pp. 247-271.

Müller J.M., Buliga, O., and Voigt, K.-I., 2018b. Fortune favors the prepared: How SMEs approach business model innovations in Industry 4.0. Technological Forecasting and Social Change, 127, pp. 2-17.

Müller, J. M., and Voigt, K. I., 2018. Sustainable industrial value creation in SMEs: a comparison between Industry 4.0 and Made in China 2025. International Journal of Precision Engineering and Manufacturing-Green Technology, 5(5), 659-670.

Müller, J.M., Maier, L., Veile, J., and Voigt, K.-I., 2017. Cooperation strategies among SMEs for implementing Industry 4.0. Hamburg International Conference on Logistics (HICL).

National Research Council, 2000. Surviving Supply Chain Integration. Surviving Supply Chain Integration. National Academies Press.

Plattform Industrie 4.0, 2017. Was ist Industrie 4.0? [online] Available at: <https://www.plattform i40.de/I40/Navigation/DE/Industrie40/WasIndustrie40/was-ist-industrie-40.html> [Accessed 6th May 2019].

Rink, C. and Wagner, S.M., 2007. Lieferantenmanagement: Strategien, Prozesse und systemtechnische Unterstützung. In W. Brenner and R. Wenger (eds.). Elektronische Beschaffung: Stand und Entwicklungstendenzen, pp. 39–62, Berlin, Heidelberg: Springer.

Schoenherr, T. and Swink, M., 2012. Revisiting the arcs of integration: Cross-validations and extensions. Journal of Operations Management, 30(1), pp. 99–115.

Stevens, G., 1989. Integrating the supply chain. International Journal of Physical Distribution & Logistics Management, 19(8), pp. 3–8.

Stewart, G., 1997. Supply chain operations reference model (SCOR): the first cross-industry framework for integrated supply-chain management. Logistics Information Management, 10(2), pp. 62–67.

Supply Chain Council., 2012. SCOR Supply Chain Operations Reference Model Revision 11.0 (11.0). Supply Chain Council.

Voigt, K. I., Müller, J. M., Veile, J. W., Becker, W., and Stradtmann, M. 2019. Industrie 4.0–Risiken für kleine und mittlere Unternehmen. In: Geschäftsmodelle in der digitalen Welt (pp. 517-538). Springer Gabler, Wiesbaden.

Voigt, K. I., Kiel, D., Müller, J. M., and Arnold, C. 2018. Industrie 4.0 aus Perspektive der nachhaltigen industriellen Wertschöpfung. In Digitalisierung im Spannungsfeld von Politik, Wirtschaft, Wissenschaft und Recht (pp. 331-343). Springer Gabler, Berlin, Heidelberg.

Voigt, K. I., Buliga, O., and Michl. K., 2017. Business Model Pioneers - How Innovators successfully implement New Business Models. Cham: Springer.

Wiengarten, F., Humphreys, P., Gimenez, C., and McIvor, R., 2016. Risk, risk management practices, and the success of supply chain integration. International Journal of Production Economics, 171(3), pp. 361–370.

Zhao, G., Feng, T., and Wang, D., 2015. Is more supply chain integration always beneficial to financial performance? Industrial Marketing Management, 45, pp. 162–172.

Zhao, X., Huo, B., Selen, W., and Yeung, J. H. Y., 2011. The impact of internal integration and relationship commitment on external integration. Journal of Operations Management, 29, pp. 17–32.

Robot-Human-Learning for Robotic Picking Processes

Mathias Rieder[1] *and Richard Verbeet*[1]

1 – Ulm University of Applied Sciences

Purpose: This research paper aims to create an environment which enables robots to learn from humans by algorithms of Computer Vision and Machine Learning for object detection and gripping. The proposed concept transforms manual picking to highly automated picking performed by robots.

Methodology: After defining requirements for a robotic picking system, a process model is proposed. This model defines how to extend traditional manual picking and which human-robot-interfaces are necessary to enable learning from humans to improve the performance of robots' object detection and gripping.

Findings: The proposed concept needs a pool of images to train an initial setup of a convolutional neural network by the YOLO-Algorithm. Therefore, a station with two cameras and a flexible positioning system for image creation is presented by which the necessary number of images can be generated with little effort.

Originality: A digital representation of an object is created based on the generated images of this station. The original idea is a feedback loop including human workers after a not successful object detection or gripping which enables robots in service to extend their ability to recognize and pick objects.

Keywords: Picking Robots, Machine Learning, Object Detection, Computer Vision, Human-Robot-Collabortaion

First received: 19.May.2019 **Revised:** 23.June.2019 **Accepted:** 26.June.2019

1 Introduction

Finding staff for carrying out logistic tasks is getting harder and harder for companies as a survey of Kohl and Pfretzschner (2017) showed. Combined with developments in engineering and Artificial Intelligence there is a trend to integrate machines into the execution of logistic tasks, either to support workers or to automate them completely (Schneider, et al. 2018). Different to tasks for transport or manufacturing standardization, it is more challenging in picking tasks because a high amount of flexibility is needed to complete these tasks. This is the main reason why there is a low level of automatization in picking processes of just 5% in warehouses, 15% are mechanized and 80% are still run manually (Bonkenburg, 2016). Fully automated picking processes, besides fully automated storage, offer several advantages: saving of space and labor cost, availability of personnel instead of robots, savings on operational cost as heating or lighting (de Koster, 2018) and facing lack of personnel in logistics.

For here discussed robots in logistics there is a suitable definition of Bonkenburg (2016) in contrast to all other robotic solutions like robotic vacuum cleaner or nursing robot: "A Robot with one or more grippers to pick up and move items within a logistics operation such as a warehouse, sorting center or last-mile".

Picking of known objects in dynamic environments by robots is a major task as shape and position of an object may change since the last visit of a robot at the object's storage location. If the position of an object is constant, e.g. for welding robots in automotive production systems, robots complete their jobs very well. Therefore, no understanding of their surroundings is necessary. But if the robot must work in cooperation with humans there are changes to the environment as, to the objects, the shelf, and the position of the objects within the shelf or their orientation. Furthermore, even the object itself can be different since the last handling process due to changing object design caused by changed package sizes or modernization of styles as it is common business. In retail there is also a constantly changing product range by introducing respectively discontinuing promotional or seasonal goods. So, a robot must adapt to this situation by object detection.

The cooperation of robots and humans is necessary because the number of objects robots can pick is very small (Schwäke, et al., 2017). A promising approach is to assign those picking orders to robots they can recognize and grip while humans pick the leftovers (Wahl, 2016). These both sections could be separated in different areas, but this would cause two major disadvantages: humans are not able to pick objects from the robots' working area, e.g. in case of a capacity bottleneck during seasonal peaks, and robots can't enlarge the number of pickable objects by working with and learning from human colleagues.

In addition to this, cooperation between robots and humans may be the answer if partial automation is desired or even required because of lack of personnel. To enable such a picking setup a process model is proposed which allows cooperation between humans and robots to guarantee robust processes and learning robots. The first step in this model is to generate the necessary data for robots' object detections. But especially for jobs in logistic environments there is a lack of data sets for training object detection systems which are essential for robot picking (Thiel, Hinckeldeyn and Kreutzfeldt, 2018). Out of these data sets the object detection system pulls "knowledge" about the objects. If data quality is low the resulting model will also be inadequate and furthermore, if there are objects not being part of the input data, they cannot be identified by the model. This data set must initially be created which means a lot of work, for comparison COCO-dataset contains 330,000 images for differentiating 80 object categories which took about 70,000 worker hours (Lin, et al., 2015). Therefore, an adaptive system is necessary whose data can represent the latest status to successfully work on picking orders.

Besides the question on how to get data for training there is a very central point mentioned by Hui (2018): "The most important question is not which detector is the best. It may not possible to answer. The real question is which detector and what configurations give us the best balance of speed and accuracy that your application needed." The two central aspects of characterizing an object detection algorithm is accuracy and speed (Hui, 2018).

Considering the heterogeneous landscape of objects combined with the variety of cameras, algorithms, impacts from surroundings like lighting and robotic and computing hardware the comparison of existing solutions is a very challenging task. This results in a need for experiments and testing. There must be a specific set of training data for each solution approach which represent the target area of the algorithm. The best way for gathering such data representing operational processes is using these processes themselves. To support the task of gathering information, external data input is needed from humans to tell the system about changes on objects as a computer system cannot reliably recognize the consequences of chances to objects for object detection and picking. There is also a need for an efficient way to collect images to train an initial object detection model which must work successful on training data before it can be used in picking customer orders in operational processes.

This leads to the question how to transform manual picking processes into highly automated ones in an efficient way ensuring operational order fulfillment. To answer this question, it is necessary to set up an object detection system as basis for robot picking evaluating which object detection algorithm is suited best for a specific logistic environment. Comparing possible algorithms, a specific training and testing data set is necessary which is not existing yet. During building up this data set, there will be further questions to be answered belonging the data set itself, e. g. how many images of each object must be recorded, how to face changes to objects and their appearance or which angles and rotations of object images are more useful to training.

Summarizing the central questions in short:

1. How to transform manually operated picking processes into highly automated ones?
2. Which algorithm(s) support object detection in logistics best?
3. What must a data set for object detection in logistics look like?
4. How can changes on the objects and within object range be handled?

Deriving from these questions the goal of the current research work is to work out a process environment that makes adaption to changes possible using human-robot-cooperation. Besides there must be an answer on how to collect images of objects on an efficient way to make a comparison of object detection algorithms possible to choose the best suitable one.

This paper is structured as follows. Section 2 reviews related work in areas of picking robots and object detection. In section 3 the chosen object detection algorithm is introduced shortly. Section 4 proposes a process model that handles the need for image data to enable object detection models including the introduction of a picture recording machine. In section 5 first results are presented and shortly discussed. Section 6 presents the conclusion. Section 7 shows further questions for research work on picking robots and object detection.

2 Related Work

For the topic of this paper there are two sections of great interest: picking robots which will do the physical job by gripping objects and object detection algorithms to determine where the robots must grip the targeted objects.

2.1 Picking Robots

For several years many research efforts have been done on flexible picking robots, e. g. for harvesting vegetables and fruits like oranges (Muscato, et al., 2005), cucumber (Van Henten, et al., 2005) or strawberry (Hayashi, et al., 2010). Current robotic applications are driven by four technology trends that enable and enhance the applicable solutions of robots in logistics. These are feet (mobility), hands (collaboration and manipulation), eyes (perception and sensors) and brains (computing power and cloud) where each trend has shown many improvements in recent years (Bonkenburg, 2016).

Nowadays there are several companies offering mobile picking robots as IAM Robotics (2019), Fetch Robotics (Fetch Robotics, 5Inc., 2019) or Magazino (Magazino GmbH, 2019) and many more supporting logistics processes by partwise process automation (Britt, 2018). But there will be more solutions supported by developments within the technology trends mentioned before, which, for example, support robots for more and more flexible gripping by tactile sensors which make robotic grippers more adaptable to their use case (Costanzo, et al., 2019). Besides better sensors, grippers are getting increasingly adaptable as the presentation of a gripper construction kit for robots by Dick, Ulrich and Bruns (2018) shows. The central component, the brain of the robot, is also being refined. Besides the constant improvement on brain's processing and architecture there is a continuous work on algorithms to extract information from sensor input to detect object positions and the object's gripping point faster and with higher accuracy (Tai, et al., 2016).

Motion as the job of moving around is solved adequate for autonomous guided vehicles since many years and works on first picking robots successfully as Magazino's Toru shows (Wahl, 2016).

To sum up: existing systems partially face the problem of picking automation and may deliver viable solutions in stable environments but there is a lack of flexibility adapting to chances in the environment.

2.2 Object Detection

For picking processes it is essential to know where the target object is located. For this job object detection algorithms determine the position of the target from sensor data - usually images from cameras - within an image that contains multiple objects (Agarwal, 2018). Semantic Segmentation, classification, localization and instance segmentation are other jobs working on images besides object detection. These tasks of Computer Vision are shown in Figure 1 outlining the differences of these.

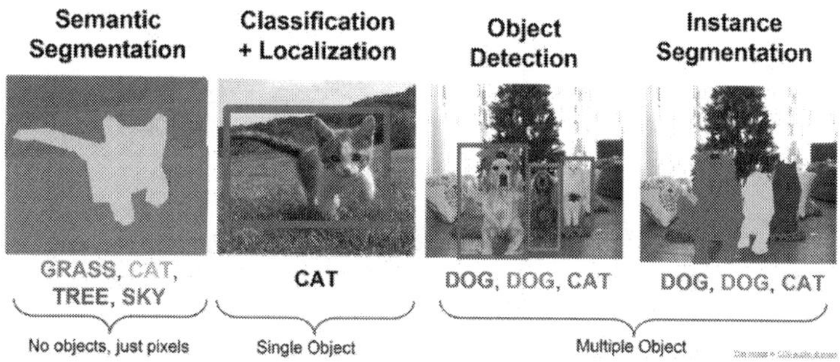

Figure 1: Comparison of semantic segmentation, classification and localization, object detection and instance segmentation (Li, Johnson and Yeung, 2017)

To improve the efficiency of object detection, Machine Learning can be used to extract information from an existing set of data to predict on unknown data (Witten, Frank and Hall, 2011). This can be applied to different domains where a great amount of data exists - the more data the better, which is the case for object detection (Domingos, 2012).

Research of recent years within the field of object detection has developed approaches based on Deep Learning, a special kind of Machine Learning.

They offer the advantage of finding features automatically, for example a neural network is taught using training data for object detection (Ouaknine, 2018a). Several algorithms for object detection were developed using Deep Learning, a comparison of these algorithms for different applications is presented by Zhao, et al. (2019). Deep Learning is used to train a neural network which later, after the training is finished, can be used for object detection tasks. But such neural networks have problems at object detection with object not being part of the training data as these tend to be identified as objects containing in the data set (Colling, et al., 2017). Machine Learning can also be used for gripping point detecting which outperforms hand-set configurations (Lenz, Lee and Saxena, 2015).

Object detection requires input data distinguished in 2D-images or 3D-information which depends on what information is available, which kind of objects must be distinguished or what accuracy is needed. If the objects look different but have identical geometric shape a combination of images and distance information may be needed, as RGB-D data, to define gripping points (Lenz, Lee and Saxena, 2015). Another option is the computation of 3D-information from 2D-images (Jabalameli, Ettehadi, and Behal, 2018).

Like other object detection algorithms YOLO is trained by images. To gather images there are different approaches like turning an object in front of a camera to take images. This idea is used for different purposes like 3D-Scanning as basis for 3D-Printing (Rother, 2017), 360-degree images for web shops (Waser, 2014) or master data and image capturing (Kraus, 2018). A similar setup to the proposed process model in this work was created by Hans and Paulus (2008). Their focus of research is on color within the images (Hans, Knopp, Paulus, 2009). Another specific setup is designed to record 360-degree-images of motor vehicles (Ruppert, 2006). A similar approach is moving the camera around an object and take images from different positions which is proposed by Zhang et al. (2016).

To compare the outcome of different algorithms meta-data is added to the data sets containing information about the set. Different data sets are used for different learning jobs like images, text or speech (Stanford and Iriondo, 2018). For an image data set this information defines the objects in the images and where within the picture the objects can be found. This enables a comparison of the output of object detection algorithms and what they should discover within the images. Latest research in object detection focuses on COCO-dataset (Common Objects in Context) which was presented by Lin, et al in 2015 including metrics measuring the performance of object detection algorithms on test images. Redmon et al. (2016) characterized the performance of their YOLO-algorithm (You Only Look Once) using different datasets (ImageNet 2012, VOC 2007, VOC 2012, Picasso, People-Art). For version 3 of YOLO there is given a comparison on COCO-dataset only (Redmon and Farhadi, 2018). YOLO is mentioned here as it is the "fastest general-purpose object detector in the literature" (Redmon, et al., 2016).

3 YOLO-Algorithm

Processing 45 images per second YOLO-algorithm can be part of a real-time object detection system (Redmon, et al., 2016). YOLO trains a convolutional neural network (CNN) with training data on a loss function (Redmon, et al., 2016). The functional principle of YOLO is shown in Figure 2 originating with the publication of Redmon et al. (2016). Within the CNN images are split into grids cells where each cell is analyzed for possible objects, marking them with a bounding box and equipping each with according confidence. Afterwards the bounding boxes of each grid cell are combined with a class probability. Each grid cell can only contain one object so the bounding box with the highest confidence is chosen. Neighboring cells respectively bounding boxes containing the same object are summarized by non-maximal suppression. In difference to other algorithms YOLO works on the whole image what makes it being so fast. But there are limitations which are detecting small objects appearing in groups or objects in new respectively unusual environments (Redmon, et al., 2016).

4 Proposal for Adaptive Generation of Learning Data

As mentioned above classifying models must deal with a changing range of products especially in businesses where product design is changed for promotional purpose. This is a major problem for a robotic picking system. It needs reliable input from the object detection system because failure in picking can be expensive causing delayed order completion or even non-fulfillment of customer orders.

This requirement leads to the need for picking processes being adaptable to changes in the environment.

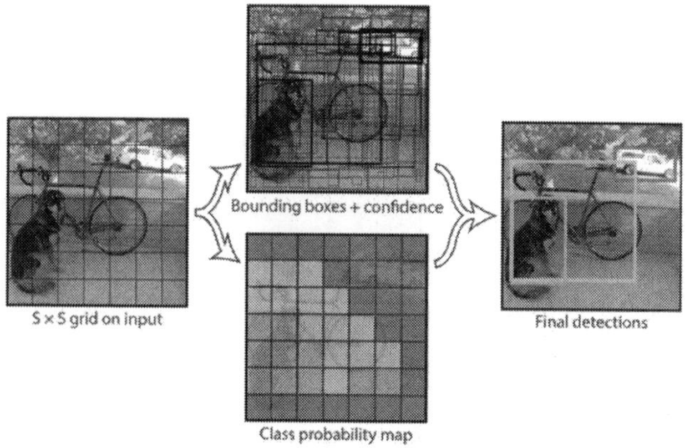

Figure 2: Working principle of YOLO-algorithm (Redmon, et al. 2016)

4.1 Process Modell

The process model consists of the two parts Learning and Operation as shown in Figure 3. Within the first part no robotic equipment is needed as it aims to build a detection model for objects which can be calculated on external computing resources. Therefore, images of the objects to differentiate must be generated. A lot of pictures are needed to calculate such a model so that images of an object from different perspectives and angles of rotation must be created. The different rotation angles and perspectives are needed as the object can appear in every orientation in a warehouse (compare Zhang, et al., 2016). For subsequent calculation of the detection model from the images they are stored in a database.

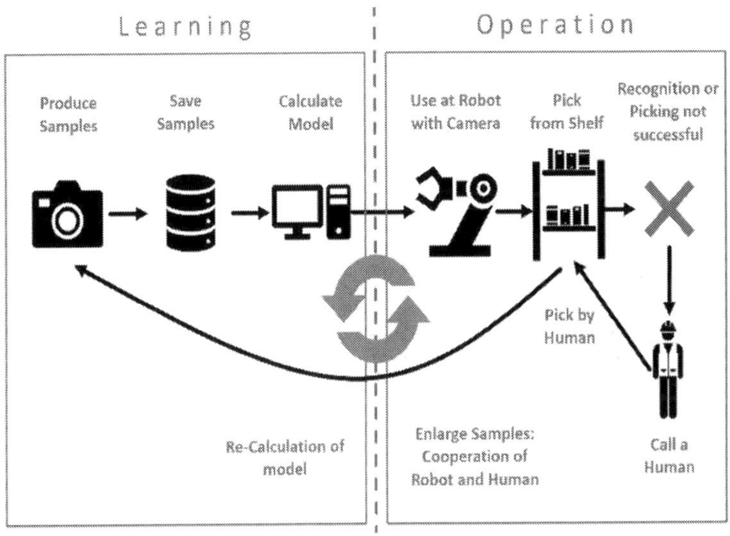

Figure 3: Process Model

If the object detection model exceeds the defined performance indicators it is used in a real picking environment which is describe by Operation in the process model. These performance indicators must be defined evaluated during testing. There the object detection model is applied to the robot control to find objects the robot must pick. An image of the target shelf is recorded by a camera mounted at the mobile robot. The model locates the target object within the images and defines grasping points from the orientation of the object and possible grasping points from the database where master data is saved. If the robot succeeds everything is fine.

If a problem occurs, e.g. the target object isn't detected in the shelf because of changes of its design or it is obscured by another object, the robots calls for a human picker. The human fulfills two important tasks.

If the object is in the shelf, he completes the order by picking the object. Furthermore, he must give feedback describing why detection was not possible according to the system's error message and, if the object is in the image, where it is located. The system uses this information to improve the detection model for the next try by including the additional images recorded in cooperation with the human picker at the shelf for model calculation. But as the calculation of such a model on a standard computer lasts several days re-calculation cannot be done in real-time on the robot as could be observed during testing.

If an object detection model performs very poor the object will be sent back to Learning: more images must be recorded with the Picture Recording Machine being introduced in the following chapter and the detection model must be trained once again.

4.2 Picturing Recording Machine

As first step to implement a setup of the introduced process model it is necessary to gather images of the different objects. As doing this manually is a time-consuming job this task is partial automated by a Picture Recording Machine which is shown in Figure 4.

Besides this the machines enables collecting images from precise orientations in a repeatable way. For gathering images on different locations, it is mobile and enables imaging of objects from different perspectives and angles of rotation. To get different angles of rotation between 0 to 360 degrees there is a turning table in the center of the machine which is driven by a stepper motor (42SHD0216-20B).

Figure 4: Picturing Recording Machine

Furthermore, the camera system mounted on the rocker is also moved by a separate stepper motor (Nema 23, 60STH86-3008B). By this, images between 0 and 90 degrees can be taken. Each motor is controlled by its own motor driver (PoStep60) triggered by a microcontroller (Raspberry Pi 3 Model B v1.2). The electronic setup is shown in Figure 5.

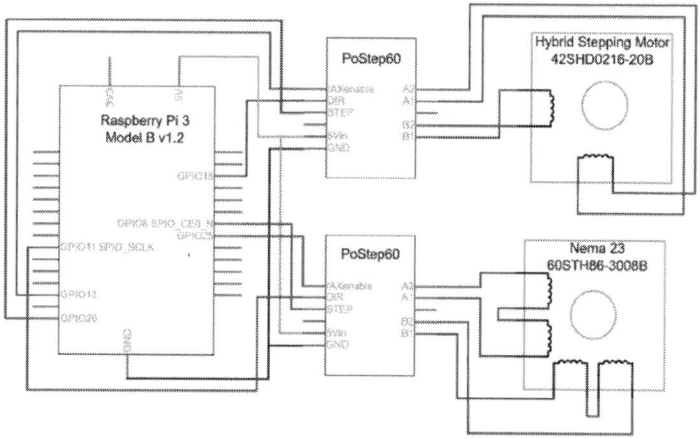

Figure 5: Wiring Diagram for the Picturing Recording Machine

This setup allows the collection of images in a semi-automatic way as there is only the change of the objects and the input of the number of images and angles to be done manually. The system gets its input by the GUI that is shown in Figure 6. This GUI is connected to the database where images and master data are stored. It is possible to use the Picture Recording Machine also for collection of master data. By clicking on the barcode-icon the object's barcode is scanned by the user and the system checks if there is existing master data and images according to this object-ID. If there is existing data, it is loaded from database if not, a new object-ID is generated. Loading data from the database also includes loading a representative view of the object in the GUI what makes comparison of existing data and the present object possible for the user of the machine.

There are options for recording images of the object as the number of images taken in one rotation and the number of perspectives. For perspectives there is the option to decide whether to include 0- or 90-degree's view or not.

The mounting on the rocker is designed to simultaneously support two different camera systems for imaging. Besides a comparison of camera-hardware it enables also comparison of different algorithms as there may be one recording 2D-data and the other one generating 3D-information.

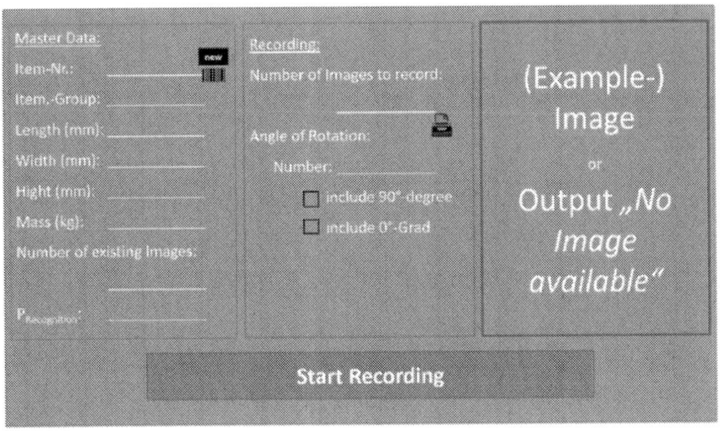

Figure 6: GUI for the Picturing Recording Machine

These images are taken by two different cameras: Microsoft Kinect One and Photoneo PhoXi 3D-Scanner M giving the option to compare training and testing of object detection algorithms on different equipment and testing the algorithm working independent on different hardware (Photoneo s. r. o., 2018; Microsoft Corporation, 2018). These ones are chosen as there will be a comparison of the industrial camera solution which comes with better features and a higher price (Photoneo) as the cheap home solution (Kinect). Both are each chosen in their category after a market research under the condition of each camera system providing color images (e.g. RGB) and depth-information.

Currently there is a Microsoft Kinect One and a Photoneo PhoXi 3D-Scanner M mounted on the rocker (Photoneo s. r. o., 2018; Microsoft Corporation, 2018).

As creation of own data sets for training is suggested by Thiel, Hinckeldeyn and Kreutzfeldt (2018) the Picture Recording Machine is an essential part of this research work and will support the follow-up steps to create an adaptive and learning environment for robots in logistics.

5 Results and Discussion

Evaluating YOLO-algorithm images of all-day office objects are taken which could be picked in a retail commissioning (coffee package, stapler, different bottles, different beverage packs, cookie pack). Figure 7 shows a first test on a manually taken image which shows the result localizing three objects within the image, giving a confidence score for each object: ice_tea (58%), water bottle (91%) and coffee (92%). The predic-tion needs 5.63 seconds which is about the time quoted by Redmon et al. and shows only predictions with confidence higher than 20% (Redmon, 2016). The complete results of the first try are shown in Table 1.

The Convolutional Neural Network for this purpose is trained with 329 training images containing one or two of the objects (254 images with one object and 75 with two). Preparation and recording these images take approximately 10 men-days. Training images are taken with Photoneo PhoXi

3D-Scanner M, testing images for evaluating object detection are shot with MS Kinect.

Even though "ice_tea" exists in 154 images (50 of them with another object) it had a bad score comparing "water bottle" and "coffee", except "stapler" which seems having not enough images or otherwise being only single in training images.

Figure 7: Testing YOLO on manually taken images

A very positive result of these test is the section of "false positives". A false positive result is an object the object detection algorithm tells it would be another object. As errors in picking are expensive this score must be low. There must be distinguished between known (part of training data) and unknown objects appearing within an image. Removing unknown objects from "false positives" gives the actual number of errors. The scores of 0% to 6% in this category resulting from only a few training images indicates YOLO to be a very promising approach.

The other way round "false negatives" tell about objects being in an image but not being detected. This is not a problem for the proposed process model as the human fallback level handles this type of error. For false negatives' score there is the chance of getting better by human feedback during operation.

Table 1: Results of testing YOLOv3 in first images

	ice_tea	**water bottle**	**coffee**	**stapler**
training images	154	100	100	50
single object	104	50	50	50
with one other object	50	50	50	0
test images	50	50	50	50
positive detections	31 (62%)	32 (64%)	46 (92%)	1 (2%)
false negatives	19 (38%)	18 (36%)	4 (8%)	49 (98%)
false positives	0 (0%)	1 (2%)	13 (26%)	1 (2%)
known objects	0 (0%)	1 (2%)	3 (6%)	1 (2%)
unknown objects	0 (0%)	0 (0%)	10 (20%)	0 (0%)
mean confidence for positive detections	69,1%	72,8%	82,7%	63,0%

To sum up the results YOLO-algorithm seems to be very promising for our purpose as there were only few images for training and testing leading to quite good results. So, there will be further research about this approach.

Supporting training for object detection the Picture Recording Machine will help by contributing efficient picture recording. First test had shown that the machine is able to record about 20 images a minute under high repeating accuracy and by systematically saving the images with according recording information (ID, angle, rotation, camera type). The number on images per minute depends on how many pictures are taken during one rotation and the number of the angles of recording. The bigger the number of images the shorter the distance to move between two recordings.

6 Conclusion

The proposed model will help to transform manual picking processes to a robotic picking system for a wide range of objects, whereby the problem of missing worker or legal restriction (working on Sundays) can be solved. For making picking in warehouses by robots possible continuous update of object data is necessary. Cooperation with humans is essential for reliable working on picking orders. In this work an approach for cooperation between robots and humans for picking processes is presented to guarantee stable output of commissioning and to support the robotic system by updating and extending their object data to increase the rate of objects being picked by robots.

The presented process model provides the following advantages which ensures robust processes as well as a learning environment for gripping robots:

1. Generation of missing image data respectively the needed number of images
2. Continuous updating of image data

3. Decoupling of generation of images, model testing and robotic installation
4. Reliable picking processes by human fallback level

Making robots learn to grip new objects with the presented process model may work slowly for a big number of objects but it guarantees stable output of the commissioning.

A part of the process model is the generation of image data of related objects which is a common problem for today's application working on object detection. Generating many images of many objects is possible by the Picture Recording Machine what tackles the problem of missing data sets. In combination with feedback from human-robot collaboration the basic data can be enriched by images from the process which makes object detection training results more stable.

Having many images also enables a comparison of different object detection algorithms which again enables to choose the best one for the specific object detection task.

7 Future Research

After gathering images of different objects there will be a closer look on which parameters affecting the output of different object detection algorithms. Therefore, tests with the recorded data set on known algorithms will be done. Another question to answer is which input has which impact on the output of an algorithm: number of images, how to handle similar looking objects, which degrees of rotation and which camera angles are more helpful than others supporting object detection. A further question to answer is if 0- and 90-degrees' view is how helpful for object detection models.

A further step towards more efficient learning could be the automated generation of coordinates of an area where the object within an image is located. Now this is done manually and very time consuming so there will be attempts for a higher degree of automation.

A very important part of the process model is the human-robot interface where information is generated that supports the learning process. There must be research on how this interface must be designed that human pickers will accept the co-working with their robotic colleagues. Besides the information the humans can give as feedback to the system must be specified that the learning system can understand what problems made detecting the object not possible.

The automation of image gathering could also enable meta-learning comparing different object detection algorithms by providing each with image data automatically and evaluating their results.

Finally, the proposed concept must be tested in a realistic field test ideally in a real commissioning of an industrial partner which has to be found.

Acknowledgements

This research work is done within Post Graduate School "Cognitive Computing in Socio-Technical Systems" of Ulm University of Applied Sciences and Ulm University. This work is part of the ZAFH Intralogistik, funded by the

European Regional Development Fund and the Ministry of Science, Research and the Arts of Baden-Württemberg, Germany (F.No. 32-7545.24-17/3/1).

Financial Disclosure

The Post Graduate School is funded by the Ministry for Science, Research and Arts of the State of Baden-Württemberg, Germany.

References

Agarwal, R., 2018. Object Detection: An End to End Theoretical Perspective - A detailed look at the most influential papers in Object Detection [online] Available at <https://towardsdatascience.com/object-detection-using-deep-learning-approaches-an-end-to-end-theoretical-perspective-4ca27eee8a9a> [Accesses 13 May 2019]

Bonkenburg, T., 2016. Robotics in Logistics - A DPDHL perspective on implications and use cases for the logistics industry, Bonn: Deutsche Post DHL Group

Britt, P., 2018. Whitepaper: 10 Robots That Can Speed Up Your Supply Chain, Framingham: Robotics Business Review

Colling, D., Hopfgarten, P., Markert, K., Neubehler, K., Eberle, F., Gilles, M., Jung, M., Kocabas, A., Firmans, K, 2017. PiRo - Ein autonomes Kommissioniersystem für inhomogene, chaotische Lager, Logistics Journal. Proceedings, DOI: 10.2195/lj_Proc_colling_de_201710_01

Costanzo, M., De Maria, G., Natale, C., Pirozzi, S., 2019. Design and Calibration of a Force/Tactile Sensor for Dexterous Manipulation, Sensors, 19(4), 966

de Koster, R. B. M., 2018. Automated and Robotic Warehouses: Developments and Research Opportunities, Logistics and Transport No 2(38)/2018, pp.33-40, DOI: 10.26411/83-1734-2015-2-38-4-18

Dick, I., Ulrich, S., Bruns, R., 2018. Autonomes Greifen mit individuell zusammengestellten Greifern des Greifer-Baukastens, Logistics Journal. Proceedings, ISSN: 2192-9084

Domingos, P., 2012. A Few Useful Things to Know about Machine Learning, Communications of the ACM, 55, 10, pp.78-87, DOI: 10.1145/2347736.2347755

Fetch Robotics, Inc., 2019. Autonomous Mobile Robots That Improve Productivity, [online] Available at <https://fetchrobotics.com/> [Accessed 12 May 2019]

Hans, W. Paulus, D., 2008. Automatisierte Objektaufnahme für Bilddatenbanken. In: Helling, Stephan; Brauers, Johannes; Hill, Bernhard; Aach, Til: 14. Workshop Farbbildverarbeitung. RWTH Aachen: Shaker. S. 143-151.

Hans, W., Knopp, B., Paulus, D., 2009. Farbmetrische Objekterkennung. In: 15. Workshop Farbbildverarbeitung. Berlin: GfAI. S. 43-51.

Hayashi, S., Shigematsu, K., Yamamoto. S., Kobayashi, K., Kohno, Y., Kamata, J., Kurita, M., 2010. Evaluation of a strawberry-harvesting robot in a field test, Biosystems Engineering, Volume 105, Issue 2, 2010, Pages 160-171, ISSN: 1537-5110,

Hui, J., 2018. Object detection: speed and accuracy comparison (Faster R-CNN, R-FCN, SSD, FPN, RetinaNet and YOLOv3), [online] Available at < https://medium.com/@jonathan_hui/object-detection-speed-and-accuracy-comparison-faster-r-cnn-r-fcn-ssd-and-yolo-5425656ae359> [Accesses 13 May 2019]

IAM Robotics, 2019. Making flexible automation a reality, [online] Available at <https://www.iamrobotics.com/> [Accessed 12 May 2019]

Jabalameli, A., Ettehadi, N., Behal, A., 2018. Edge-Based Recognition of Novel Objects forRobotic Grasping, arXiv:1802.08753v1 [cs.RO]

Kohl, A.-K., Pfretzschner, F., 2018. Logistimonitor 2018 Der Wirtschaftszweig in Zahlen - Ergebnisse einer Expertenbefragung von Statista und der Bundesvereinigung Logistik (BVL) e.V.. Bremen

Kraus, W., 2018. Digitale Prozesse im Großhandelsunternehmen: Logistik 4.0 - Roboter im Warenlager, Stuttgart, 09 October 2018. Stuttgart: Fraunhofer IPA

Lenz, I., Lee, H., Saxena, A., 2015. Deep Learning for Detecting Robotic Grasps, The International of Robotics Research, 34(4-5), pp.705-724

Li, F.-F., Johnson, J., Yeung, S., 2017. Detection and Segmentation, CS231n: Convolutional Neural Networks for Visual Recognition,Stanford University, [online] Available at < http://cs231n.stanford.edu/slides/2017/cs231n_2017_lecture11.pdf> [Accessed 10 June 2019]

Lin, T.-Y., Maire, M., Belongie, S., Bourdev, L., Girshick, R., Hays, J., Perona, P., Ramanan, D., Zitnick, C. L., Dollár, P., 2015. Microsoft COCO: Common Objects in Context, Computer Vision and Pattern Recognition, arXiv:1405.0312v3 [cs.CV]

Magazino GmbH, 2019. Intelligente Robotik und Lagerlogistik, [online] Available at <https://www.magazino.eu/> [Accessed 12 May 2019]

Microsoft Corporation, 2018. Kinect für Windows. [online] Available at: <https://developer.microsoft.com/de-de/windows/kinect> [Accessed 11 De-cember 2018]

Muscato, G., Prestifilippo, M., Abbate, N., Rizzuto, I., 2005. A prototype of an orange picking robot: past history, the new robot and experimental results, In: Industrial Robot: An International Journal, Vol. 32 Issue: 2, pp.128-138, DOI: 10.1108/01439910510582255

Ouaknine, A., 2018. Review of Deep Learning Algorithms for Image Semantic Segmentation, [online] Available at < https://medium.com/@arthur_ouaknine/review-of-deep-learning-algorithms-for-image-semantic-segmentation-509a600f7b57> [Accessed 1 May 2019]

Photoneo s. r. o., 2018. PhoXi® 3D-Scanner M. [online] Available at: <https://www.phtoneo.com/prduct-detail/phoxi-3d-scanner-m/?lang=de> [Accessed 11 November 2018]

Redmon, J., 2016. YOLO: Real Time Object Detection, [online] <https://github.com/pjreddie/darknet/wiki/YOLO:-Real-Time-Object-Detection> [Accessed 07 June 2019]

Redmon, J., Divvala, S., Girshick, R., Farhadi, A., 2016. You Only Look Once: Unified, Real-Time Object Detection. Computing Research Repository (CoRR), arXiv:1506.02640 [cs.CV]

Redmon, J., Farhadi, A., 2018. YOLOv3: An Incremental Improvement. arXiv:1506.02640 [cs.CV]

Rother, H., 2017. 3D-Drucken...und dann? - Weiterbearbeitung, Verbindung & Verede-lung von 3D-Druck-Teilen, München: Carl Hanser Verlag

Ruppert, W., Patentanwälte Freischem, 2006. Verfahren zur Aufnahme digitaler Abbildungen. Köln. EP 1 958 148 B1 (active).

Schneider, J., Gruchmann, T., Brauckmann, A., Hanke T., 2018. Arbeitswelten der Logistik im Wandel: Automatisierungstechnik und Ergonomieunterstützung für eine innovative Arbeitsplatzgestaltung in der Intralogistik. In: B. Hermeier, T. Heupel, S. Fichtner-Rosada, S., ed. 2018. Arbeitswelten der Zukunft. Wiesbaden: Springer Fachmedien GmbH. pp.51-66

Schwäke, K., Dick, I., Bruns, R., Ulrich, S., 2017. Entwicklung eines flexiblen, vollautomatischen Kommissionierroboters, Logistics Journal. Proceedings, ISSN: 2192-9084

Stanford, S., Iriondo, R., 2018. The Best Public Datasets for Machine Learning and Data Science - Free Open Datasets for Machine Learning & Data Science, [online] Available at < https://medium.com/towards-artificial-intelligence/the-50-best-public-datasets-for-machine-learning-d80e9f030279> [Accessed 13 May 2019]

Tai, K., El-Sayed, A.-R., Shahriari, M., Biglarbegian, M., Mahmud, S., 2016. State of the Art Robotic Grippers and Applications, Robotics, 5, 11; DOI: 10.3390/robotics5020011

Thiel, M., Hinckeldeyn, J., Kreutzfeldt, J., 2018. Deep-Learning-Verfahren zur 3D-Objekterkennung in der Logistik. In: Wissenschaftliche Gesellschaft für Technische Logistik e. V., 14. Fachkolloquium der WGTL. Wien, Austria, 26-27 September 2018, Rostock-Warnemünde: Logistics Journal

Van Henten, E.J., Van't Slot, D.A., Hol, C.W.J., Van Willigenburg, L.G., 2009. Optimal manipulator design for a cucumber harvesting robot, Computers and Electronics in Agriculture, Volume 65, Issue 2, 2009, Pages 247-257, ISSN 0168-1699, DOI: 10.1016/j.compag.2008.11.004.

Wahl, F., 2016. Pick-by-Robot: Kommissionierroboter für die Logistik 4.0. Future Manufacturing. 2016/5. Frankfurt am Main: VDMA Verlag. pp.16-17

Waser, G., 2014. 360-Grad-Bilder: eine Attraktion für Webshops. Marketing & Kommunikation, 01/2014

Witten, I.H., Frank, E., Hall, M.A., 2011. Data mining - Practical machine learning tools and techniques, 3 ed., Amsterdam: Morgan Kaufmann Publishers Inc.

Zhao, Z.-Q., Zheng, P., Xu, S.-t., Wu, X., 2019. Object Detection with Deep Learning: A Review, arXiv:1807.05511v2 [cs.CV]

Zhang, H., Long, P., Zhou, D., Qian, Z., Wang, Z., Wan, W., Manocha, D., Park, C., Hu, T., Cao, C., Chen, Y., Chow, M., Pan, J., 2016. DoraPicker: An Autonomous Picking System for General Objects, arXiv:1603.06317v1 [cs.RO]

II.
Innovation and Technology Management

Digital Logistics, Strategic Cognitive Readiness and Employee Training

Thomas Neukirchen[1] *and Matthias Klumpp*[1]

1 – FOM University of Applied Sciences

Purpose: We consider long-term structural implementation risks connected to digitalization in production and logistics. Similar to considerations regarding automation, we identify areas of potential corporate benefits and corresponding necessary adjustments in task descriptions and employee behavior.

Methodology: Case study using an interactive process analysis and annotations method (interaction room) for a digitalization effort in a German power plant component company, including employees in digital developments early on. Survey of the concept of strategic cognitive analysis for logistics digitalization (systematic literature research).

Findings: Analysis of the concept of strategic cognitive readiness is given for logistics digitalization and related to training requirements. This is connected to the presented case study which results in (i) augmented and digitalized process (ii) employee training requirements regarding strategic cognitive readiness.

Originality: The concept of strategic cognitive readiness, originating from a military context, had been widely identified as a core variable with regards to successful automation prior to related widespread adoption. Nonetheless, recent research shows that measures aimed at effectively integrating employees, if taken, have fallen short in many cases.

Keywords: Digitalization, Training, Cognitive Readiness, Case Study

First received: 16.May.2019 **Revised:** 01.Jun.2019 **Accepted:** 04.June.2019

1 Introduction

Digital logistics requires new perspectives and analyses regarding cognitive and psychological risk prevention due to the increased mental workload in automated systems. The aim of this threefold interdisciplinary approach is to arrive at efficient digitalization measures, co-developed with employees who remain adaptable in the sense of strategic cognitive readiness (Grier, 2012) while sustaining their psychological health and wellbeing. This is achieved by numerous activities intersecting with the three mentioned leverage points, such as process analysis and digitalization strategy, risk assessments, process optimization and competence development, workshops for leadership, cooperation and self-management, introduction of a digital error culture etc. This article discusses the foundations for the approach, gives an outline of project activities and assesses the joint approach (employee-centered digitalization and training) to the digitalization of a particular process (offer preparation) in a company (German power plant components). The approach is motivated by the following context: A number of challenges connected with digitalization in logistics and production exhibit similarities to considerations voiced during the advent of widespread automation in the 1970s and 1980s (Thackray, 1980; Bainbridge, 1983; Cummings, Gao and Thornburg, 2016; Frey and Osborne, 2017). Given that, contemporary and recent work both highlight that a number of issues, however accurately forecast, have not been prevented, nor mitigated. Thus, the question for useful lessons for digitalization drawn from analogous (unresolved) automation issues arises.

In this contribution, we address this question regarding possible consequences of digitalization for employees in logistics and production. For meaningful focus we narrow down our exploratory research to psycho-social matters and team-work, and with a view in terms of a reoccurring concept called 'cognitive readiness'. This term, which is explained below, ties together many important human-resources issues connected with automation shortcomings as well as digitalization risks. For different applications, definitions of the concept have to be adapted adequately and differentiation with respect to timescales and planning horizons need attention, as these surely deviate from the original military context of the term. As a condition for an optimal (cognitive) work performance, a balance of mental workload and stress is widely seen as prerequisite for maximal adaptability (Hancock and Warm, 1989; Guznov, Matthews and Warm, 2010), while an imbalance is likely to impair performance and, if persisting, to lead to profound psychological harm. On the other hand, a low workload and stress level is associated with boredom and detachment from tasks (Cooke, et al. 2010). Symptoms of the latter, such as impairments of short-term memory, perceptual focus, and attention seem as undesirable as those of an overwhelming workload that may even prevent any engagement in a task (Ewing and Fairclough, 2010; Paulaus, et al., 2009). Between these non-optimal states, an 'optimal' work experience and engagement have long since been established (Csikszentmihalyi, 2000). However, the flow-concept may not accurately incorporate changes to logistics work induced by pervasive digitalization (Lucke and Rensing, 2014; Zijm and Klumpp, 2017). Digitalization of processes can induce efficiency gains, of course, while the direction of changes in both quality and quantity of workload and stress are generally not a given. To ensure logistics digitalization takes place in a sustainable fashion on both sides (productivity and human resources), we propose a blend of measures on different timescales, intended to integrate employees into digitalization efforts as soon as possible to ensure both efficient use and preempt inadequate workload developments soon. Further, efficiency of use and prevention of psychological harm are covered with training workshops and risk assessments. The latter is conducted twice at least: during the digitalization process, that is, before a measure takes full effect and is to be used widely. Then, after particular processes have undergone digitalization to their full extent, another risk assessment will be scheduled to aid in prevention of long-term psychological risks and gain data on significant changes in the type of workload and its requirements induced by digitalization (Rintala and Suolanen, 2005; Vuori, Helander and Okkonen, 2019).Thus we will be able to systematically describe areas of potential for individual and corporate benefit and changes to be expected, induced by digitalization and constituting 'work 4.0' in logistics and production.

The specific research questions for this article are: Given that digitalization is viewed to induce substantial changes to logistics work environments (added complexity, density, collaboration etc.) and the corresponding employee capabilities (e.g. skills, knowledge, abilities, motivation), which conceptual recommendations can be given for the design of (team-) work and process design? The concept of strategic cognitive readiness, originating from a military context, subsumes important psychological components, many of which are likely to be either required or affected by digitalized Work 4.0 environments. We thus analyze the concept and explore its applicability in a company case study regarding integrative process analysis of a proposal preparation process for a German power plant component company operating in an efficient lot size one environment. Thus, further research questions for this article are: Which process steps/elements can be modified by digitalization in a way that employees are involved in the development of digitalization measures themselves? How does the concept of strategic cognitive readiness inform the participation of logistics employees in digitalization efforts? These recommendations will be given of a real industry case involving complex, labor-intensive and communication-heavy tasks with, and this merits emphasis, iteration factor one, meaning project structures largely eluding comprehensive standard procedures.

The remainder of the article is structured as follows: Section 2 defines 'strategic cognitive readiness' and gives an overview of its role in an automation and digitalization context. Relevant issues for digitalization and work 4.0 are derived. Section 3 elaborates on the method used to elicit areas of potential and significant process gaps. This method (scoping with annotations, interaction room (Book, Gruhn and Striemer, 2017)) involves relevant employees in a bottom-up process of identifying shortcomings in given company processes. This serves e.g. to develop and implement new software such as idea management apps with participation of the very end-users from a conceptual phase onward, thus maximizing usage likelihood. Further, this section presents an implementation roadmap for digitalization measures containing psychological risk-assessments during the initial phases of requirements definition as well as after implementation. This is meant to ensure efficient use of measures by employees and maintenance of psychological health in view of risks identified (i) during an initial interview phase with employees and (i) drawn from the literature. The specific company case presented in section four draws from both (i) and (ii) as well as the annotation method cited above. Section five concludes.

As we will be working towards a specific applied case to appear further below, we will assume throughout this contribution that the process a company will observe is complex in the sense that no efficiency can be gained would any single stakeholder override team decisions with their respective specialized knowledge, contracts are fixed before all make-or-buy decisions can be made (e.g. for manufacturing) and repetition factor is one, thus generally limiting the accuracy of history-based cost calculations.

2 'Cognitive Readiness' and its Role in Logistics Automation and Digitalization

'Cognitive readiness' has been used to refer to a concept from a military context (Walsh and Shingledecker, 2006). One working definition used by the Institute for Defense Analyses reads as "Cognitive readiness is the mental preparation (including skills, knowledge, abilities, motivations, and personal dispositions) an individual needs to establish and sustain competent performance in the complex and unpredictable environment of modern military operations" (Morrison and Fletcher, 2002). The authors identify ten psychological components for the concept in its original context: Situation awareness, memory, transfer of training, metacognition, automaticity, problem solving, decision-making, mental flexibility and creativity, leadership, and emotion. Decision-making, for instance, is there defined as the act of choosing a tactic or strategy, often primed by patterns learned beforehand. This is quite central to the understanding of cognitive readiness we will pursue, as it underlines that some aspects of this quality are accessible to systematic training while others are inherently individual.

Readiness is defined via distinction from effectiveness. The latter refers (in the current context as well as in the original military terminology) to some evaluation of an outcome of an individual's or team's effort towards an operational goal. The former is a predictor or predictive measure of effectiveness. Cognitive readiness is further determined by cognitive variables such as intelligence and personality, motivation, beliefs etc. Given conditions like or analogous to complexity, dynamics and scarcity of resources and strong time constraints, one may accept that cognitive readiness as a requirement applies to any individual or team having to react quickly to unpredictable and rapidly emerging issues (Morrison and Fletcher, 2002). Different accounts on the term cognitive readiness take into account time-scales and planning horizons, leading, e.g., to differentiations such as operational - tactical - strategic cognitive readiness (Grier, 2012)). While the above definition by Fletcher corresponds to operational cognitive readiness, strategic cognitive readiness as defined for a military context by Grier (2012) appears most fitting for the treatment of (logistics) work environments since it adequately captures the quality of most tasks and their time horizon: "An individual's potential to perform assigned planning and organizational duties in the complex and unpredictable environment of military operations" (p.404). We explore the concept for a logistics work context, the military connotation does not apply here. Still, conditions necessary to enable employees to remain in the state of strategic cognitive readiness are potentially affected by pervasive digitalization measures. Further, efficient use, monitoring, and adaptation of these pervasive digitalization measures requires just the kind of optimal cognitive performance described by strategic cognitive readiness. In a sense, it is a reverse and preventive operating concept to organizational blindness/myopia (Knudsen, 2011).

2.1 Strategic Cognitive Readiness as a Work-related Concept

For any work context that requires prompt and adequate reaction to some emerging challenge, one may define prevention measures as training requirements. This applies even more to slightly less dramatic but usually more common scenarios where both non-routine situations occur and these may be characterized as abnormal, thus eluding any sole reliance on rule-based behavior (Sklet, 2006; Kluge, 2014). Opposed to the latter is knowledge-based behavior. This adds a less volatile dimension to the concept of cognitive readiness in a work context: Critical thinking and decision making are universal abilities which can be (recurrently) trained to a degree such that skill may be executed by individuals or teams under pressure, facing abnormal issues (Kluge, 2014). For many logistics and production work cases, time surely is essential. However, not so much on a scale of instances as in the military context but rather as one dimension of resource constraint. Still, for sophisticated tasks and complex operations, such as development of power plant components, the most adequate parts of heavily fragmented and widely distributed knowledge need to be both available and used. Classically, this would be a logistics problem in itself and digitalization is meant to dispose of that need exactly. Still, information quality and distribution can be critically influenced by quality of digitalization measures, e.g. as in our case, an efficient digital idea management for highly complex logistics and construction projects with iteration factor one. Kluge (2014) characterizes challenges posed by abnormal or non-routine situations as such that necessitate reliance on skills which remained unused for longer periods of time, and in the specific case of abnormal situations, such that are ambiguous and may contain major threats to a system while response time is limited. Extending these properties of cognitive readiness to requirements for a digitized system, e.g. intelligent digital idea management, the time dimension may lose some of its relevance with respect to the case we will treat further below. This, however, is ideally just due to formerly hybrid processes being seamlessly digitized. More essential for digitalization adding to a process with repetition factor one would be the timely availability of all relevant knowledge a company owns (e.g. construction specifications, granular time budgets) to gain most exact cost calculations, for instance. Work by Cummings, Gao and Thornburg (2016) or Thackray (1980), for instance, has identified numerous aspects of cognitive readiness and their interconnections. Those from automation and digitalization contexts specifically relevant to digital logistics work and employee training as discussed in this article are given in Table 1 along with representative sources. There (Table 1) we use terms and definitions as given in (Fletcher, 2004) and list literature which reports on these and analogously defined aspects relevant for digitalization contexts, respectively. In the case discussion in section four we relate workshop outcomes to these concepts to show worthwhile leverage points for employee training.

Table 1: Strategic cognitive readiness aspects relevant in automation and digitalization contexts

Components (Fletcher, 2004)	**Descriptions and sub-components**
Situation Awareness	Perception of environmental elements and events with respect to time or space, the comprehension of their meaning, and the projection of their future status (Cummings, Gao and Thornburg, 2016; Endsley, 2016; Gorman, et al., 2016)
Memory, Transfer of Training	Active, reconstructive ability to recall and recognize in the current operational situation patterns that will lead to likely solutions (Fletcher, 2004); ability to apply what is learned in one context to a different performance context. It can be measured by the ability to select and apply procedural knowledge gained in one context to another ("low-road" transfer) or by the application of principles abstracted from a set of contexts ("high-road" transfer) to another (Perkins and Salomon, 1992; Simoni and Bastian, 2018)
Metacognition	Metacognitive knowledge, regulation, experiences (Flavell, 1979); metacognitive awareness, including declarative, procedural, and conditional knowledge (Jacobs and Paris, 1987); metacognitive regulation skills (planning, monitoring, evaluating, (Schraw,

Components (Fletcher, 2004)	**Descriptions and sub-components**
	1998), problem solving, executive management and strategic knowledge (Zohar and Barzilai, 2013)
Automaticity	Ability to do things without occupying the mind with the low-level details required, allowing it to become an automatic response pattern (Bargh, Chen and Burrows, 1996; Bargh, 2014)
Problem Solving	Mental process: Use of generic or ad hoc methods in a structured manner aimed a solution of problems (Hayes, 1981; Schacter, Gilbert and Wegner, 2011, p.376)
Decision-Making	Cognitive process leading to the selection of a belief or a course of action among several alternative possibilities, might result in action, preference formation (Kahneman and Tversky, 1982) or choice from alternatives(Bellman and Zadeh, 1970; Bazerman and Moore, 2013)

2.2 Related Current Issues in Digital Logistics Work

Digital work developments in logistics and supply chain management encompass a multitude of factors, concepts and technologies. Major steps in this direction can be recognized in assembly and picking processes with the merger or formerly separated human and robot work areas. In current new settings, cobots work together in mixed teams with humans. At the same time, also artificial intelligence applications start to support blue- as well as white-collar workers. Many applications have been implemented like pick-by-voice or pick-by-light picking systems in intralogistics, as well as digital augmented systems in production or distribution logistics enabling new business model in platform economy systems (Klumpp and Ruiner, 2018a; Lang, et al., 2019). In such systems, the human worker is directed and steered by automated commands from the warehouse management or logistics software system in order to increase operational and picking efficiency. This brings new forms of capabilities and requirements for human workers, but also helps them to substitute physical with control tasks (Fletcher, et al., 2019). It also puts significant strain on the human cognitive system as not only a large number of commands by voice or digital light signals have to be computed and enacted, but also a semi-automated feedback mechanism has to be adhered to, in most cases scanning the picked goods with a barcode scanner in order to allow the computer system to document and control workers activities. In many cases, the question of human and cognitive workload has prevented the use of further automated systems as workers are not capable of dealing with too complex digital systems while implementing physical tasks in production and material handling. Furthermore, this leads to the question of a social sustainability of human work in logistics and supply chain management: If workers are not sufficiently motivated and integrated, the long-term success and integration of digital solutions in logistics work systems is endangered (Klumpp, Neukirchen and Koop, 2019; Klumpp and Zijm, 2019; Lang, et al., 2019; Oeshmy, et al., 2019).

Moreover, further digital developments are expected within the transportation sector. For example autonomous driving will bring deeply changed workplaces about – driving will no longer be a physical task in any sense it used to be like for the actual steering, shifting and navigation. Such tasks are automated and transferred to artificial intelligence entities. Instead, human operators will remain in the driver seat with supervision and strategy tasks like pilots are used to have since many decades (Demmel, et al., 2019; Fraedrich, et al., 2019; Klumpp, 2019). Therefore, situational awareness (traffic situation), customer contact and communication as well as strategic planning (platooning with other vehicles or safety preparation and inspection) will take over the working time of driving personnel instead of physical tasks. This will also require new competencies for the workers as for example technical insight, know-when (e.g. to interrupt automated systems) instead of know-how as well as customer and quality management approaches. This leads to the research and business practice requirement for logistics to develop measurement and mitigation concepts regarding mental workload in digital work settings (Arvan, et al., 2019; Emmanouilidis, et al., 2019). Furthermore, this might also open up questions of regulation and law making in the political arena in order to mitigate risks and fears connected to automation and AI applications in transportation and logistics (Klumpp and Ruiner, 2018b).

3 Materials and Methods

The case study has been conducted as an interactive process analysis in a German industry company (power plant components). The detailed methodology of workshops, annotations, interactive analysis and discussion is given below and in (Book, Grapenthin and Gruhn, 2014; Book, Gruhn and

Striemer, 2017). The literature studied for this article and specifically for the concept of strategic cognitive readiness (Fletcher, 2004; Grier, 2012) can be organized hierarchically, as a few widely accepted definitions have been presented. These refer to and consist of the components as listed in section 2, thus widely accepted definitions for these common components have been researched. Criteria for the inclusion of a definition have been number of citations and, if applicable, use in pertinent specialist dictionaries (e.g. VandenBos, 2015).

3.1 Interactive Process Analysis

The specific case analysis is outlined in section four. There we consider a German industry case to exemplify the concepts and ideas for early employee integration in digitalization projects. The company is a typical German machinery producer about 150 years old and connected to the electrical power generation plant industry. As specific components are constructed, produced and built into power plants worldwide, production and logistics processes are tailored towards an efficient lot size one environment. Regulation and service requirements are very high as also the nuclear power plant sector is addressed. In this section we consider one component of the interaction room method used to analyze a business process, in the case example preparation of an offer. Here, precision and completeness do not matter as much as arriving at a common understanding of desired typical procedures, data, interfaces etc. for all stakeholders. Company efforts regarding digitalization have been described along two dimensions as follows (Book, Gruhn and Striemer, 2017), p.64). It is helpful to keep these in mind when particular measures are being discussed further on, such as the concept of experimental spaces: Digital capabilities means that potential of digitalization is explored and researched systematically, resulting in a crisp strategy for areas/processes and ways in which digitalization investments will take place. This requires knowledge of products, distribution channels and detailed insights into customer demands. To be successful, a clear idea of available and useful technology and adequate contexts is indispensable, e.g. to automate interfaces and to integrate real-world objects. Leadership capabilities refers to knowledge and acceptance of the fact that new digitalization solutions have an emergent character. Necessarily, this knowledge extends to stakeholders regarding absence of precise predictability of IT projects. The important trade-off here is between keeping focus on the digitalization goals at hand and leaving space and means for experimentation.

The method we refer to (interaction room, (Book, Gruhn and Striemer, 2017)) offers procedures for all aspects of software development and digitalization, addressing, all stakeholders, objectives, interfaces etc. with a variety of "sub-rooms" and canvases, each of which can be put to use for single aspects such as objects, processes, integration etc. All these can be integrated into the whole comprehensive development endeavor, or, in the consulting-like context we use it here, particular "rooms" (such as the one for scoping) and canvases may be employed. We do so with adapting the process canvas to one particular company process. This represents the case study referred to earlier in this text. In section 4, this specific application will be detailed, while in this section a general description of the method is to follow. The whole interaction room idea is aptly described in (Book, Grapenthin and Gruhn, 2014; Book, Gruhn and Striemer, 2017) by its inventors. For the current purpose, we need not be concerned with the complete mapping procedure of a project since our company case presents us with an established process in need of improvement. Thus, in the language of the framework, a filled process canvas would be given already - we may assume a process map as given. The interaction room literature defines a sensible notation for the process canvas. However, in practice, one needs to consider conventions particular to the company case, if only for the sake of efficient communication. Thus the notation we give here will have been overridden in parts in the documentation of the company case in section four. One should keep in mind that this usually is a cooperative, interdisciplinary process in which a coach and a number of domain experts (e.g. controlling, finance, engineering, accounting, and marketing) partake. Because of this, adherence to UML-prescriptions (Fowler, 2010) always just follows intuitive modelling. Annotations are used to aid in expressing all matters that go beyond the limits of common process models. It is crucial to actually have stakeholders (usually employees) act this procedure out together by placing annotation symbols on a process map and discussing these subsequently. Crucial aspects such as value drivers, hidden cost potentials and risks will be specifically located on the process map and it is ensured that all stakeholders achieve common knowledge (Milgrom, 1981) on all critical issues. Additionally, the procedure reveals blind spots caused by complexity of a process. During a session, stakeholders receive physical symbols representing annotations to be placed on a process map visible to all participants. This way of adding natural language to a formal process model can help unravel a project's challenge and pitfalls early on instead of overlooking them with rigorous formalism. An annotation example would be a note attached on the process map with a "?"-symbol, representing uncertainty, which is categorized as a risk driver (contrasting e.g. value drivers). This annotation represents high risks for the current setup and its implementation and its analysis is prioritized. The ensuing discussion would have to address at least the questions for a definition of this source of uncertainty and required preventive action. The process analysis in section four illuminates both notation and the use/evaluation of annotations.

3.2 Digitalization Implementation with Psychological Risk-Assessment

If one wants to employ a risk assessment which ensures legal certainty, adhering to standards is a promising option. Using ISO 10075 as guidelines (Demerouti, et al., 2012), the following cornerstones for a risk assessment for psychological workload can be derived. There a tripartite structure is

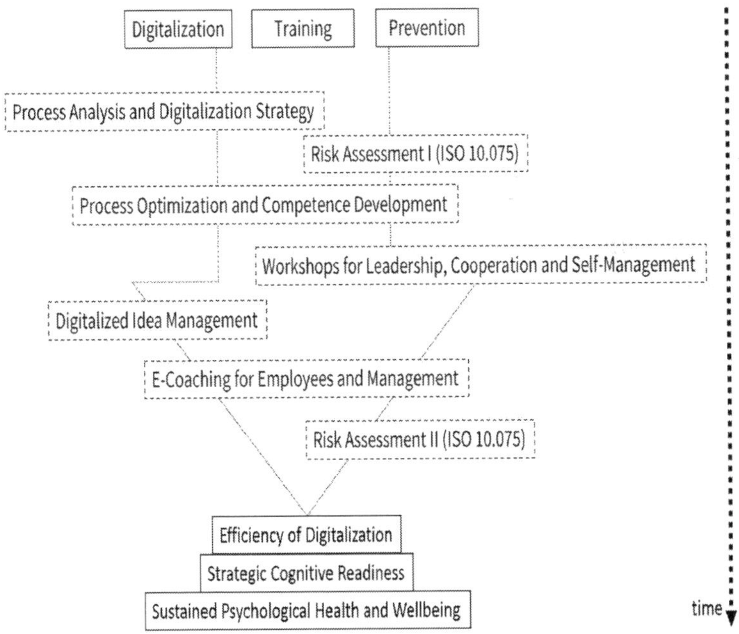

Figure 1: Analysis, optimization and evaluation steps scheduled for the research project; risk assessments (I, II) may be iterated.

given for measurements of work load, with the aim of considering aspects affecting health, safety, wellbeing, and productivity.

Basically, psychological stress and strain in risk assessments of digitized jobs have so far been insufficiently addressed. Although there is a standard for measuring mental workload (ISO 10.075), there is currently no adequate and proven instrument for measuring mental workload at the workplace, taking into account the specifics of digitized work. Therefore, a cross-sector requirement analysis tool for measuring mental stress and strain in digitized work systems should be used for practical application. In the following, this and the digitized business processes to be observed here are to be outlined (Figure 1 shows a chart of the procedure). The steps bracketed by the risk assessments are considered as an iterative procedure, that is, they can be run through several times as needed.

The core idea with the aim of understanding the specific situation in a single company is that of a "laboratory idea" or “experimental space”. In particular, protected learning spaces, employees, together with executives and works councils as experts in an integrated project approach, are actively involved in the development of digital design solutions. The variety of perspectives and the combination of top-down and bottom-up approaches should ensure greater efficiency, effect, and acceptance of the solutions.

4 Digital Logistics and Employee Training - Case

4.1 Proposal Preparation and Core Sub-Processes

The overarching goal for the case research activities is one particular value creation process, here it is preparation of an offer regarding a product. Leveraging digitalization for improved pricing implies improvements in the quality of calculations and pricing precision. Further, it means making better use of existing (decentralized) knowledge. Three core tasks have been identified to this end, each to be mapped to its individual effects and their contribution to the overall goal.

Table 2: Tasks and Desired Effects in the Observed Offer Preparation Process

Value Creation Process: Pricing/Offer Preparation (Lot Size One)	
Task	*Effect*
Digitalization and indexing of archives, integrative process analysis	Digital availability of experience-based knowledge, e.g. construction documents
Digitized time and attendance recording in construction	Transparency with mobile and granular recording, precise time budget allocation
Upgrades to idea management, including interfaces	Efficient idea management through integration, improved error culture

4.2 Annotations and Target-Performance Comparison

The most basic assumption is the trade-off between the optimal realization of a complex system involving human-computer interaction and efficient allocation of resources, including time. A comprehensive understanding of sufficiently complex matters is not attainable with maximum efficiency. This plain fact calls for abstraction and decision criteria for retention and omission of elements deemed important/negligible, respectively. The procedure we chose to employ is oriented towards value generation, which explains why the case described in this article focuses on sufficiently value-adding processes (Bergman, King and Lyytinen, 2002; Boehm and Huang, 2003; Grapenthin, et al., 2013). The crucial aim is to have relevant stakeholders arrive at a common understanding of a process and its identified shortcomings. There exist obvious and less obvious hindrances to this: These can originate from specializations and individual qualification, variations in documentation style between departments or inefficient prioritization of critical issues. Some of this can be explained by inherent redundancies and ambiguities of natural language: Here the trade-off is between the inefficiencies of natural language - available to all stakeholders - and formal, specialized languages (Pohl and Rupp, 2009; Book, Grapenthin and Gruhn, 2012; 2014). A systematic categorization of common understanding (explicit vs. implicit, true vs. false) has been given by Fricker and Glinz (2013). Table 3 highlights the categories of understanding that shall be discerned during the procedure.

During the process analysis workshop, annotations for the following issues were used: Value drivers, complexity, uncertainty, automation, dependence, manual task, need for training, media disruption. The remainder of this section will elaborate on each annotation (numbered 1-21) from the workshop session. The process is centered on a (desired) software solution for a calculation loop ('tool' in the following) that would ideally contain interfaces to customers, engineering, work preparation, ERP, and controlling.

Table 3: Categories of Common Understanding (implicit/explicit, relevant/irrelevant, true/false; Fricker and Glinz, 2013)

	Categories of common understanding	
Relevant, but Unknown/Unnoticed	Implicit, True, Relevant	Explicit, True, Relevant
	Implicit, False, Relevant	Explicit, False, Relevant
	Implicit, false, Irrelevant	Explicit, False, Irrelevant

Descriptions of exemplary improvement potentials are given in the following for each annotation in order to show the power of the implemented annotation method in digitalization projects: (1), manual task: According to employees in customer service, requests need to be handled since people need to interact, e.g. to unravel implicit requirements. (2), manual task: feedback channel, e.g. for completed projects, evaluation requires expert knowledge. (3), media disruption for historic reasons. (4), dependence: Mandatory verification of data generated by employees from assembly, complicated interface between current tool and proposal. (5), value driver: Calculation is a major value driver in general. Mutual dependence between flexibility and precision; major downside risk associated with design errors. (6), uncertainty: Lot size one environment limits the use of historic data. Company uses historically 'grown' database. (7), automation: Potential for automation, currently several interfaces between different types of software. (8), automation and media disruption: media disruption exists for feedback from engineering to tool, manual exchange of Excel-sheets. (9), need for training: Engineering employees need training to be able to use different file formats. (10), automation: Several different systems in use, fragmented across departments - employees express desire to consolidate this. The general link with strategic cognitive readiness is explained below. For example, the following process steps 11-19 require situation awareness on the assembly level (blue collar employees) and the component metacognition, specifically metacognitive knowledge, declarative and procedural knowledge on behalf of the employees in administration and engineering: (11), media disruption and automation: Interface between tool and work preparation does not exist. Manual transfer of data from Excel sheets to Access database. Reliance on historical data for scheduling. (12), uncertainty: Sub-processes in work preparation is not transparent, record of processes is fragmented rather than comprehensive, as is use of tools/software. (13), value drivers: Proposed digital tool would lower fragmentation of processes and increase transparency of work preparation, thus increasing adherence to schedules and quality. (14), need for training, to enable de-fragmentation of sub-processes and tool usage. (15), value drivers and automation: Time recording is not automated, which could increase precision decisively. (16), manual task: Assembly consist of diverse manual tasks which are mapped to time budgets via manual time recording using barcode scanning. (17 and 18), uncertainty: Mapping and recording of time budgets is currently highly non-transparent and imprecise. While product data sheets and schedules are extremely detailed, time recording is currently much less precise. (19), value driver: Which at-tributes do need recording, regarding their contribution to revenue generation? (20 and 21), uncertainty, dependence: Currently, no active cost monitoring during a project; dependence between tool and controlling.

All of these sites of digitalization potential have been unraveled during workshops with employees involved in the process by having them freely annotate process steps with very limited number of annotations. Beyond the detailed suggestions, it is evident that knowledge of this central process has been fragmented and explicit awareness of the listed shortcomings and potential risks was not given before the annotation procedure. In fact, it took some effort to get these employees to actually conduct the annotation process, as everyone was convinced of the sufficiency of their knowledge and efficiency of the current process (Hart, 1991). This highlights the need for training regarding awareness of such shortcomings and, ideally, aiming at enabling employees to conduct this type of inquiry independently. With reference to components of strategic cognitive readiness, situation awareness, while being more aptly replaced by sense making as requirement for the process analysis, is part of the requirements for assembly workers, whose activities are mapped to annotations 11-19. Thus, the component metacognition, and more specifically metacognitive knowledge, declarative and procedural knowledge with respect to sub-processes in assembly work and with respect to the whole process at the abstract level discussed in the workshop have been a main requirement. The annotation process itself requires individuals to plan, monitor, evaluate and revise their own thinking processes and products (Hartman, 2013). Automaticity, that is, the ability to do things without occupying the mind with the low-level details required, allowing it to become an automatic response pattern (Bargh, 2014), is connected to this in two ways: First, participants needed to be aware of assembly processes taking place in this manner to evaluate, which of these have potential for automation. Second, automaticity of social behavior (self-reported) played a role in participants' willingness to consider taking part in the process analysis a valuable activity. Problem solving, surely, is a requirement of the whole process analysis, while more specifically, the generic method (annotations and structured discussion) applied by participants to the problem (process analysis). For the choice of possible process improvements and their ranking, individual choices and individual and collective preferences had to be formed (decision-making). Specifically, decision-making procedures inherent to the annotation analysis or implied by it (e.g. the suggestive character of the annotation 'automation') needed to be made explicit in the discussion ensuing the annotation procedure, to achieve a preference order of process improvement measures supported by all stakeholders.

4.3 Synthesis: The Role of Employee Training in Fostering Strategic Cognitive Readiness

Necessary conditions to enable employees to remain in the state of strategic cognitive readiness (Grier, 2012) are affected by pervasive digitalization, and efficient use of digitalization measures requires just the kind of optimal cognitive performance described by strategic cognitive readiness. It is sensible to integrate efforts toward development and sustenance of it as soon as possible on a digitalization project timeline, preferably beginning with conceptualization and requirements definition phases, before even a single line of code has been written. Employee training ensures sustainability of the described state, and psychological risk assessments add to this as well as contributing to avoidance of adverse effects of digitalization on employee health and well-being. Idea Management (Oldham and Cummings, 1996) intersects with the process discussed in the case example and appears to be a good leverage point for employee-centered digitalization since its processes embody a company's error culture, and, by extension, its communication culture as a whole (Hand, 2016; Crama, Sting and Wu, 2019).

5 Discussion and Outlook

Since only one case study was implemented, applicability and transferability of the findings are limited. However, in viewing the annotations approach to digitalization process analysis in light of the strategic cognitive readiness requirements, a few general recommendations for logistics employee training can be derived: It is advisable to involve employees in (bottom-up) digitalization of processes as early on as possible to gain acceptance and an apt picture of implicit and explicit requirements (Book, Grapenthin and Gruhn, 2014). It is recommended to train employees in the annotations method in order to enable companies to iteratively assess and maintain requirements and process knowledge. Thus to this point in the research project, triangulation with the practical results is still insufficient and it will be crucial to gather further data from additional case studies in order to establish triangulation (Mangan, Lalwani and Gardner, 2004). This will take place over the course of the research project DIAMANT (2018-2021) when more company cases are examined and iterations of case studies like the one presented in this article take place. Applicability and transferability of current findings is quite limited, which is also due to limited ability of the authors to disclose company information. A more detailed process description, including a detailed process diagram would likely make the company recognizable as their (iteration factor one) product is quite unique. Later publications further down the project timeline are likely to present more comprehensive and detailed results.

Financial Disclosure

We gratefully acknowledge funding from the German Federal Ministry of Labor and Social Affairs (DIAMANT, EXP.00.00014.18).

References

ACM (Hg.), 2012. Proceedings of the ACM SIGSOFT 20th International Symposium on the Foundations of Software Engi- neering (FSE 2012).

Arvan, M.; Fahimnia, B.; Reisi, M.; Siemsen, E., 2019. Integrating human judgement into quantitative forecasting methods: A review. In: Omega 86, S. 237–252.

Bainbridge, L. (1983. Ironies of Automation. In: Automatica 19 (6), pp. 775–779.

Bargh, J. A., 2014. The four horsemen of automaticity: Intention, awareness, efficiency, and control as separate issues. In: Robert S. Wyer, JR. und Thomas K. Srull (Hg.): Handbook of Social Cognition, Second Edition. Volume 1: Basic Processes Volume 2: Applications. 2nd ed. Hoboken: Taylor and Francis.

Bargh, John A.; Chen, Mark; Burrows, Lara, 1996. Automaticity of social behavior. Direct effects of trait construct and stereotype activation on action. In: Journal of Personality and Social Psychology 71 (2), pp. 230–244. DOI: 10.1037/0022-3514.71.2.230.

Bazerman, Max H.; Moore, Don A., 2013. Judgment in managerial decision making. 8. ed. Hoboken, NJ: Wiley.

Bellman, R. E.; Zadeh, L. A., 1970. Decision-Making in a Fuzzy Environment. In: Management Science 17 (4), B-141-B-164. DOI: 10.1287/mnsc.17.4.B141.

Bergman, M.; King, J. L.; Lyytinen, K., 2002. Large-Scale Requirements Analysis Revisited: The need for Understanding the Political Ecology of Requirements Engineering. In: Requirements Engineering 7 (3), pp. 152–171.

Boehm, B.; Huang, L. G., 2003. Value-Based Software Engineering: A Case Study. In: Computer 36 (3), pp. 33–41.

Book, M.; Grapenthin, S.; Gruhn, V., 2012. Seeing the Forest and the Trees: Focusing Team Interaction on Value and Effort Drivers. In: ACM (Hg.): Proceedings of the ACM SIGSOFT 20th International Symposium on the Foundations of Software Engi- neering (FSE 2012), pp. 1–4.

Book, M.; Grapenthin, S.; Gruhn, V., 2014. Highlighting Value and Effort Drivers Early in Business and System Models. In: H. Fujita und A. Selamat (Hg.): Intelligent Software Methodologies, Tools and Techniques: 13th International Conference, SoMeT 2014. Cham: Springer, pp. 211–222.

Book, M.; Gruhn, V.; Striemer, R., 2017. Erfolgreiche agile Projekte: Pragmatische Kooperation und faires Contracting. Wiesbaden: Springer Vieweg.

Book, Matthias; Gruhn, Volker; Striemer, Rüdiger, 2017). Erfolgreiche agile Projekte. Pragmatische Kooperation und faires Contracting. Berlin: Springer Vieweg (Xpert.press).

Cooke, N.; Cummings, M. L.; Hancock, P.; Marras, W.; Warm, J, 2010. Bored beyond belief: How automation is boring us to distraction. In: Proceeedings of the Human Factors and Ergonomics Society 54th Annual Meeting, pp. 758.

Crama, Pascale; Sting, Fabian J.; Wu, Yaozhong, 2019. Encouraging Help Across Projects. In: Management Science 65 (3), pp. 1408–1429. DOI: 10.1287/mnsc.2017.2988.

Csikszentmihalyi, M., 2000. Happines, Flow, and Economics Equality. In: American Psychologist 55 (10), pp. 1163–1164.

Cummings, M. L.; Gao, F.; Thornburg, K. M., 2016. Boredom in the Workplace: A New Look at an Old Problem. In: Human Factors 58 (2), pp. 279–300.

Demerouti, E.; Fergen, A.; Glaser, J.; Herbig, B.; Hofmann, A.; Nachreiner, F. et al. 2012. Psychische Belastung und Beanspruchung am Arbeitsplatz : inklusive DIN EN ISO 10075-1 bis -3 // Psychische Belastung und Beanspruchung am Arbeitsplatz. Inklusive DIN EN ISO 10075-1 bis -3. Unter Mitarbeit von Andrea Fergen. 1. Auflage. Berlin: Beuth; Beuth Verlag GmbH (Forum Arbeitsschutz).

Demmel, S.; Gruyer, D.; Kurkhardt, J. M.; Glaser, S.; Larue, G.; Orfila, O.; Rakotonirainy, A., 2019. Global risk assessment in an autonomous driving context: Impact on both the car and the driver. In: IFAC-Papers 51 (34), pp. 390–395.

Emmanouilidis, C.; Pistofidis, P.; Bertoncelj, L.; Katsouros, V.; Fournaris, A.; Koulamas, C., 2019. Enabling the human in the loop: Linked data and knowledge in industrial cyber-physical systems // Enabling the human in the loop. Linked data and knowledge in industrial cyber-physical systems. In: Annual Reviews in Control. DOI: 10.1016/j.arcontrol.2019.03.004.

Endsley, Mica R., 2016. Toward a Theory of Situation Awareness in Dynamic Systems. In: Hum Factors 37 (1), pp. 32–64. DOI: 10.1518/001872095779049543.

Ewing, K. C.; Fairclough, S. H, 2010. The effect of an extrinsic incentive on psychophysiological measures of mental effort and motivational disposition when task demand is varied. In: Proceeedings of the Human Factors and Ergonomics Society 54th Annual Meeting, pp. 259–263.

Flavell, John H., 1979. Metacognition and cognitive monitoring. A new area of cognitive-developmental inquiry. In: American Psychologist 34 (10), pp. 906–911. DOI: 10.1037/0003-066X.34.10.906.

Fletcher, J. D., 2004. Cognitive Readiness: Preparing for the Unexpected. Institute for Defense Analyses. Alexandria, Virginia. Online verfügbar unter https://apps.dtic.mil/docs/citations/ADA458683, zuletzt geprüft am 10.05.2019.

Fletcher, Sarah R.; Johnson, Teegan; Adlon, Tobias; Larreina, Jon; Casla, Patricia; Parigot, Laure et al., 2019. Adaptive automation assembly. Identifying system requirements for technical efficiency and worker satisfaction. In: Computers & Industrial Engineering. DOI: 10.1016/j.cie.2019.03.036.

Fowler, Martin, 2010. UML distilled. A brief guide to the standard object modeling language. 3. ed, 16. printing. Boston, MA: Addison-Wesley.

Fraedrich, E.; Heinrichs, D.; Bahamonde-Birke, F. J.; Cyganski, R., 2019. Autonomous driving, the built environment and policy implications. In: Transportation Research Part A: Policy and Practice 122, pp. 162–172.

Frey, Carl Benedikt; Osborne, Michael A., 2017. The future of employment. How susceptible are jobs to computerisation? In: Technological Forecasting and Social Change 114, pp. 254–280. DOI: 10.1016/j.techfore.2016.08.019.

Freytag, M.; Kotzab, H.; Pannek, J. (Hg.), 2017). Dynamics in Logistics. Berlin: Springer.

Fricker, Samuel; Glinz, Martin, 2013. On shared understanding in software engineering. DOI: 10.5167/uzh-88170.

Fujita, H.; Selamat, A. (Hg.), 2014. Intelligent Software Methodologies, Tools and Techniques: 13th International Conference, SoMeT 2014. Cham: Springer.

Gorman, Jamie C.; Cooke, Nancy J.; Pedersen, Harry K.; Winner, Jennifer; Andrews, Dee; Amazeen, Polemnia G. (2016. Changes in Team Composition after a Break. Building Adapative Command-and-Control Teams. In: Proceedings of the Human Factors and Ergonomics Society Annual Meeting 50 (3), pp. 487–491. DOI: 10.1177/154193120605000358.

Grapenthin, S.; Book, M.; Gruhn, V.; Schneider, C.; Völker, K., 2013. Reducing complexity using an interaction room – An experience report. In: ACM Special Interest Group on the Design of Communication Conference (SIGDOC 2013).

Grier, R. A., 2012. Cognitive Readiness at the Tactical Level: A Review of Measures. In: Proceeedings of the Human Factors and Ergonomics Society 55th Annual Meeting, pp. 404–408.

Guznov, S.; Matthews, G.; Warm, J.: Team member personality, performance, and stress in Roboflag synthetic task environment. In: Proceeedings of the Human Factors and Ergonomics Society 54th Annual Meeting, pp. 1679–1683.

Hancock, P. A.; Warm, J. S.: A dynamic model of stress and sustained attention. In: Human Factors 31, pp. 519–537.

Hand, Martin (2016. Making Digital Cultures: Routledge.

Hart, Paul't, 1991. Irving L. Janis' Victims of Groupthink. In: Political Psychology 12 (2), pp. 247. DOI: 10.2307/3791464.

Hartman, Hope J., 2013. Developing Students' Metacognitive Knowledge and Skills. In: Hope J. Hartman (Hg.): Metacognition in learning and instruction. Theory, research, and practice. Dordrecht, Boston: Kluwer Academic Publishers (Neuropsychology and cognition, v. 19), pp. 33–68.

Hartman, Hope J. (Hg.), 2013. Metacognition in learning and instruction. Theory, research, and practice. Dordrecht, Boston: Kluwer Academic Publishers (Neuropsychology and cognition, v. 19).

Hayes, J. R., 1981. The Complete Problem Solver. Philadelphia, PA: The Franklin Institute.

Jacobs, Janis E.; Paris, Scott G., 1987. Children's Metacognition About Reading. Issues in Definition, Measurement, and Instruction. In: Educational Psychologist 22 (3-4), pp. 255–278. DOI: 10.1080/00461520.1987.9653052.

Kahneman, Daniel; Tversky, Amos, 1982. The Psychology of Preferences. In: Sci Am 246 (1), pp. 160–173. DOI: 10.1038/scientificamerican0182-160.

Kluge, A., 2014. The acquisition of knowledge and skills for taskwork and teamwork to control complex technical systems. A cognitive and macroergonomics Perspective. Dortrecht: Springer.

Klumpp, M., 2018. Automation and artificial intelligence in business logistics systems: human reactions and collaboration requirements. In: International Journal of Logistics Research and Applications 21, pp. 224–242.

Klumpp, M., 2019. Artificial Intelligence Applications. In: W. H. M. Zijm, M. Klumpp, A. Regattieri, S. Heragu, Henk Zijm, Matthias Klumpp und Sunderesh S. Heragu (Hg.): Operations, Logistics and Supply Chain Management. Lecture Notes in Logistics. // Operations, logistics and supply chain management. Cham: Springer (Lecture Notes in Logistics book series), pp. 637–662.

Klumpp, M.; Neukirchen, T.; Koop, W., 2019. Mobile Learning and Human-Artificial Cooperation in Logistics. In: H. Proff (Hg.): MOBILITÄT IN ZEITEN DER VERÄNDERUNG. Vergangenheit gegenwart zukunft. [S.l.]: GABLER, pp. 371–382.

Klumpp, M.; Ruiner, C., 2018. Digitalization and Work Organization in New Urban Food Delivery Systems. In: International Journal on Food System Dynamics 9 (5), pp. 399–408.

Klumpp, M.; Ruiner, C., 2018. Regulation for Artificial Intelligence and Robotics in Transportation, Logistics, and Supply Chain Management: Background and Developments. In: Network Industries Quarterly 20 (2), pp. 3–7.

Klumpp, Matthias; Zijm, Henk, 2019. Logistics Innovation and Social Sustainability. How to Prevent an Artificial Divide in Human-Computer Interaction. In: J Bus Logist 29 (7), p. 733. DOI: 10.1111/jbl.12198.

Knudsen, Morten, 2011. Forms of Inattentiveness. The Production of Blindness in the Development of a Technology for the Observation of Quality in Health Services. In: Organization Studies 32, 7), pp. 963–989. DOI: 10.1177/0170840611410827.

Lang, S.; Kota, M. S. S. D.; Weigert, D.; Behrendt, F., 2019. Mixed reality in production and logistics: discussion the application potentials of Microsoft HoloLens. In: Procedia Computer Science, 149), pp. 118–129.

Lucke, U.; Rensing, C.: A Survey on Pervasive Education. In: Pervasive and Mobile Computing. Special Issue on Pervasive Education, Elsevier.

Mangan, John; Lalwani, Chandra; Gardner, Bernard, 2004. Combining quantitative and qualitative methodologies in logistics research. In: Int Jnl Phys Dist & Log Manage 34 (7), pp. 565–578. DOI: 10.1108/09600030410552258.

Milgrom, Paul, 1981. An Axiomatic Characterization of Common Knowledge. In: Econometrica 49 (1), p. 219. DOI: 10.2307/1911137.

Morrison, J. E.; Fletcher, J. D., 2002. Cognitive Readiness. Institute for Defense Analyses. Alexandria, Virginia (IDA Paper, P-3735), zuletzt geprüft am 30.04.2019.

Oeshmy, D. E.; Makdisi, J.; da Silva, E. H. D. R.; Angelis, J., 2019. Managing human errors: Augmented reality systems as a tool in the quality journey. In: Procedia Manufacturing (28), p. 24–30.

Oldham, G. R.; Cummings, A., 1996. EMPLOYEE CREATIVITY. PERSONAL AND CONTEXTUAL FACTORS AT WORK. In: Academy of Management Journal 39 (3), pp. 607–634. DOI: 10.2307/256657.

Paulaus, M. P; Potterat, E. G.; Tyler, M. K.; VanOrden, K. F.; Bauman, J.; Momen, N.; et al., 7. A neuroscience approch to optimizing brain resources for human performnace in extreme environements. In: Neuroscience and Biobehavioral Reviews 33, pp. 1080–1088.

Perkins, D. N.; Salomon, G., 1992. Transfer of learning. In: International Encyclopedia of Education 2, pp. 6452–6457.

Petkovic, T.; Puljiz, D.; Markovic, I.; Hein, B., 2019. Human intention estimation based on hidden Markov model motion validation for safe flexible robotized warehouses. In: Robotics and Computer-Integrated Manufacturing (57), pp. 182–196.

Pohl, Klaus; Rupp, Chris, 2009. Basiswissen Requirements Engineering. Aus- und Weiterbildung zum "Certified Professional for Requirements Engineering" ; Foundation Level nach IREB-Standard. 1. Aufl. Heidelberg: dpunkt-Verl.

Proff, H. (Hg.), 2019. MOBILITÄT IN ZEITEN DER VERÄNDERUNG. Vergangenheit gegenwart zukunft. [S.l.]: GABLER.

Rintala, Niina; Suolanen, Sanna, 2005. The Implications of Digitalization for Job Descriptions, Competencies and the Quality of Working Life. In: Nordicom Review 26 (2), pp. 53–67. DOI: 10.1515/nor-2017-0258.

Schacter, Daniel L.; Gilbert, Daniel Todd; Wegner, Daniel M., 2011. Psychology. 2. ed. New York, NY: Worth.

Schraw, Gregory, 1998. Promoting general metacognitive awareness. In: Instructional Science 26 (1/2), pp. 113–125. DOI: 10.1023/A:1003044231033.

Simoni, Carla de; Bastian, Claudia C., 2018. Working memory updating and binding training: Bayesian evidence supporting the absence of transfer. In: Journal of experimental psychology. General 147 (6), pp. 829–858. DOI: 10.1037/xge0000453.

Sklet, S., 2006. Safety Barriers: Definition, Classification, and Performance. In: Journal of Loss Prevention in the Process Industries 19 (5), pp. 494–506.

Thackray, R. I., 1980. Boredom and Monotony as a Consequence of Automation. A Consideration of the Evidence Relating Boredom and Monotony to Stress. FAA Office of Aviation Medicine (FAA Office of Aviation Medicine Reports, FAA-AM-80-1).

VandenBos, Gary R., 2015. APA dictionary of psychology (2nd ed.). Washington: American Psychological Association.

Vuori, Vilma; Helander, Nina; Okkonen, Jussi, 2019. Digitalization in knowledge work. The dream of enhanced performance. In: Cogn Tech Work 21 (2), pp. 237–252. DOI: 10.1007/s10111-018-0501-3.

Walsh, W. J.; Shingledecker, C., 2006. An Analysis of a Joint and Expeditionary Mindset. United States Army Research Institute for the Behavioral and Social Sciences. Arlington, VA.

Wyer, Robert S., JR.; Srull, Thomas K. (Hg.), 2014. Handbook of Social Cognition, Second Edition. Volume 1: Basic Processes Volume 2: Applications. 2nd ed. Hoboken: Taylor and Francis.

Zijm, W. H. M.; Klumpp, M., 2017. Future Logistics. What to Expect, How to Adapt. In: M. Freytag, H. Kotzab und J. Pannek (Hg.): Dynamics in Logistics. Berlin: Springer, pp. 365–379.

Zijm, W. H. M.; Klumpp, M.; Regattieri, A.; Heragu, S.; Zijm, Henk; Klumpp, Matthias; Heragu, Sunderesh S. (Hg.), 2019. Operations, Logistics and Supply Chain Management. Lecture Notes in Logistics. // Operations, logistics and supply chain management. Cham: Springer (Lecture Notes in Logistics book series).

Zohar, Anat; Barzilai, Sarit, 2013. A review of research on metacognition in science education. Current and future directions. In: Studies in Science Education 49 (2), pp. 121–169. DOI: 10.1080/03057267.2013.847261.

SmartAirCargoTrailer – Autonomous Short Distance Transports in Air Cargo

Benjamin Bierwirth[1]*, Ulrich Schwanecke*[1]*, Thomas Gietzen*[1]*,*

Daniel Andrés Lopéz[1]*, Robert Brylka*[1]

1 – Hochschule RheinMain

Purpose: SmartAirCargoTrailers improve short-distance air cargo transports within airport premises. They create real-time-transparency between the partners in the transport chain and thereby reduces waiting times while at the same time increasing equipment utilization. An autonomous truck with swap bodies facilitates these short-distance transports.

Methodology: The SmartAirCargoTrailer system consist of an autonomous truck with swap trailers which is controlled by a cloud platform. To allow for real-time-transparency all shipments loaded onto the trailer are detected by a camera-based system.

Findings: To enable reliable scannings of AirwayBill numbers (AWBs) a system of multiple cameras was developed. Challenges arose from the big variationin shipment size and scanning in motion. While the truck could be triggered automatically based on time or filling level, also human interaction had to be integrated.

Originality: The system introduces a combined push-pull-algorithm to optimize the utilization of the autonomous trucks. The camera-based barcode scanning allows for shipment identification without interrupting the loading process. Autonomous driving in the mixed traffic environments of the landside airport premises is another innovative part of the project.

Keywords: Air Cargo, Camera-based Barcode Detection, Neural Networks, Autonomous Transports

First received: 19.May.2019 **Revised:** 27.May.2019 **Accepted:** 14.June.2019

1 Introduction

Air transport is an important part of global supply chains, particularly in view of decreasing lead times. Air cargo accounts for one third of the global cargo value exchange while at the same time representing less than 1% of the cargo volume (Shepherd, Shingal, and Ray, 2016). Global eCommerce is the main driving factor of future growth.[1] At present, the market for smaller packages and express shipments is being dominated by the leading Courier, Express and Parcel service providers (CEP). In an environment of increasing distance and decreasing lead time requirements, faster modes of transport have become more and more important. To meet those customer requirements and to better compete with CEP-providers, the industry, represented by the International Air Transport Association (IATA), aims at a significant acceleration of the overall transport time (IATA, 2018).

While increasing the speed of air transport seems to be neither economical nor technically unambitious, acceleration of the overall lead time within processes on the ground deserve all attention. Especially at hub airports, the links between continental forwarder networks and intercontinental airline networks provide potential for improvement. Typically, physical exchanges of shipments between forwarder hubs and air cargo handling facilities of designated airlines are executed via road transport. The covered distance is often less than a few kilometers. As a result, the process of delivering or receiving shipments between forwarder hub and air cargo handling facilities takes hours whereas the actual driving activity accounts for

[1] For further information on global supply chains and the impact of digitization on freight forwarding refer to Lehmacher, 2015 and Dietrich and Fiege, 2017.

a few minutes only. The purpose of the forwarder hub is to consolidate single shipments to achieve a better weight and volume mix and reduced kg-pricing for bigger shipments. As many shipments are last-minute-deliveries (or pick-ups from the shipper site) this consolidation is based on dynamic and experienced-based decision making. As the consolidated air cargo shipments have to be delivered prior to a Latest Acceptance Time (LAT) forwarders tend to deliver at the latest possible time to maximize the consolidation potential. However, in peak times the congestion of the air cargo handling facilities generates the decision problem regarding the right cut-off-time at the forwarder facility: Apart from the LAT and the transport distance the delay caused by the congestion has to be taken into account.

The aim of the SmartAirCargo trailer is to increase the transparency of the process by realizing a demand-driven transport concept. The autonomous transports are planned for August and September 2019. So far the basic functions and integration of camera-based barcode reading and the cloud platform are in workable conditions while the preparatory work on the autonomous transport is almost finished.

In the following, the paper gives a short review of relevant literature regarding air cargo processes, camera-based barcode identification, and autonomous transports (Chapter 2). Then, the air cargo supply chain is introduced in more detail (Chapter 3). Chapter 4 consists of the goals and a concept overview. Preliminary and expected results, but also limitations are detailed in chapter 5. At the end an outlook is given.

2 Literature Review

In this section we give a brief overview of the relevant and latest relevant publications While the section of air cargo provides an general overview of the area of application, the following sections highlight the current status in the area of autonomous vehicles in freight transport and camera-based barcode reading. The literature review on air cargo is the basis for the description of the air cargo supply chain and its challenges while camera-based barcode reading and autonomous vehicles are two central conceptual elements of the SmartAirCargoTrailer.

2.1 Air Cargo

Most publications related to air transport and airport operation focus either on passenger processes, airside ground handling or on network design. Feng et al. provide a literature review of air cargo operation studies (Feng, Li, and Shen, 2015). Lange has identified an increase in departure delays caused by cargo operations (Lange, 2019). Liu et al. specify this fact by developing a model to calculate costs of flight delays caused by late package deliveries (Liu, Yin, and M. Hansen, 2019). The airport classification defined by Mayer can be used to identify cargo hubs (Mayer, 2016). Selinka et al. provide an analytical solution for truck handling at air cargo terminals (Selinka, Franz, and Stolletz, 2016) while Azadian et al. formulate an algorithm to optimize air cargo pick-up and deliveries (Azadian, Murat, and Chinnam, 2017). Brandt and Nickel focus on the planning problem of air cargo (F. Brandt and Nickel, 2018). Bierwirth and Schocke have analyzed the status quo and challenges of digitization in the air cargo supply chain

(Bierwirth and Schocke, 2017). In summary, it can be stated that handling agents have to face various challenges heavily influencing costs and performance.

2.2 Autonomous Vehicles

In intralogistics, the use of AGVs (Automated Guide Vehicles) is common since decades (Schmidt, 2018 and Ullrich, n.d.). Applications in transport are rare as the use on public roads has some challenges regarding liability and regulations (Bartolini, Tettamanti, and Varga, 2017 and Hey, 2019). The AGVs of Container Terminal Altenwerder (CTA) in Hamburg are used for container transport between operate quay and staging area. Technology-wise they are centrally managed in a closed environment without other vehicles or humans beings around and they navigate with the help of transponders in the ground (Ranau, 2011). Today camera- or radar-based systems are applied and under further development (Dixon, 2018 and Meinel, 2018)

The interaction between human and AGVs or robots is under development. While research and experimental applications in warehouse and production environment focus on full automation (Bonini et al., 2015), flexible AGVs (Wurll, 2016) and interaction with workers (Neubauer and Schauer, 2017) and do not require fixed guiding infrastructures, the steering of autonomous trucks for road transport is more complex and therefore requires a different approach. Especially the navigation and docking or maneuvering of truck-trailer combinations is challenging.[2]

---

[2] For the docking process further information can be found in Clarembaux et al., 2016, David and Manivannan, 2014. For the special requirements on navigation refer to Stahn, Stark, and Stopp, 2007 and Kober, Huber, and Oberfell, 2017.

The current state and future visions for autonomous driving can be found in (Bäumler and Kotzab, 2017, Clausen and IVI, 2018 and Flämig, 2016). An overview of AGVs already applied can be found at (Kückelhaus, Zeiler, and Niezgoda, 2014). Pipp, Reiners, and Roesgen (2018) describes applications of autonomous driving in an airport environment. In (Parreira and Meech, 2011) autonomous haulage systems are discussed, but so far technical and regulative restrictions hinder the realisation. While platooning is conceptually designed (Kunze et al., 2011) and tested to increase efficiency on highways, autonomous trucks are deployed in on-site applications delivering goods from one building to another or in moving swap bodies (A. Brandt, 2018 or KAMAG, 2019).

2.3 Camera-based Barcode Reading

In general, camera-based barcode reading can be divided into two tasks: *detection* and *decoding*. Most work focuses on only one of these two tasks. For example, there are many commercial and free software libraries for decoding barcodes, such as *QS-Barcode SDK*, *QR-Code SDK*, *ZXing*, or *ZBar* to name a few. All these libraries can reliably decode a barcode if it can be seen in an image of sufficient quality, i.e. size, sharpness and orientation, otherwise they fail.

The size of the image area in which the barcode is visible is perhaps the most important aspect of camera-based barcode decoding. If the minimum thickness of a single bar of a barcode cannot be displayed correctly, the barcode cannot be decoded. Thus, a sufficient local resolution must be guaranteed by using a proper camera setup. Another important factor in camera-based barcode decoding, which can be corrected algorithmically to a certain extent, is the sharpness of the image containing the barcode. Beside defocusing, motion of the barcode recorded by the camera is an important reason for blur. The alignment of a captured barcode in an image is also critical, as most barcode decoding algorithms expect horizontally oriented barcodes, which cannot be guaranteed in practice. A reliable camera-based barcode reading system suitable for practical use must therefore be able to cope with the three problems just mentioned. Thus, beside reliable barcode localization, the deblurring of the image regions containing the barcodes and their proper horizontal alignment are important pre-processing steps before the actual decoding.

State of the art localization methods such as (Gallo and Manduchi, 2011, Sörös and Flörkemeier, 2013, Creusot and Munawar, 2015, Yun and Kim, 2017 or Namane and Arezki, 2017), detect 1D or 2D barcodes from blurry, low contrast and low resolution images. However, all these methods require that the barcodes are very present in the image, i.e. the distance between camera and barcode is small and the barcode is aligned almost parallel to the camera image. In addition, they focus only on localizing barcodes, but do not correct their orientation or sharpen the image, both of which, as we have seen, are important for decoding. Katona and Nyúl, 2013 developed an algorithm to address some of these drawbacks by using morphological operations, but also does not consider small resolutions. More recent approaches based on deep learning proposed a barcode localization pipeline (see D. K. Hansen et al., 2017 and Zhang et al., 2018). The authors used a specialized neural network to locate barcodes and then rotate them to align them horizontally. However, The proposed approaches do not cope with motion blur. In (Yahyanejad and Ström, 2010) the authors presented a method for deblurring barcode images that are mainly affected by translational motion blur. A more recent neural network-based approach to sharpen barcode images was presented by Kupyn et al., 2018.

3 Air Cargo Supply Chain

For this paper, we focus on the processes between forwarder and air cargo handling agents. For a better understanding of the overall context we describe a generic air cargo supply chain from shipper to consignee first (see also Figure 1). We then highlight the relevant processes and describe them in more detail.

The air cargo supply chain starts with a shipper, usually a company, as air cargo is mostly used in the Business-to-Business (B2B) context. Based on a contract between shipper and consignee a forwarder is contracted by one of the parties based on the INCOTERMS agreed. The shipment is picked up by the forwarder and brought to a local forwarder facilities. From there, it is consolidated with other air cargo shipments and transported to the forwarder hub at the airport. There, all air cargo shipments with the same destination (airport-wise) will be consolidated. Part of the consolidation process at the forwarder hub is also the booking of airline capacity. As air cargo pricing is varying and multiple carriers and routings are available for almost all destinations, bookings are done at the latest possible point – ideally when all cargo is known and already at the hub. Then, shipments will be labeled with the airway bill (AWB) number according to a standardized format defined by IATA.

The consolidated shipments will then be transported to the airline air cargo handling facility. Typically, only home-base carriers operate their own facilities while most airlines use handling agents offering their services to multiple airlines. Transport from forwarder to airline, therefore, typically consists of multiple shipments to multiple destinations of multiple airlines. It is the handling agent who receives and consolidates all shipments of all forwarders for one flight. This process is called ”build-up”. The handling agent has his own cut-off-time for the transport from the handling facility to the plane. Depending on the airport and permissions, this is done by the handling agent or a ground handling agent. The cargo is then transported to the destination, sometimes with an additional handling process at the airline hub. After arrival at the destination airport the cargo follows the same processes only in reverse order. Most important is the customs clearance process as the shipments can only be processed any further after clearance. Usually, the forwarder will focus on a limited number of urgent shipments to be picked up from the handling agent immediately after clearance while the rest of the shipments will be processed later.

The physical exchange of shipments is accompanied by the exchange of transport documents. Although IATA and many companies try to digitize documents only slightly more than 50% of all shipments are transported with an electronic AWB (IATA, 2019). The handling agents try to improve transparency by implementing truck slot booking systems where the incoming or outgoing AWBs have to be listed. But for short distance transports those slots are booked either with no AWB information or just-in-time.

For shipments requiring special handling such as Dangerous Goods, Pharmaceuticals or Perishables, additional physical and documentary processes apply. However, this lies outside the focus of this paper.

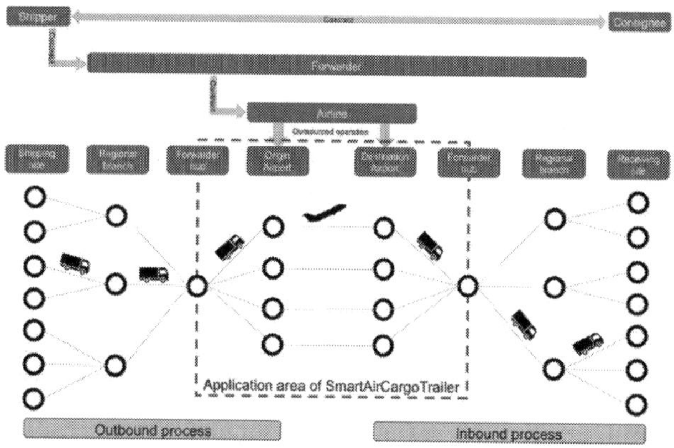

Figure 1: Simplified air cargo supply chain and application area of the SmartAirCargoTrailer.

4 Methodology

4.1 From Gap Analysis to Goals

In chapter 2, the main deficits in the air cargo supply chain have already been identified, so the main goal of the presented concept can be easily deducted as the optimization of short distance transports in the air cargo supply chain. In more detail, this goal can be broken down into several sub goals:

- To increase truck and truck driver utilization and load factors
- To optimize the planning processes and to avoid of peaks by increasing the transparency
- To reduce congestion at the handling agent facilities
- To simplify import processes

4.2 Concept Overview

The concept of the SmartAirCargoTrailer rests on the following three elements which are illustrated in Figure 2:

1. Air cargo shipments which are ready to be transported to the next supply chain partner (outbound: the handling agent; inbound: the forwarder) will be loaded directly onto a trailer. The loading or unloading will be surveyed and steered by a camera-based barcode reading system.
2. The information about the shipments loaded onto the trailer will be shared between the parties involved, typically the forwarder and the handling agent. A cloud-based platform is serving as the enabling technology.
3. Transports will be executed by using trailers or swap-bodies with an autonomous truck managed by the cloud platform.

The shipment identification process is based on the existing AWB-Labels which usually provide a 1D barcode. The use of more modern identification technologies such as Radio Frequency IDentification (RFID) would be preferable, but the global scope and the volatile network structures circumvent the introduction of RFID tags early in the supply chain. What is more important, as machinery parts belong to the most shipped category, are the difficulties with too much surrounding metal that would also lead to detection rates well below 100%, but would require an extra process (tagging each piece). Labeling the piece with the AWB number is always necessary as not all facilities around the globe would be equipped with RFID readers. As the AWB-labeling is done after the booking, which – as mentioned above – comes after virtually consolidating the shipments, this happens shortly before these are transported to the handling agent. Therefore, it can be assumed that most labels will be placed on the outside of a pallet and can be decoded by a camera from one side or the top. Instead of consolidating the shipments in the warehouse in a dedicated staging area, the individual shipments are loaded directly onto the trailer. When identified, the information is sent to the cloud-based platform. The camera system is designed so that it can be integrated into the trailer and is compatible with all warehouses and facilities. The cloud-based platform receives the information from the camera system. Additionally, the consolidation plan will be provided by the forwarder. The system has to provide information to the operative staff about the loaded shipments and the number of pieces and missing shipments or pieces. The transport itself can then be automatically triggered by multiple events, which will be detailed below. Using an autonomous truck is possible as the transport is done on-airport premises in an infrastructure which is known and where many of todays obstacles to autonomous driving like pedestrians or cyclists may rarely interfere. As the truck only moves the swap bodies or trailers, waiting times are irrelevant.

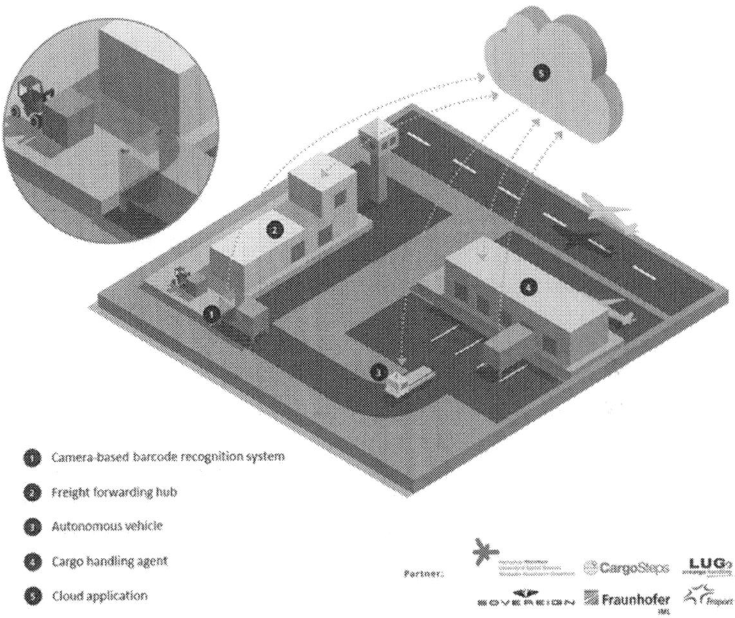

Figure 2: Elements and partners of the SmartAirCargoTrailer (Source: Sieke and Mehrtens, 2019).

5 Preliminary and Expected Results

As the testing of the autonomous vehicle will start this August, only the preparatory results are shown.

5.1 Process

As defined in the concept section, first the overall process with different use cases (or event-triggers) was developed. Up to 17 participants are involved in the air cargo supply chain from shipper to departure. At the beginning of a loading process, the transportation plan has to be uploaded to the system to ensure that the system can be identified correctly, i.e. intended AWB numbers with the right number of pieces as well as to identify and warn if wrong AWBs are loaded onto the trailer. To ensure compliance with safety and security regulation the swap body has to be closed and sealed before it is picked up by the AGV. After transport it has to be ensured that the swap body will be opened only after the AGV has finished its docking process. As for security regulation the intactness of the seal has to be checked.

Today's regulations require that a person presents the shipment related documents and that the identity and security status of this person, the driver, is verified. In the future, the projects propose the use of electronic seals which are also managed by the platform to lock the swap body or trailer and transport the documents inside. As the AGV and the status of the seal would be under 100% surveillance, as all motions would be tracked, this would enhance security. Any manipulation could be detected immediately.

With regard to triggering autonomous transports the following events have been defined:

- The basic push process is triggered by a full swap body. The loading status, or to be precise the empty loading meters, are monitored with an additional camera or an array of ultrasound detectors. To prevent a disruption of operation at the loading facility the AGV will be triggered before 100% capacity will be reached.
- Another trigger-event can be derived from the LAT of one of the shipments loaded. If one of the shipments comes close to its LAT,

the transport will be triggered and even if the loading is not complete, the operative staff would be requested to close the swap body. In this case a manual intervention – a rejection or approval – by the transport manager is necessary. The sending facility decides if they want to finish loading first and risk missing the LAT or if they would modify their transport plan.

- A pull request from the receiving side can be sent if the load of the shipment onto the trailer would be needed to optimize their workflow and processes although the trailer still has capacity left and also time-wise a transport is not necessary. This could be the case when heavy shipments which are handled first in the build-up process would be already processed at the forwarder, but as capacity in the trailer is still available and more shipments of the same forwarder are currently being handled in the forwarders facility, the forwarder intends not to transport those shipments. The handling agent on the other side would profit from those shipments as the build-up could start and other shipments which are at the handling agent's warehouse could be processed afterwards. Although this generates an additional transport, the overall process benefits from early delivery of the shipments (as build-up already started the peak load after LAT is significantly lower).
- A manual pull request could also be sent from the receiving side if it seems appropriate or beneficial to bring in available shipments. This could be triggered by free manpower or free truck docks at the receiving side, esp. if they expect upcoming peaks.
- For highly time-critical shipments such as urgently awaited imports an on-demand transport can be triggered automatically

once the shipment has been registered as being loaded onto the trailer.

- While the push of a trailer is triggered by reaching maximum capacity it could be that the cargo loaded is not time-critical and at the same time the receiving side is under peak load conditions or expects a peak soon. In this case, the receiving side could advise the AGV to transport the swap body to a staging area. As a consequence, the forwarder profits from being able to continue its operations while the handling agent can focus on time-critical shipments.
- To optimize work load, the transport manager of the handling agent could delay a transport that is generated by LAT if the manager decides that the shipments could be handled later, for example if the LAT-critical shipment is a small pieces which will be processed late in the build-up process. This would lead to fewer transports.

Generally speaking, to understand the roles and tasks of the different parties involved in transports seems to be of utmost importance:

- The operative warehouse staff should have a mobile device which shows them which shipments should be loaded, how many pieces have been loaded and wether any mistakes have occurred. They need options to modify the number of pieces detected (in case the barcodes of some pieces could not be read or any other detection error occurred).
- The transport manager is responsible for initially creating the transport list, but also for any updates and modifications. In case

a transport request is generated, the transport manager has the possibility to intervene. The transport manager could also send out a pull request.

- For the setup with multiple parties and AGVs an AGV manager is needed. The responsibilities of the AGV manager would be to monitor the AGV operation and solve conflicts, for example if not all transport requests could be served due to a limited number of operating AGVs.

5.2 Cloud-based Platform

The cloud-based platform links the elements of the SmartAirCargoTrailer concept, mainly the camera system, the autonomous vehicle and the user frontend. The architecture was developed based on the process and role models. The system is designed as a stand-alone solution, as previous experience with digitization and interfaces in the air cargo supply chain (see Bierwirth and Schocke, 2017 for details) showed that IT integration would be challenging and time-consuming. The web-based frontends would have to run on mobile devices as well as on stationary computers. While the camera system should provide AWB numbers and the responding number of pieces, the interface with the autonomous vehicle has to exchange status information as well as transport orders. For export or outbound shipments a transport plan (according to the consolidation) has to be provided by the forwarder.

The core of the platform is the algorithm which automatically manages the autonomous vehicle based on multiple triggers detailed above. As the automatic triggers are time-driven the algorithm has to take into account the process time of the vehicle operation. To increase or maximize the vehicle utilization the vehicle may perform other transport orders and has to drive to the forwarder facility first. After the swap body is closed and a starting signal is given – at least to ensure the closing of the swap body – the transport is performed and the swap body arrives at the handling agent. So depending on the current position of the AGV the total transport time has to be considered to ensure that the transport arrives prior to LAT. The platform consists of a server application (backend) and a web application which serves as the frontend. The cloud application platform Heroku is used to provide a highly scalable infrastructure. Programming is done using Python and the framework Django. The platform uses a Postgresql database, while the transport orders are managed with the help of a worker process. The interfaces with the AGV and the camera system use a REST API.

The web frontend is programmed in responsive HTML5 using the JavaScript Framework Angular.js and websocket connections which enables instant updates. The web site offers a login to ensure security and user role selection. The current status of the transportation plan is displayed for the operative staff. Single entries can be modified by the user in case of reading errors. Completed shipments will be indicated in green, incomplete shipments with an orange indicator and wrong shipments shown in red. Shipments which are missing completely are shown in grey.

To generate input data prior to finishing the interface with the camera system a web-app for Android was developed. Based on the basic barcode reading capabilities of smart phones AWB numbers could be scanned and the information processing could be tested. User role models, IT security and specific processes, e.g. to ensure that the swap body is closed, will be added in later stages.

5.3 Autonomous Transport

For autonomous transports we decided to use swap bodies in combination with swap body trans- porters upgraded with AGV capabilities (see Figure 3). The AGV was already successfully tested at closed cross-docking facilities. The guidance of the vehicle is GPS- and camera-based so no modification of infrastructure is needed.

Figure 3: AGV swap body truck (Source: KAMAG, 2018).

The internal airport roads were originally designed to allow for the operation of all ground handling equipment with a lane with of 5 m, the movement of the swap bodies with an additional width due to the down-folded legs will be possible. As airport roads are private roads they allow the operation of vehicles without legal permission (Straßenverkehrszulassung) and a license plate with the approval of the airport operator only. The speed limits on these roads is 30 km/h. The area is connected to the public road system via two gates. Although the area is accessible to everyone who wants to enter, the registration procedure or the need of an airport ID ensures that the roads are not used by misguided passenger cars or through traffic.

The final system test is planned for several weeks, but prior to that the autonomous vehicle has to be adapted to the special conditions of the airport premises. Prior to that, the routing will be tested with a conventional swap body transporter. The swap body will contain the camera system which then will be battery-powered. As the sites of the handling agent and the forwarder were those of the project partners, the preferable routing had to be chosen taking into account that both intersections and turns or changing lanes and possible interferences with normal traffic do create additional problems. Occasionally, trucks and other vehicles stop or even park at the roadside which can pause problems. Figure 4 shows the chosen route which contains two right and two left turns where the right of way has to be given. To gain experience and avoid peak traffic hours the tests with the AGV will not be conducted neither during shiftover times (7-9 am and 4-6 pm) nor during cargo traffic peak times.

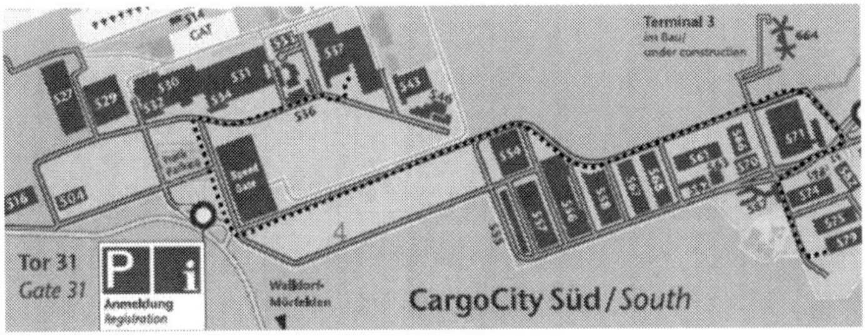

Figure 4: Chosen AGV-routing (Source: Own representation based on Fraport, 2019).

5.4 Camera-based Barcode Reading

The two biggest challenges in decoding barcodes using conventional camera systems during air cargo loading are the large field of view that needs to be covered and the rather difficult lighting conditions in the vicinity of the loading ramp. The latter often lead to blurred images. In the following, we describe our camera setup, with which we try to ensure that we can capture as many barcodes as possible. Furthermore, we present our approache to deal with the large field of view and blurred image regions containing barcodes.

Figure 5 shows the proposed camera-setup, which fits the typical dimensions in the surrounding of a loading ramp. The load is on a standard pallet moving through an archway with a size of approximately 3m 2.8m into a trailer. Our camera arrangement consists of cameras at the corners and upper center of the archway to cover all sides of the pallet. All cameras are hardware synchronized, which guarantees that alle images are taken at the same time. This allows to filter out barcodes that are seen simultaneously in different camera images.

Despite a well-thought-out optical setup, based on several synchronized cameras, the reliable recognition of the barcodes in the camera images and, above all, their subsequent decoding poses a major challenge. To meet these challenges, our camera-based system for reading barcodes essentially consists of the three successive stages (see also Figure 6).

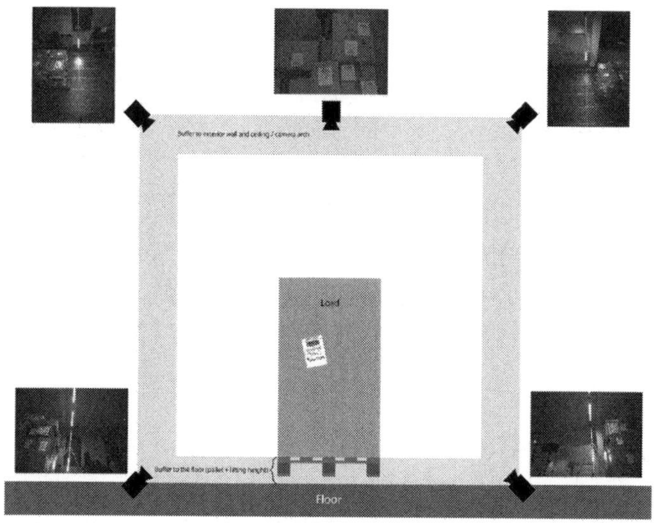

Figure 5: Camera arch equipped with synchronized cameras.

1. *barcode detection*, where image regions that contain barcodes are marked,
2. *image region enhancement*, where the image regions containing barcodes may be sharpened and, above all, aligned horizontally, and finally
3. *barcode decoding*, where the detected and aligned barcode is decoded.

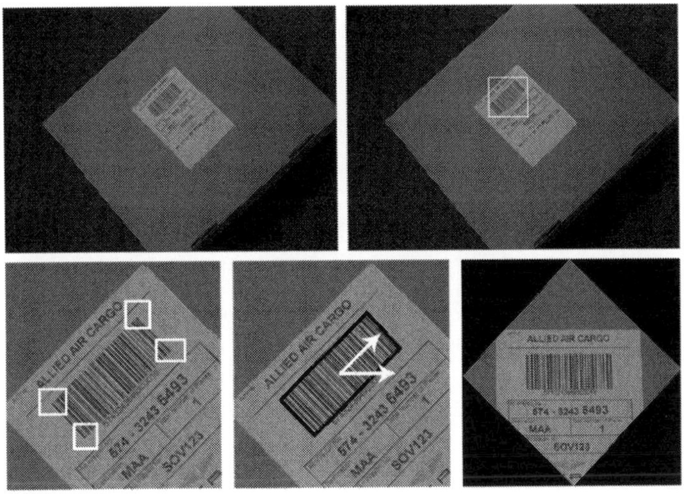

Figure 6: Top row: Original camera image and barcode detected with *Barcode-YOLO*. Bottom row: Corners of the barcode detected with *Corner-YOLO*, determined orientation, and aligned image region that is used to decode the barcode.

The aim of the first stage of our system is to recognize all relevant barcode candidates in each of the camera images (see Figure 7). This step is based on the real-time object detection system YOLO (see Redmon, Divvala, et al., 2016 and Redmon and Farhadi, 2017), a neural network that outputs bounding boxes for objects it detects in real-time. Since YOLO is a deep neural network, it requires a lot of training data to work properly. However, there are pre-trained variants of YOLO that can be adapted to specific problems. We have customized such a pre-trained network for the task of barcode detection using 300 manually labelled images of barcodes. In the following we will refer to this network as Barcode-YOLO.

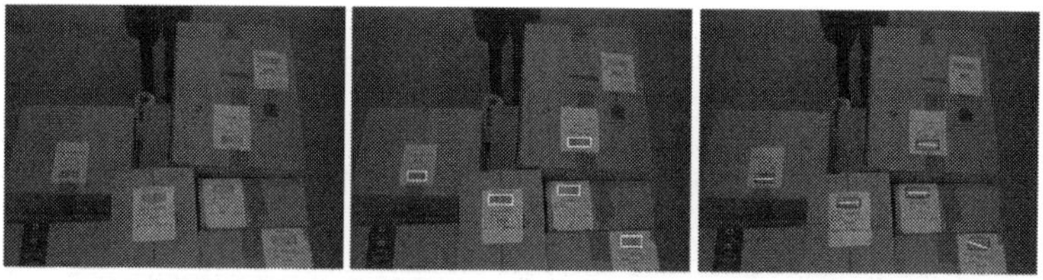

Figure 7: From left to right: Original camera image, detecting relevant barcodes (ignoring others) and decoded barcodes.

In the second step we estimate the alignment of the detected barcodes (see Figure 6). For this step we use another YOLO network (which we call *Corner-YOLO*) that we have trained to recognize the corners of a barcode. As illustrated in Figure 6 the input for *Corner-YOLO* is a cropped area of the original image based on the prediction boxes of *Barcode-YOLO*. Creating a rotated minimum rectangle over the centers of the detected corners of the boxes, we can estimate the rotation of the barcodes.

The last step is the decoding of the barcodes. For this step we use the 1D/2D barcode image processing library ZXing[3]. As already mentioned, the decoding rate can be further improved by first deburring the rotated image area containing the barcode. It turned out that neural network-based approaches such as the one presented in (Kupyn et al., 2018) are not suitable for deblurring as they provide visually very appealing results but hallucinate false lines leading to incorrect barcodes. Thus, algorithms such as the one presented in (Yahyanejad and Ström, 2010) should be used for deblurring.

---

[3] ZXing speaks "zebra crossing". You can find it at: https://github.com/zxing

5.5 Limitations

Although the overall concept is applicable to air cargo supply chains in general, some limitations have to be considered.

- The concept is only applicable for cargo hub airports where short distance transports between on-airport forwarder hubs and handling agent facilities exist. Furthermore, the concept requires designated truck doors for swap bodies or trailers during loading processes which are longer than at present. As for the handling agents, their number is usually limited, in most cases less than 10, but sometimes even below 5. As for the forwarders the concept has been designed to serve the 5 to 10 biggest players on a specific airport. If the number of truck doors is identified as a limiting factor the concept of AGV based swap body positioning at cross docking facilities could be applied.
- The camera system is trained and developed based on the IATA AWB standard with some optional modification by the forwarder. As the labels are not 100% standardized, the system would have to be slightly modified and trained to read new partners labels.
- A mobile internet connection is necessary to ensure near-time communication with the platform and to provide timely feedback to the operator while loading. The information of the identified shipments has to appear on the mobile device within seconds so the operator can modify or rescan the shipments if errors occur.
- The autonomous transport relies on the operation of an unregistered vehicle within a gated airport area with almost no through-traffic and no pedestrians or cyclists. To transfer the results to

other airports additional requirements may have to be considered. The system is designated for airports with multiple facilities, forwarder hubs and handling agents. At some airports there is either one common building or buildings are connected so transports are done with forklifts or tow tractors. The basic principles of the cloud-based platform, the steering algorithm, and the camera-system can be used, but other types of trailers and AGVs have to be used.

- The autonomous transport may encounter difficulties in certain road conditions (e.g. snow or heavy rains) which are unknown as of today.

6 Managerial Implications and Outlook

The SmartAirCargo trailer concept is meant to improve short distance transports. While the outbound operation of the forwarder remains unchanged, handling agents gets additional information about upcoming shipments and options to optimize workload at the truck gates and in the build-up process by pulling useful shipments from the forwarder. The inbound process is accelerated by pushing all shipments which passed customs to the forwarder. By means of an automatic trigger urgent shipments can be processed without any delay.

Overall the systems helps to overcome the LAT-driven export peaks and to come to a more demand- driven supply chain optimized process flow with more dynamic LATs.

The use of swap bodies (or trailers) increases truck utilization as loading or unloading does not require a truck and its driver to be present any more. By using an AGV, manpower costs could be eliminated. In view of the current truck driver shortage the deployment of AGVs seems to offer a most adequate perspective. The camera-system ensures the completeness of shipments by providing a piece count and minimizes the chances of loading wrong shipments. In combination with the platform the workload of the transport manager is reduced as the system ensures punctual delivery.

The main investments required to make such a concept operational stem from the AGV and the camera system. On the other hand, the following cost savings can be achieved:

- Manpower costs (mostly truck drivers). As hub airports operate 24/7 – even if a night curfew exist – for the continuous operation of one truck a minimum of 5 drivers are needed.
- Quality costs which consist of penalties for missed flights, rebooking or paying for unused capacity, cost of last-minute actions to deliver missing pieces, cost of searching or even cost for unjustified claims.

With the achieved transparency, currently existing buffer times can be reduced which makes it possible to shorten lead times, so cargo could either be picked up later or delivered earlier. Also the throughput of the facility can be raised.

Attention ought to be given to the following issues for future research and developments:

- Air cargo transports should be tested with trailers as they provide more capacity, particularly mega-trailers with less limitations regarding shipment size. Even as trailers are used in closed areas they would have to be connected to the braking and electric systems of the tow truck. Using an AGV would require an automatic connection for both systems.
- The power supply for the camera-system in the trailer has to be changed, as the current battery-based power supply is not capable of a 24/7 operation. Two options exist: Either the system is charged by a connection to the truck while being transported or the charging takes places while being connected to a truck gate.
- Overall, it would be preferable to change the AGV to a e-mobility version. As total payload is seldom used, the speeds are rather low and as the total distance travelled is low only a limited battery capacity is needed while charging options could be provided at the facilities.
- An integration with cargo community systems (CCS) and the warehouse management systems (WMS) has to be developed, to make use of truck slot booking systems, but also to integrate the information flow to make sure that transport plans are automatically uploaded and updated from the WMS.
- The security of the outbound process has to be developed in more detail with the regulation being modified, to allow for a non-human transport and handover of secured shipments. Therefore electronic seals have to be added to the system and security-related alerts and monitoring has to be programmed. The approval

from the responsible authority, the Luft- fahrtbundesamt (LBA), is required prior to a day-to-day-operation.

- A method of automated image processing meant to detect any damages has to be developed to be used for a better claims management.
- The camera system could also be used to verify dimensions.
- Once data from regular operations is available the transport planning algorithm can be optimized with AI and predictive analytics.[4]

---

[4] Kappel et al., 2019 describes potential use cases for AI in autonomous driving.

Acknowledgements

The authors would like to thank our project partners Sovereign Speed, CargoSteps, Fraunhofer IML, Fraport, LUG aircargo handling, KAMAG and the associated partners Dachser and QuickCargoService.

Financial Disclosure

This project (HA project no. 575/17-57) is funded in the framework of Hessen ModellProjekte, financed with funds of LOEWE – Landes-Offensive zur Entwicklung Wissenschaftlich-ökonomischer Exzellenz, Förderlinie 3: KMU-Verbundvorhaben (State Offensive for the Development of Scientific and Economic Excellence).

References

Anyz, T. and R. Touzani, 2019. SmartAirCargoTrailer - Project deliverables.

Azadian, F., A. Murat, and R. B. Chinnam, 2017. An unpaired pickup and delivery problem with time dependent assignment costs: Application in air cargo transportation. In: European Journal of Operational Research 263.1, pp. 188–202.

Bartolini, C., T. Tettamanti, and I. Varga, 2017. Critical features of autonomous road transport from the perspective of technological regulation and law. In: Transportation Research Procedia 27, pp. 791– 798.

Bäumler, I. and H. Kotzab, 2017. Intelligent transport systems for road freight transport—an overview. In: Dynamics in Logistics. Springer, pp. 279–290.

Bierwirth, B. and K.-O. Schocke, 2017. Lead-time optimization potential of digitization in Air Cargo. In: Proceedings of the Hamburg International Conference of Logistics (HICL). epubli, pp. 75–98.

Bonini, M., D. Prenesti, A. Urru, and W. Echelmeyer, 2015. Towards the full automation of distribution centers. In: 2015 4th International Conference on Advanced Logistics and Transport (ICALT). IEEE, pp. 47–52.

Brandt, A., July 2018). DB Schenker tests automated "Wiesel" swap-body transporter. URL: https:// www.deutschebahn.com/en/presse/press_releases/DB-Schenker-tests-automated-Wiesel- swap-body-transporter-3183398.

Brandt, F. and S. Nickel, 2018. The air cargo load planning problem - a consolidated problem definition and literature review on related problems. In: European Journal of Operational Research.

Clarembaux, L. G., J. Pérez, D. Gonzalez, and F. Nashashibi, 2016. Perception and control strategies for autonomous docking for electric freight vehicles. In: Transportation Research Procedia 14, pp. 1516– 1522.

Clausen, U. and M. K. IVI, 2018. Automatisiertes Fahren. In: Digitalisierung. Springer, pp. 385–411.

Creusot, C. and A. Munawar, 2015. Real-Time Barcode Detection in the Wild. In: 2015 IEEE Winter Conference on Applications of Computer Vision, pp. 239–245.

David, J. and P. Manivannan, 2014. Control of truck-trailer mobile robots: a survey. In: Intelligent Service Robotics 7.4, pp. 245–258.

Dietrich, A. and F. Fiege, 2017. Digitale Transformation des Speditionsgeschäfts umfasst mehr als Spedition 4.0. In: Wirtschaftsinformatik & Management 9.3, pp. 36–45.

Dixon, R., 2018. Trends in der Automobil-Sensorik. In: Automobil-Sensorik 2. Springer, pp. 17–28.

D. W. Hukins, A. Hunter, and A. M. Korsunsky. Vol. 2229. International Association of Engineers. Newswood Limited, pp. 501–507.

Feng, B., Y. Li, and Z.-J. M. Shen, 2015. Air cargo operations: Literature review and comparison with practices. In: Transportation Research Part C: Emerging Technologies 56, pp. 263–280.

Flämig, H., 2016. Autonomous vehicles and autonomous driving in freight transport. In: Autonomous driving. Springer, pp. 365–385.

Fraport, 2019. Frankfurt Airport Map. URL: https://maps- frankfurt. com/ img/ 0 / frankfurt- airport-train-station-map.jpg.

Gallo, O. and R. Manduchi, 2011. Reading 1D Barcodes with Mobile Phones Using Deformable Tem- plates. In: IEEE Transactions on Pattern Analysis and Machine Intelligence 33.9, pp. 1834–1843.

Hansen, D. K., K. Nasrollahi, C. B. Rasmusen, and T. B. Moeslund, 2017. Real-Time Barcode Detection and Classification using Deep Learning. In: Proceedings of the 9th International Joint Conference on Computational Intelligence - Volume 1: IJCCI, INSTICC. SciTePress, pp. 321–327.

Hey, T., 2019. Die außervertragliche Haftung des Herstellers autonomer Fahrzeuge bei Unfällen im Straßenverkehr. Springer.

IATA, 2018. IATA Cargo Strategy. URL: https://www.iata.org/whatwedo/cargo/documents/ cargo-strategy.pdf.

IATA, 2019. e-awb international monthly report - march 2019. URL: https : / / www . iata . org / whatwedo/cargo/e/eawb/Documents/e-awb-monthly-report-r17.pdf.

KAMAG, 2018. KAMAG Wiesel AGV. URL: https://www.tii-group.com/communication/news- overview/detail/getarticle/News/detail/save-the-date-10-july-2018.html.

KAMAG, 2019. E-Wiesel AGV. URL: https://www.kamag.com/products/logistics-transporters/e- wiesel-agv.html.

Kappel, M., E. Krune, M. Waldburger, and B. Wilsch, 2019. Die Rolle der KI beim automatisierten Fahren. In: Künstliche Intelligenz. Springer, pp. 176–193.

Katona, M. and L. G. Nyúl, 2013. Efficient 1D and 2D Barcode Detection Using Mathematical Morphol- ogy. In: Mathematical Morphology and Its Applications to Signal and Image Processing. Ed. by C. L. L. Hendriks, G. Borgefors, and R. Strand. Berlin, Heidelberg: Springer Berlin Heidelberg, pp. 464–475.

Kober, W., R. Huber, and R. Oberfell, 2017. Vehicle Reference Lane Calculation for Autonomous Vehicle Guidance Control. In: Automated Driving. Springer, pp. 141–158.

Kückelhaus, M., K. Zeiler, and D. Niezgoda, Dec. 2014. Self-driving vehicles in logistics.

Kunze, R., R. Ramakers, K. Henning, and S. Jeschke, 2011. Organization and operation of electronically coupled truck platoons on german motorways. In: Automation, Communication and Cybernetics in Science and Engineering 2009/2010. Springer, pp. 427–439.

Kupyn, O., V. Budzan, M. Mykhailych, D. Mishkin, and J. Matas, 2018. DeblurGAN: Blind Motion Deblurring Using Conditional Adversarial Networks. In: 2018 IEEE Conference on Computer Vision and Pattern Recognition, CVPR 2018, pp. 8183–8192.

Lange, A., 2019. Does cargo matter? The impact of air cargo operations on departure on-time per- formance for combination carriers. In: Transportation Research Part A: Policy and Practice 119, pp. 214–223.

Lehmacher, W., 2015. Globale Supply Chain: Technischer Fortschritt, Transformation und Circular Econ- omy. Springer-Verlag.

Liu, Y., M. Yin, and M. Hansen, 2019. Economic costs of air cargo flight delays related to late package deliveries. In: Transportation Research Part E: Logistics and Transportation Review 125, pp. 388–401.

Mayer, R., 2016. Airport classification based on cargo characteristics. In: Journal of Transport Geography 54, pp. 53–65.

Meinel, H. H., 2018. Radarsensors and autonomous driving—yesterday, today and tomorrow!. In: e & i Elektrotechnik und Informationstechnik 135.4-5, pp. 370–377.

Namane, A. and M. Arezki, July 2017. Fast Real Time 1D Barcode Detection From Webcam Images Using the Bars Detection Method. In: World Congress on Engineering 2017. Ed. by S. I. Ao, L. Gelman,

Neubauer, M. and O. Schauer, 2017. Human factors in the design of automated transport logistics. In: International Conference on Applied Human Factors and Ergonomics. Springer, pp. 1145–1156.

Parreira, J. and J. Meech, 2011. Autonomous haulage systems–justification and opportunity. In: International Conference on Autonomous and Intelligent Systems. Springer, pp. 63–72.

Pipp, T., P.-A. Reiners, and J. von Roesgen, 2018. Driverless Vehicles: Stand der Technik und Anwen- dung am Fallbeispiel Flughafen. In: Mobilität 4.0–neue Geschäftsmodelle für Produkt-und Dienstleis- tungsinnovationen. Springer, pp. 203–244.

Ranau, M., 2011. Planning approach for dimensioning of automated traffic areas at seaport container terminals. In: Handbook of terminal planning. Springer, pp. 179–193.

Redmon, J., S. K. Divvala, R. B. Girshick, and A. Farhadi, 2016. You Only Look Once: Unified, Real-Time Object Detection. In: 2016 IEEE Conference on Computer Vision and Pattern Recognition, CVPR 2016, pp. 779–788.

Redmon, J. and A. Farhadi, 2017. YOLO9000: Better, Faster, Stronger. In: 2017 IEEE Conference on Computer Vision and Pattern Recognition, CVPR 2017, pp. 6517–6525.

Schmidt, T., 2018. Innerbetriebliche Logistik. Springer.

Selinka, G., A. Franz, and R. Stolletz, 2016. Time-dependent performance approximation of truck handling operations at an air cargo terminal. In: Computers & Operations Research 65, pp. 164–173.

Shepherd, B., A. Shingal, and A. Ray, 2016. Value of Air Cargo: Air Transport and Global Value Chains.

Sieke, H. and L. Mehrtens, 2019. SmartAirCargoTrailer - Project deliverables.

Sörös, G. and C. Flörkemeier, 2013. Blur-resistant joint 1D and 2D barcode localization for smart- phones. In: MUM. Ed. by M. Kranz, K. Synnes, S. Boring, and K. V. Laerhoven. ACM, 11:1–11:8.

Stahn, R., T. Stark, and A. Stopp, 2007. Laser scanner-based navigation and motion planning for truck-trailer combinations. In: 2007 IEEE/ASME international conference on advanced intelligent mechatronics. IEEE, pp. 1–6.

Ullrich, G., 2013. Fahrerlose Transportsysteme: Eine Fibel-mit Praxisanwendungen-zur Technik-für die Planung.

Wurll, C., 2016. Das Bewegliche Lager auf Basis eines Cyber-physischen Systems. In: Handbuch Industrie 4.0: Produktion, Automatisierung und Logistik, pp. 1–28.

Yahyanejad, S. and J. Ström, 2010. Removing motion blur from barcode images. In: pp. 41–46. DOI: 10.1109/CVPRW.2010.5543258.

Yun, I. and J. Kim, 2017. Vision-based 1D barcode localization method for scale and rotation invariant. In: TENCON 2017 - 2017 IEEE Region 10 Conference, pp. 2204–2208.

Zhang, H., G. Shi, L. Liu, M. Zhao, and Z. Liang, 2018. Detection and identification method of medical label barcode based on deep learning. In: 2018 Eighth International Conference on Image Processing Theory, Tools and Applications (IPTA), pp. 1–6.

A Quantitative Assessment of the Collaborative Logistics Benefits

Camillo Loro[1], Riccardo Mangiaracina[1], Angela Tumino[1], Alessandro Perego[1]

1 – Politecnico of Milan

Purpose: The purpose of this analysis is to study the Collaborative Logistics process, its implementation within the Consumer Electronics and Household Appliances Supply Chain and to assess the benefits stemming from it.

Methodology: To achieve the objectives the Activity-Based-Cost logic has been used. Each single defined activity has been associated to its most appropriate cost drivers. Moreover, an analytic simulation model has been built and implemented through a simplified representation of the Supply Chain composed of three actors: a retailer, a logistics operator and a manufacturer.

Findings: The benefits stemming from the Collaborative Logistics project are equal to 35.51%. The majority of Supply Chain benefits comes from the reduction in Stock-Out costs and Inventory Management costs.

Originality: The analytical model has been developed to foster the diffusion of this process by showing the potential benefits achievable through its implementation.

Keywords: eSupply Chain Collaboration, Collaborative Logistics, Consumer Electronics and Household Appliances Industry, Digitalisation

First received: 22.May.2019 **Revised:** 12.June.2019 **Accepted:** 17.June.2019

1 Introduction

In recent years, the adoption of innovative technologies in Supply Chain Management is accelerating faster and faster, changing the competitive paradigm. The needs of the final customers are getting more sophisticated, and this drives the companies to develop innovative products in ever-shorter time at affordable prices. To deal with these market changes, companies could implement collaborative projects both upstream and downstream in the Supply Chain, creating a competitive advantage. Indeed, these projects enhance the efficiency and effectiveness, such as production or inventory costs reduction and service level improvement. In particular, the Italian Consumer Electronics and Household Appliances Industry is characterised by a heterogeneous product portfolio and a wide product variety in terms of value and size. In the Italian market the most of companies are small and medium enterprises, the Supply Chain is rather concentrated in the upstream levels and extremely fragmented at the retail channel level. On the one hand, this complexity represented an initial barrier to establish collaboration among all the partners; on the other hand, the implementation of innovative technologies was necessary to create a competitive advantage. In this industry the use of e-Supply Chain Collaboration solutions is still sporadic: Consignment Stock between Manufacturers and Suppliers, Continuous Replenishment Programs (CRP) and after-sales support solutions between Manufacturers and Retailers are the most common. Recently, new collaborative processes have been emerged, such as the Collaborative Logistics project, able to centralise in a unique digital solution strategic and operational information in order to better integrate the actors of the chain. Based on these new technologies, there is room for improvement both in the efficiency and in the effectiveness of logistics activities (Cervello and Triventi, 2018).

The paper is organised as follows: the literature review presents evidence from the extant knowledge, the research questions and methodology defines the research questions and describes the adopted methodology, the findings section introduces the results, and conclusions and further development summarise the contribution and limitations of the model which can be addressed through future works.

2 Literature review

In the current scenario of the digital economy, the use of Internet can be a key factor for the enhancement of the Supply Chain Management (SCM) (Tan et al., 2002), thanks to the exchange of the information in real time and to the launch of collaborative project between the different business partners, thus creating the e-Supply Chain Management (e-SCM) (Gimenez and Ramalhinho, 2004). An e-Supply Chain can be described as a strategy of business where the e-commerce helps to optimise the processes, reduce the time production, give a faster and efficient response to the final customers (Luo et al., 2001). The advantage of the internet is that it allows the integration of the various systems of the different actors of a specific supply chain (Braglia and Frosolini, 2014), but the enterprise needs to invest a lot of resources, management commitment (as well as the commitment of the partners), time and energy and to handle the organizational changes to obtain real benefits from such solution (Pant et al., 2003).

With the adoption of integrated system within the organisation, companies could achieve different benefits such as: the developing of a cooperative business climate and an interactive approach to the supply chain, the capacity to foresee and react to demand fluctuation, and to improve the relationship between the actors of the supply chain and the consequent creation of strategic partnership (Luo et al., 2001). Other types of benefits are: the optimization of the internal and external operations, the response to the market in real-time (Pant et al., 2003), the cost savings (due, for examples, to the reduction of inventory level, procurement costs, and cycle time) and the increasing in the profitability (Akyuz and Rehan, 2009).

In this context, even the notion of the Supply Chain Collaboration (SCC) has gained the attention of many authors. The Supply Chain Collaboration arises when two or more firms share the responsibility of exchanging common planning, management, execution, and performance measurement information (Anthony, 2000) in order to accomplish a "synchronized Supply Chain" (Anderson and Lee, 1999). The reason is that integration with Supply Chain partners is needed to provide a value-add for the clients, with a consequent increase in satisfaction level and sales (Cao and Zhang, 2011). Moreover, thanks to the agreements with partners, a company can increase efficiency, reducing wastes, decreasing costs and achieving in this way a potential continuous flow (Narus end Anderson, 1995). Furthermore, partners have realised that they could extend the collaboration agreement to different activities with the purpose of enlarging the efficiency and the effectiveness of their processes and that the advancement in technologies have helped to deal with different activities, which imply the communication with partners (Cascio and Montealegre, 2016). Thanks to technological improvements partners have increased the quality and the type of information to exchange and have been able to extend the relationship with partners over different types of inter-organizational processes (Cao and Zhang, 2011). One of these projects is the Collaborative Logistics that builds on the existing relationship between trading partners (Manufacturers and Retailers) by extending this relationship to one or more service providers (Logistics Operators), act as the collaborative "hub" with their logistics base (Kaveh and Khosravi Samani, 2009). The Collaborative Logistics process can decrease the costs, and increase supply chain efficiency, in addition to make trading partners more flexible to manage the consumer demand fluctuation (Kaveh and Khosravi Samani, 2009).

The adoption of this solution is taking attention in the Italian Consumer Electronics and Household Appliances Supply Chain (eInvoicing & eCommerce B2b Observatory, 2018), and in the literature there is a lack of analysis regarding the benefits derived from the sharing of forecasting data stemming from the exchange of information between manufacturers, logistics operators and specialised retailers, enabled by the ICT-driven in the order-to-payment cycle integration (Cervello and Triventi, 2018). In fact, the communication of such data increases the visibility of each actor along the entire Consumer Electronics and Household Appliances Supply Chain (Cervello and Triventi, 2018).

3 Research questions and methodology

Given the identified gap, this paper attempts to contribute to the extant literature by proposing a model that aims at identifying which activities are involved into the implementation of a Collaborative Logistics project, how the processes change with respect to the traditional approach and how the adoption of this new approach can generate benefits for all the actors in the Supply Chain.

To reach this objective, the following research questions were identified:

RQ1: *How is the Collaborative Logistics project performed? What are the involved activities and the benefits stemming from its implementation? How can their costs be modelled?*

RQ2: *What are the benefits of adopting Collaborative Logistics in the Consumer Electronics & Household Appliances industry and how are they distributed among the triad of actors (i.e. Manufacturer, Retailer and Logistics Operator)? What are the benefits for the entire Supply Chain?*

The employed methodology encompasses specific activities, belonging to two different phases. The first step is characterised by the empirical analysis. It aims at defining the Collaborative Logistics model. In particular, 56 potential companies have been initially contacted. Finally, 12 out of 56 companies (4 Retailers, 4 Manufacturers and 4 Logistics Operators) have been selected, basing of the level of digitalisation and the role in the Supply Chain, and the implementation of the specific project we analysed to gather information through the interview. Thus, this phase was represented by the description of the process and the identification of all the activities performed by different actors. In particular, two different scenarios have been studied: AS-WAS, before the adoption of the Collaborative Logistics project; AS-IS, after the adoption of the Collaborative Logistics project. In the second phase, the model has been structured from an analytic perspective, in order to quantify the benefits coming from the adoption of a Logistics Collaborative project, through the Activity Based Methodology. The benefits have been computed with respect to three actors: Retailer, Manufacturer and Logistics Operator. Furthermore, the benefits of the overall Supply Chain have been quantified. Five differential costs elements between the two scenarios have been considered: Administrative costs, Inventory Management costs, Stock-Out costs, Penalty costs and Transportation costs. The simulation has taken into account a simplified Supply Chain made up of three companies. It was performed through Excel.

3.1 General hypotheses

The model represents a simplified representation of the reality, some general hypotheses need to be formulated. The hypotheses of the model try to be as adherent to reality as possible, so that the generated output can give companies trustful insights.

The cost modelling is based on two categories of hypotheses:

- The Demand profile;
- Inventory management.

With reference to the time, a time horizon equal to one year is applied, considering 52 weeks. Furthermore, in the simulation a time bucket of one week is used. The demand presents a strong seasonality, on the basis of what emerged from Statista website, filtering the Italian Household Appliances and Consumer Electronics sector in the year 2017. In particular, the Seasonality Index (S.I.) – for what concerns the demand seen by the retailer - is equal to the ratio between the actual monthly demand (Dt) and the average monthly one (μ_D) (equation 1). The demand follows a standard normal distribution and the considered variables are independent one to each other. The seasonality of the demand profile influences the computation of all the parameters and the variables (equations 2, 3, 4 and 5). The Table 1 lists all the variables of the seasonality index.

Table 1: Demand seasonality components

Variable	**Description**	**Unit of measure**	**Source**
$\boldsymbol{S.I._t}$	Seasonality index	–	Formula result
$\boldsymbol{D_t}$	Demand of the week t	[items]	Simulation computation
$\boldsymbol{D_{t;T+LT}}$	Demand over the Replenishment Period and the Replenishment Lead Time of the *week t*	[items]	Simulation computation
$\boldsymbol{SS_{t;T+LT}}$	Safety Stock facing the Demand and Lead Time Variability	[items]	Simulation computation

Variable	Description	Unit of measure	Source
	over the Replenishment Period and the Replenishment Lead Time of the *week t*		
AT_t	Maximum Inventory Level for the *week t*	[items]	Formula results

Below, all the formulas are reported:

$$S.I._t = \frac{D_t}{\mu_D} \tag{1}$$

$$D_t = (\mu_D + z_t \cdot \sigma_D) \cdot S.I._t \qquad with\ z_t \sim N(0;1) \tag{2}$$

$$D_{t;T+LT} = \mu_D \cdot (T + LT) \cdot S.I._t \tag{3}$$

$$SS_{t;T+LT} = k \cdot \left(\sqrt{(T+LT) \cdot \sigma_D^2 + \mu_D{}^2 \cdot \sigma_{LT}^2}\right) \cdot S.I._t \tag{4}$$

$$AT_t = D_{t;T+LT} + SS_{t;T+LT} \tag{5}$$

This implies that the inventory level is not "static" but adherent to the real demand trend. The Retailer demand does not change from the AS-WAS to the AS-IS situation because it is always influenced by the variability of the final customer, so the collaboration or the technology used do not influence it. Instead upstream, the Manufacturer in the AS-WAS faces the variability both of the final customer and the Retailer. The Manufacturer demand forecasting is less accurate than the Retailer one because it is influenced by a "double" variability: that of the Retailer and that of the final consumer. Historical data do not come solely from the Manufacturer, but also by the Retailer. This affect all the subsequent activities of the model, including the production planning and the inventory management. The huge variability is smoothed in the AS-IS because it comes only by the final customer thanks to the visibility ensured by the platform.

The inventory planning model chosen is the Periodic Review. The orders are placed every T (Time, order interval) that in the simulation is equal to one week, both for the Manufacturer and the Retailer. The implementation of a fixed replenishment time period requires the existence of an Availability Target (AT) to be reached and a variable purchasing quantity (Q_t) which is computed on the basis of the inventory level (IL_t) (equation 6).

The availability target (AT) (equation 7) takes into consideration both the average demand, during the fixed replenishment period (T) and the replenishment lead time (LT), and the demand variability taken into account with the safety stock (SS).

In particular, the average demand (D_{T+LT}) (equation 8) and the safety stocks (SS_{T+LT}) (equation 9) are computed as.

In the Table 2 below all the variables of the inventory level componenets are described.

Table 2: Inventory level components

Variable	Description	Unit of measure	Source
T	Fixed Replenishment Period	[weeks]	Interviews
LT	Replenishment Lead Time	[weeks]	Interviews
Q_t	Purchasing Quantity	[items]	Formula result
AT	Availability Target	[items]	Formula result
IL_t	Inventory Level	[items]	Simulation computation
D_{T+LT}	Average Demand over the Replenishment	[items]	Simulation computation

Variable	Description	Unit of measure	Source
	Period and the Replenishment Lead Time		
SS_{T+LT}	Safety Stock facing the Demand and Lead Time Variability over the Replenishment Period and the Replenishment Lead Time	[items]	Simulation computation
μ_D	Average Weekly Demand	[items/weeks]	Simulation computation
k	Function of the Service Level	–	Simulation computation
σ_D^2	Demand Variance	[items^2]	Hypothesis
σ_{LT}^2	Lead Time Variance	[weeks^2]	Hypothesis

Below, all the formulas concerning the inventory level are reported:

$$Q_t = AT - IL_t \tag{6}$$

$$AT = D_{T+LT} + SS_{T+LT} \tag{7}$$

$$D_{T+LT} = \mu_D \cdot (T + LT) \tag{8}$$

$$SS_{T+LT} = k \cdot \sqrt{(T + LT) \cdot \sigma_D^2 + \mu_D{}^2 \cdot \sigma_{LT}^2} \tag{9}$$

Obviously, different demands are considered with respect to different classes. The Stock Keeping Units (SKUs) are divided into three classes (see Table 3 for details) on the bases of the inventory turnover index, collected in Statista website. Class A is made by the high-rotating SKUs, class B by the medium-rotating and class C by the low-rotating. The assignment of the SKUs to the classes follows the marginal increase methodology. This items classification holds true for both the Retailer and the Manufacturer.

Table 3: Division in classes

Classes	**Items**	**Volume**
Class A	1%	50%
Class B	50%	40%

Classes	Items	Volume
Class C	49%	10%

3.2 Cost modelling

After the definition of the general hypotheses, the cost structure has been modelled so that all the costs drivers are included.

The costs categories considered within the simulation are:

- Administrative costs;
- Inventory management costs;
- Stock-out costs;
- Transportation costs;
- Penalty costs.

Table 4: Cost categories for each actor

	Adminis-trative Costs	Inventory Management Costs	Stock Out Costs	Trans-porta-tion Costs	Pen-alty Costs
Manu-facturer	Impacted	Impacted	Im-pacted	Not-im-pacted	Im-pacted

	Administrative Costs	Inventory Management Costs	Stock Out Costs	Transportation Costs	Penalty Costs
Logistics Operator	Impacted	Not-impacted	Not-impacted	Impacted	Impacted
Retailer	Impacted	Impacted	Impacted	Impacted	Not-impacted
Entire Supply Chain	Impacted	Impacted	Impacted	Impacted	Impacted

In Table 4, the categories considered for each actor and for the whole Supply Chain perspective are synthesized.

In order to distinguish the costs related to different actors, some indexes are used: "M" refers to the manufacturer, "R" to the retailer, "L" for the Logistics Operator. Moreover, as the items are divided in three classes, the subscript "i" is used to specify them and "t" refers to the time period (one week in this case).

The Supply Chain costs function (equation 10) is given by:

$$Total\ \ costs = \sum_{i=M,L} Penalty\ \ costs_i + \sum_{i=M,R} (Inventory\ \ management\ \ costs_i + Stock - out\ costs_i) + \sum_{i=R,L} Transportation\ \ costs_i + \sum_{i=M,R,L} Administrative\ \ costs_{i_i} \quad (10)$$

The Manufacturer costs function (equation 11) is given by:

$$Total\ costs = Penalty\ costs + Administrative\ costs + Inventory\ management\ costs + Stock - out\ costs \quad (11)$$

The Retailer costs function (equation 12) is equal to:

$$Total\ costs = Administrative\ costs + Transportation\ costs + Stock - out\ costs + Inventory\ management\ costs \quad (12)$$

The Logistics Operator costs function (equation 13) is given by:

$$Total\ costs = Administrative\ costs + \ Transportation\ costs + Penalty\ \ costs \quad (13)$$

For all the costs categories, all the formulas and the specifications are reported. For the administrative cost the main components are illustrated in the Table 5 and the related formulas in the equation 14.

Table 5: Administrative cost components

Variables	Description	Unit of measure	Source
T_{proc}	Delivery Processing Time	[hour]	Interview
$c\,l_h$	Administration Hourly Labor Cost	[€/hour]	https://www.fisco-etasse.com
C_{paper}	Paper cost	[€/sheet]	eCommerce website
$sheets$	Number of sheets per order	[sheet/order]	Hypothesis
C_{space}	Space Occupation Cost	[€/m^3]	Hypothesis
$space_{sheets}$	Space Occupied by Sheets	[m^3/sheet]	Hypothesis
N_o	Number of deliveries per Week	[order/week]	Interview

$$\text{Administrative Cost} = \left[\mathrm{T_{proc}} \cdot \mathrm{c\,l_h} + \mathrm{C_{paper}} \cdot \mathrm{sheets} + \mathrm{C_{space}} \cdot \mathrm{space_{sheets}} \cdot \mathrm{sheets}\right] \cdot \mathrm{N_o} \quad (14)$$

For the inventory management cost, formula and components (as example are reported cost and components referring to the manufacturer case) are illustrated in the Table 6 and the related formulas in the equation 15, equation 16, and equation 17.

Table 6: Inventory management cost components

Variables	**Description**	**Unit of measure**	**Source**
r	Inventory Carrying Cost Rate	[€/(€*year)]	Financial website
$c_{u;M}$	Manufacturing Unitary Cost	[€/item]	Consumer Electronics and Household Appliances websites
k_M	Manufacturer Function of the Service Level	–	Simulation computation

Variables	Description	Unit of measure	Source
CS_M	Manufacturer Average Cycle Stock	[items]	Formula result
SS_M	Manufacturer Average Safety Stock	[items]	Formula result
T_M	Manufacturer Fixed Replenishment Period	[weeks]	Hypothesis
LT_M	Manufacturer Replenishment Lead Time	[weeks]	Interview
${\sigma_{D_{M;i}}}^2$	Manufacturer Demand Variance for the Product Class i	[items^2]	Hypothesis
$\mu_{M;i}$	Manufacturer Average Weekly Demand for the Product Class i	[items/weeks]	Interview

Variables	Description	Unit of measure	Source
$\sigma^2{}_{LT;M}$	Manufacturer Lead Time Variance	[weeks^2]	Hypothesis
$CS_{M;t;i}$	Manufacturer Cycle Stock for the Product Class i in the week t	[items]	Formula result
$S.I._{M;t}$	Manufacturer Seasonality Index of the week t	–	Formula result

$$Inventory\ Management\ Cost_M\ (IMC_M) = r \cdot c_{u;M} \cdot (CS_M + SS_M) \quad (15)$$

$$Avg.Safety\ Stock_M = \frac{k_M \cdot \sum_{t=1}^{52} \sum_{i=1}^{3} \left(\sqrt{(T_M + LT_M) \cdot \sigma_{D_{M;i}}{}^2 + \mu_{M;i}{}^2 \cdot \sigma^2{}_{LT;M}} \right) \cdot S.I._{M;t}}{52} \quad (16)$$

$$Average\ Cycle\ Stock_M\ (CS_M) = \frac{\sum_{t=1}^{52} \sum_{i=1}^{3} CS_{M;t;i} \cdot S.I._{M;t}}{52} \quad (17)$$

For the stock-out cost, formula and components (as example are reported cost and components referring to the manufacturer case) are illustrated in the Table 7 and the related formulas in the equation 18 and equation 19.

Table 7: Stock-out cost components

Variables	**Description**	**Unit of measure**	**Source**
SL_R	Retailer Service Level	[percentage]	Interview
$D_{R;\, yearly}$	Yearly Retailer Demand	[items]	Interview
$p_{u;\, R}$	Retailer Unit Price	[€/item]	Statista
$c_{u;\, R}$	Retailer Unit Cost	[€/item]	Simulation computation
SL_M	Manufacturer Service Level	[percentage]	Interview
$D_{M;\, yearly}$	Yearly Manufacturer Demand	[items]	Interview
$p_{u;\, M}$	Manufacturer Unit Price	[€/item]	Statista

Variables	Description	Unit of measure	Source
$c_{u;\,M}$	Manufacturer Unit Cost	[€/item]	Simulation computation

$$Stock - out\ costs_M = \ (1 - SL_M) \cdot D_{M;\,yearly} \cdot (p_{u;\,M} - c_{u;\,M}) \quad (18)$$

$$D_{M;\,yearly} = \sum_{t=1}^{52} \sum_{i=1}^{3} D_{R;\,t;\,i} \quad (19)$$

For the transportation cost, formula and components are illustrated in the Table 8 and the related formulas in the equation 20, equation 21, and equation 22 for the Retailer, and equation 23 for the Logistics Operator.

Table 8: Transportation cost components

Variables	Description	Unit of measure	Source
cl_h	Hourly haulers Labor Cost	[€/hour]	https://www.fisco-etasse.com
AS	Average Speed of Truck	[km/hour]	Ministero delle Infrastrutture e dei Trasporti

Variables	Description	Unit of measure	Source
l	Diesel liters consumed	liters	Simulation computation
$\#Deliveries$	Number of Deliveries	–	Simulation computation
ton	Weight of goods transported	ton	Simulation computation

For Retailer:

$$Tranportation\ costs = 0{,}75 \cdot Transportation\ share + 0{,}25 \cdot Diesel\ share \tag{20}$$

$$Transportation\ share_R = \#Deliveries \cdot \left[\sum_{t=1}^{52}\sum_{i=1}^{3}\frac{Km}{ton}\cdot\left(\frac{€}{km \cdot ton}\right)\right] \tag{21}$$

$$Diesel\ share = \frac{€}{l}\cdot\left(km \cdot \frac{l}{km}\right) \tag{22}$$

For Logistics Operator:

$$Transportation\ costs_L = \frac{Km}{AS}\cdot\ cl_h \tag{23}$$

Eventually, for the penalty cost, formula and components are illustrated in the Table 9 and the related formulas in the equation 24 and equation 25 for the Manufacturer, and equation 26 and equation 27 for the Logistics Operator.

Table 9: Penalty cost components

Variables	Description	Unit of measure	Source
$percentage\ of\ OTD$	Logistics Operator's On-Time-Delivery	[percentage]	Interview
0,2	Penalties Payed by the Logistics Operator in case of Consignments not Delivered on Time	[euro]	Hypothesis
0,2	Penalties Payed by the Manufacturer	[euro]	Hypothesis
$percentage\ of\ OTIF$	Manufacturer's On-Time-In-Full	[percentage]	Interview

Variables	Description	Unit of measure	Source
P_M	Purchase price paid by the Retailer to the Manufacturer	[euro]	Hypothesis

For Manufacturer:

$$Penalty\ Costs_M = \#Deliveries \cdot [(1 - percentage\ of OTIF) \cdot 0{,}2 \cdot P_M \quad (24)$$

$$OTIF\ (percetange) = \frac{number\ of\ deliveries\ OTIF}{total\ number\ of\ deliveries} \cdot 100 \quad (25)$$

For Logistics Operator:

$$Penalty\ costs_L = (1 - percentage\ of\ OTD) \cdot 0{,}2 \cdot Transportation\ costs_R \quad (26)$$

$$OTD = \frac{Orders\ delivered\ on\ time}{total\ orders\ shipped} \quad (27)$$

4 Findings

The Italian Consumer Electronics and Household Appliances Industry is characterised by high product variety and high fragmentation at the downstream side of the Supply Chain. This sector is attending the rising of a new collaborative model accessible via B2b Web Portals: Collaborative Logistics. The reason why this model is rising in the matter of Supply Chain Management is that this type of collaboration integrates in one single solution manufacturers, logistics operators and retailers operational and strategic information. This is possible since the technology supporting this program allows partners to quickly share data on a centralised platform, which enhances visibility on that specific information needed by partners. Moreover, the main technology used between manufacturers and their suppliers is B2b Portals for the development of automatic replenishment and Consignment Stock solutions.

Analysing more in depth the Collaborative Logistics project, the model description has the aim of analysing the logistics processes of Retailer, Manufacturer and Logistics Operator, identifying the main activities and understanding how they change when passing from the AS-WAS to the AS-IS scenario. In particular, the AS-WAS scenario considers the Supply Chain before the adoption of the Collaborative Logistics project. While the AS-IS scenario takes into consideration the process after the adoption of this solution. The process is divided into different phases. The first step for all the activities is the forecast of the annual demand. In the AS-WAS scenario, each actors of the chain made the forecast as a single entity, on the basis of the order history of the immediate customers. This creates great variability due to the information distortion. In the AS-IS model, the demand forecast is made in an aggregated way, given that each actor shares online information across the whole Supply Chain (including third Parties Logistics). Thanks to the new technology, it is possible to communicate and coordinate the demand and the production with the other players in the Supply Chain. This significantly reduces the variability and the inefficiencies, leading to a reduction of the total Supply Chain lead time. Another difference between the AS-WAS and AS-IS situation is the way in which the request for reservation of the delivery slot is made. The main functionality of the platform is the one related to the booking of delivery slots, made possible by sharing the agendas of the different depots. This phase changes very greatly in the two situations. The phases majorly impacted by the implementation of the collaborative model (in the AS-IS situation) and in particular by the technology are: creation of the appointment request, made by the Logistics Operator; management of the appointment request, made by the retailer. The technology on which the portal is based (e.g., a Cloud solution) optimizes this process by shortening booking times and thus speed up the physical delivery of goods. Therefore, by eliminating all traditional methods such as telephone/email, and any other means that can be used to send data in an unstructured format, it is possible to speed up the booking process downstream. Summarizing, the exploitation of the technology creates a single, shared platform accessible from every device, for an integrated management of inter-company processes: it supports a single standard method for managing bookings and, more generally, delivery activities through a shared tool within the entire Supply Chain for Household Appliances and Consumer Electronics industry.

The benefits of using the Collaborative Logistics portal are several and transversal for each category of users accessing the platform. The positive contribution of the platform is not limited to purely administrative activities (e.g., management of documentation related to the delivery process), but also has an impact on typically operational activities (e.g., loading/unloading of the vehicle). This project allows to improve both the supply side (efficiency) and the demand-side (effectiveness) issues. By combining these two elements, the project helps adjusting Supply Chain processes to better meet demand without neglecting efficiency. In addition, having identified a single standard and structured communication language enables better communication between the parties, which can be immediately understood and automated. The combination of these features allows to improve the efficiency and effectiveness of the process, thanks to a better quality of data (proven, for example, by the reduction of errors in data due to data entry) and a higher productivity of the resources used that can be freed from the execution of low value-added activities to devote themselves to other. The shared data allows to co-manage the customer and so creates implicitly a value for everyone.

In detail, the benefits for the Manufacturer related to the project implementation are the higher visibility on the products present at the Retailer PoS, with the opportunity to better program the internal operations to satisfy the retailer requirements and, if present, to incur in less penalties, the arrangement, with the other partners, of the quantities and the time windows of the deliveries in order to touch the goods as low times as possible, the optimization of the production planning, the increasing of orders frequency and accuracy and reduction of orders size, and the general reduction of logistics costs.

While the benefits for the Logistics Operator can be synthesised in the following way: the possibility to plan the delivery in advance, with a reduction of waiting time for loading at the depot and unloading at the Point of Sale, and the reduction of the number of trips, due to a major aggregation of different manufacturers.

The main benefits for the Retailer are, instead, the higher service level to the market, due to the certainty of delivery date from the Manufacturer, and the possibility of claiming about problems by using the online platform.

Finally, through the platform, it is also possible to monitor the performances by some Key Performance Indicators (KPIs).

Here are some examples: the appointment response: e.g., percentage of response, percentage of expired response, the degree of acceptance of Collaborative Logistics: e.g., percentage of acceptance, percentage of rejection, the exception monitoring: e.g., percentage of advances, number of reprogramming, and the monitoring deliveries (e.g., percentage of On-Time Deliveries).

According to the interviewed companies, different implementation barriers are present, such as: the internal organizational inertia to the change, for all the three considered actors, due to the employees' fear of moving on a new solution, very different from the traditional one. To overcome this barrier, is fundamental to involve all the human resources at any level and pass from a vertical organization to a transversal, cross-managed one, the necessity of educating the human resources to use the new technology, through a training program, with the aim of having skilled people, the Retailer resistance to stock and demand information sharing with the Manufacturer, and the difficulties in coordinating the logistics and the commercial functions.

4.1 Quantitative benefits

The cost structure has been modelled so that all the costs drivers are included in the simulation. The costs categories considered within the simulation are:

- Administrative costs;
- Inventory management costs;

- Stock-out costs;
- Transportation costs;
- Penalty costs.

Once the model has been initialised, it is possible to analyse the obtained results. The outputs refer to the benefits that each actor could achieve by adopting the Collaborative Logistics project.

The Manufacturer could achieve high benefits in implementing the Collaborative Logistics project. After its introduction, it faces a reduction of 11.25% in the total costs, which corresponds to an absolute value of € 5,880,787.24. In this way, it is easier to understand which the cost components are having the highest impact on the manufacturer benefits. The Inventory Management costs face the highest absolute variation from the AS-WAS to the AS-IS. However, the highest incidence value is covered by the Penalty costs. Consequently, even if their percentage variation is not the highest one (+25.16%), they have a huge effect on the total benefits due to their incidence on them (47.54% in the ASWAS and 67.05% in the AS-IS). On the opposite side, the Administrative costs have the biggest percentage variation (-93.07%) from the AS-WAS to the AS-IS situation, but they do not heavily affect the total benefits, as they cover a small portion of the total costs (0.48% in the AS-WAS and 0.04% in the AS-IS).

The Logistics Operator could achieve high benefits in implementing the Collaborative Logistics project equal to a reduction of 9.74% in the total costs, which corresponds to an absolute value of € 283,927.70. The reduction in the other costs leads to an overall benefit, even if the total costs reduction is the lowest compared to the Manufacturer and the Retailer. The costs undergoing the highest variations, thanks to the Collaborative Logistics project, are the Administrative costs. The percentage variation of the Transportation costs is -11.5%, they have a huge effect on the total benefits due to their incidence on them (60.71% in the AS-WAS and 59.54% in the AS-IS). Moreover, the Administrative costs have the biggest percentage variation (-72.39%) from the AS-WAS to the AS-IS situation, but they do not heavily affect the total benefits, as they cover a small portion of the total costs (7% in the AS-WAS and 2.4% in the AS-IS).

The Retailer benefits are the highest compared to the Manufacturer and Logistics Operator ones. The Retailer achieves a reduction in the total cost of -17.24%, corresponding to an absolute value of € 24,554,784.31. The costs undergoing the highest variations, thanks to the Collaborative Logistics project, are the Administrative and Stock-out costs. Moreover, the increase in the Inventory Management Costs is a consequence of the reduction of the bullwhip effect. The inventory benefits are not equally distributed among the Manufacturer and the Retailer; only the Retailer faces the avoidable costs of inventory management and, to make it accept the strategy, a costs' sharing assurance is required. The most impacting costs over the total benefits are the Transportation costs. The Transportation costs are reduced because there is an optimization of the trips that allows to saturate better the trucks and given that the number of trucks used is tied to the cost, consequently the total costs are reduced. Therefore, this high percent variation has a huge impact on the total benefits as the Transportation costs cover a big portion of them (75.88% in the AS-WAS and 80.01% in the AS-IS).

After having explained the benefits for each actor of the triad, it is fundamental to analyze the improvements under the entire Supply Chain perspective. Considering the Supply Chain as a whole, the benefits are still positive. In the AS-WAS situation, each player acted as an individual entity and, as it is clear from the data shown above, the goal of everyone was to minimize its own costs rather than the Supply Chain ones. The consequence was inefficiency due to higher Supply Chain costs, in particular, in terms of both Inventory and Stock-Out. There is a reduction of 35.51% in the total costs, which corresponds to an absolute value of € 23,686,761.21. The highest benefits for the Supply Chain come from the Stock Out Costs reduction (- € 12,210,000.00). In particular, the Transportation costs are composed by the Transportation share and the Diesel share, but since the Transportation share represents a cost for the Retailer and an income for the Logistics Operator, they are not differential considering the entire Supply Chain perspective. Moreover, the total Supply Chain Inventory Management costs mainly decrease because of the Manufacturer, this is due to the implementation of the PULL logic instead of the PUSH. Thanks to the collaboration, the inventories shift to the Retailer, enabling the chain to cope with the demand variation. This implies that one player in the Supply Chain bears more risks than the other players along the Supply Chain. As concerns the Administrative costs, they represent a very small part of the total costs, but they are the only ones that decrease for each of the three actors.

5 Conclusions and further development

This paper has the main aim to provide an effective practical tool for managers to understand the main benefits of the Collaborative Logistics project. The analytical model has been developed to foster the diffusion of this process by showing the potential benefits achievable through its implementation. Moreover, the impact for scholars is the analytical contributions, since there is a lack of quantitative assessment stemming from the Collaborative Logistics in the current literature. The work sheds light on a new process that is becoming more and more important in the Italian context and that has not been deeply studied by scholars yet.

Nevertheless, some limits are present. First of all, just three actors along the chain have been considered: Retailer, Manufacturer and Logistics Operator. An important hint for future researches could be to study the inclusion of other actors upstream, as an example the manufacturers' supplier. Furthermore, a simplification has been introduced within the analytical model: only a one-to-one relationship has been studied among the three actors.

It would be interesting to simulate a real Supply Chain with the introduction of multiple actors at different echelons and discovering how their interactions could change the results. Furthermore, better results could be achieved in more efficient way, programming Excel macros to simulate the interactions among different Supply Chain actors. Finally, even other analytic tools could be used in future researches to reach higher precision and complexity level, such as S.I.R.I.O., which is a simulation tool that allows to assess the cost differential between different hypotheses. It allows to minimise the overall costs for the chain and understand how costs vary as the variables varies.

Acknowledgments

We want to thank Giulia Cervello and Simone for their support in the performance of the present research. We also thank the companies interviewed for the active support in this study, providing the necessary data to obtain the appraisals presented.

References

Akyuz, G. A., & Rehan, M. 2009. Requirements for forming an 'e-supply chain'. International Journal of Production Research, 47(12), 3265-3287.

Anderson, D. L., & Lee, H. 1999. Synchronized supply chains: the new frontier. Achieving supply chain excellence through technology, 2, 12-21.

Anthony, T. 2000. Supply chain collaboration: success in the new internet economy. Achieving supply chain excellence through technology, 2, 41-44.

Braglia, M., and Frosolini, M. 2014. An integrated approach to implement project management information systems within the extended enterprise. International Journal of Project Management, 32(1), 18-29.

Cao, M., & Zhang, Q. 2011. Supply chain collaboration: Impact on collaborative advantage and firm performance. Journal of operations management, 29(3), 163-180.

Cascio, W. F., & Montealegre, R. 2016. How technology is changing work and organizations. Annual Review of Organizational Psychology and Organizational Behavior, 3, 349-375.

Cervello G., & Triventi S. 2018. A quantitative assessment of the collaborative logistics benefits in the consumer electronics and household appliances industry. Available at: <https://www.politesi.polimi.it/handle/10589/144441> [Accessed 22 May 2019].

Giménez, C., & Ramalhinho, H. 2004. E-supply chain management [: review, implications and directions for future research. Universitat Pompeu Fabra.

Kaveh, N., & Khosravi Samani, N. 2009. How collaborative logistics management increases supply chain efficiency.

L'eSupply Chain Collaboration in Italia, eInvoicing & eCommerce B2b Observatory, Politecnico of Milan, 2018. [online] Available at: <https://www.osservatori.net/it_it/pubblicazioni/l-esupply-chain-collaboration-in-italia> [Accessed 14 June 2019].

Luo, Y., Wirojanagud, P., & Caudill, R. J. 2001. Network-based optimization and simulation of sustainable e-supply chain management. In Proceedings of the 2001 IEEE International Symposium on Electronics and the Environment, pp. 185-190.

Narus, J. A., & Anderson, J. C. 1995. Using teams to manage collaborative relationships in business markets. Journal of Business-to-Business Marketing, 2(3), 17-46.

Pant, S., Sethi, R., & Bhandari, M. 2003. Making sense of the e-supply chain landscape: an implementation framework. International Journal of Information Management, 23(3), 201-221.

Tan, X., Yen, D. C., and Fang, X. 2002. Internet integrated customer relationship management a key success factor for companies in the e-commerce arena. Journal of Computer Information Systems, 42(3), 77-86.

Algorithm for Situation-dependent Adaptation of Velocity for Shuttle Based Systems

Thomas Kriehn[1], Franziska Schloz[2], Robert Schulz[2], Markus Fittinghoff[1]

1 – Heilbronn University

2 – University of Stuttgart

Purpose: Shuttle based storage and retrieval systems (SBS/RS) are suitable for applications which require a high throughput. Many times, however, the maximum performance of SBS/RS is not required. For example, when customers initiate a large number of retrieval requests at a specific time, or when a large number of storage requests enter the system at fixed times due to scheduled inbound deliveries. This article presents and discusses an algorithm that is based on closed-loop-control.

Methodology: A situation-dependent adaptation of the velocity to the currently required throughput or the number of currently awaiting orders requires an algorithm which needs to be implemented in the control of the SBS/RS. A simulation model of a SBS/RS will be introduced, which contains the control of the shuttle carriers and elevators as well as a model for calculating the energy requirement.

Findings: The results of this paper is the quantified energy saving by the application of the algorithm for situation-dependent adaption of velocity for SBS/RS

Originality: To our knowledge this is the first paper that introduces a situation-dependent adaption of velocity for SBS/RS.

Keywords: Shuttle Based Systems, Simulation, Energy Consumption, Velocity Adaption

First received: 19.May.2019 **Revised:** 28.May.2019 **Accepted:** 17.June.2019

1. Introduction

SBS/RS consists of one or more shuttle carriers, at least one elevator, a rack structure and a control system (VDI-2692, 2015). Systems with aisle- and tier-captive shuttle carriers are often used for high troughput demands. Tier-to-tier shuttle carriers cannot change the aisle, but the tiers. In such cases, the shuttle carrier uses the elevator to change tiers. This usually leads to throughput-reducing waiting times (Kriehn, 2017). It is also possible to combine both systems. The combination is same as a tier-captive SBS/RS with an additional elevator for shuttle carriers.

Aisle- and tier-captive SBS/RS use a shuttle carrier for each tier, which cannot leave the tier. The shuttle carrier and the elevator use buffer locations at each tier to store or retrieve totes. As a result, the horizontal and vertical transport is largely decoupled from one another. Accordingly, aisle- and tier-captive SBS/RS can often achieve a higher throughput than tier-to-tier SBS/RS (Kriehn, 2017).

Figure 1 shows an aisle- and tier-captive, tier-to-tier and a combined SBS/RS.

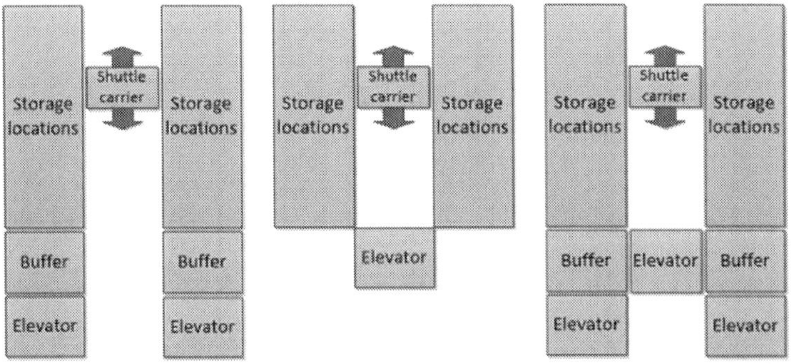

Figure 1: Tier-captive (left), tier-to-tier (middle) and combined (rigtht) SBS/RS, top view

The energy consumption of a SBS/RS can be decreased by reduced velocity of the shuttle carriers. But if the velocity is reduced permanently, the throughput also decreases. Therefore, the velocity should be lowered or increased depending on the situation. By reducing the velocity, no or only a small reduction of the throughput and the retrieval request processing time should occur. Therefore, the velocity should be adapted to the current situation in the SBS/RS. With only a few retrieval requests, the velocity should be relatively low, with many requests, the velocity should be relatively high. This paper presents an algorithm for that purpose.

We compare the energy consumption (calculated in Joule), the troughput (measured with a simulation model in totes per hour) and the retrieval request processing time (in seconds) with and without applying the algorithm for situation-dependent adaptation of velocity. The energy consumption of an SBS/RS consists of the energy consumption for elevators, shuttle carriers and tote handling attachments as well as the constant (standby) energy consumption. The throughput of an SBS/RS is the number of totes that are stored or retrieved in a certain time, e.g. in one hour. The retrieval request processing time is the time that a retrieval request spends in the SBS/RS. The time starts as soon as the retrieval request is received by the SBS / RS, and it ends as soon as the corresponding tote leaves the SBS/RS.

This paper is structured as follows: Chapter 1 gives an introduction. Chapter 2 contains an explanation of the existing literature on papers that deal with energy consumption for SBS/RS (2.1), the description of the analytical model for the calculation of the energy consumption (2.2) and the simulation model (2.3) as well as the algorithms for situation-dependent adaptation of velocity for shuttle carriers. (2.4). Chapter 2 ends with the results received by the applied algorithm (2.5). Chapter 3 contains a summary of the paper.

2. Models and algorithms

2.1 Literature

In (Ekren, 2018; Eder, 2017; Borovinsek, 2017; Akpunar, 2016; Lerher, 2016) models for calculating the energy consumption of SBS/RS are shown. In (Ekren, 2018; Borovinsek, 2017; Akpunar, 2016; Lerher, 2016) the power of the elevators and shuttle carriers are calculated with friction coefficients and multiplied by the travel time. This leads to the calculation of the energy consumption. With these analytical models, it is possible to show the potential in decreasing the velocity of shuttle carriers. With appropriate analytical models for calculating the throughput, e.g. in (Eder, 2017; Lerher, 2016), can also be shown how the achievable throughput decreases by lowering the velocity. It is also possible to measure the throughput with a simulation model, as we do in this paper.

In (Eder, 2017) the energy consumption is calculated on the basis of the law of conservation of energy. Coefficients of friction for horizontal movement are taken into account, but only for traveling with constant velocity. Our approach is the calculation of the energy consumption on the basis of the balance of forces. We consider for our model coefficients of friction for the entire route (acceleration, traveling with constant velocity, deceleration). The friction coefficient thus also increases the consumption of energy during the acceleration and reduces the recuperation of energy during the deceleration.

In this paper the reduction of energy consumption without or with very little loss of throughput is shown by applying the developed algorithm. The algorithm decreases and increases the velocity of the shuttle carriers depending on the number of retrieval requests. To our knowledge there is no other situation-dependent adaptation of the velocity of shuttle carriers published.

2.2 Analytical model to calculate energy consumption

The energy consideration includes the following motion processes:

- Movement lift,
- movement shuttle carrier, and
- movement tote handling attachement.

The movements can be divided into two movement patterns:

- Movement with only acceleration and deceleration (case 1) and
- movement, with acceleration, deceleration and a constant velocity component (case 2).

Figure 2 shows the movement sequence for case 1.

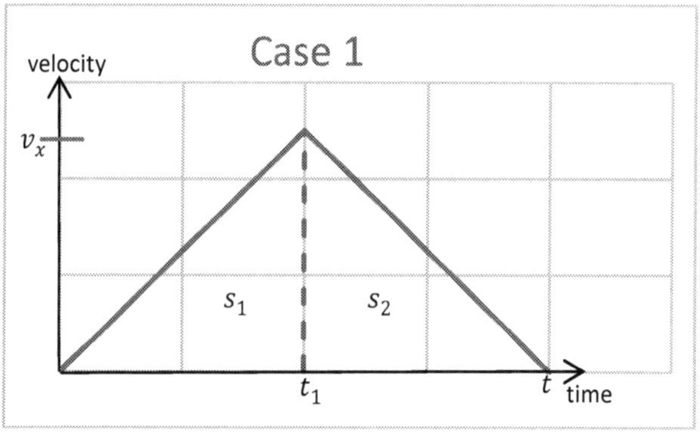

Figure 2: Movement sequence, case 1

s_1 is the distance traveled from the beginning to the end of the acceleration. s_2 is the distance traveled from the beginning to the end of the deceleration. The length of the distance s_1 has the numerical value of the area below the function line to t_1. t_1 is the time to reach v_x. t is the time required for the movement, from the beginning of the movement to the standstill of the mass. If acceleration and deceleration are identical, then s_1 and s_2 are equal ($s_1 = v_x \cdot \frac{t_1}{2} = s_2$). For different acceleration and deceleration values is $s_2 = v_x \cdot \frac{t_2 - t_1}{2}$. v_x is the achieved velocity. v_x is less than or equal to the maximum achievable velocity v.

Figure 3 shows the movement sequence for case 2.

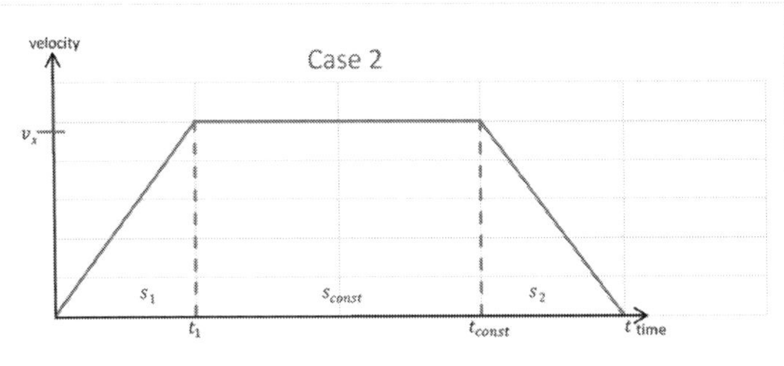

Figure 3: Case 2

s_{const} is the distance traveled at constant velocity.

Calculation of the critical distance to make a case distinction

Case 1 occurs when the distance to be traveled is less than or equal to the distance at which the maximum velocity is reached. Case 2 occurs when the distance to be traveled is greater than the distance at which the maximum velocity is reached. The distance required to reach the maximum velocity (and then immediately start with the deceleration) is referred to as the critical distance s_{crit}.

The critical distance is calculated as follows:

$$s_{crit} = \frac{a_1}{2}(t_1)^2 + \frac{a_2}{2}(t - t_1)^2$$

$$s_{crit} = \frac{a_1}{2}\left(\frac{v}{a_1}\right)^2 + \frac{a_2}{2}\left(\frac{v}{a_2}\right)^2$$

$$s_{crit} = \frac{1}{2}v^2\left(\frac{1}{a_1} + \frac{1}{a_2}\right) \tag{1}$$

a_1 denotes the acceleration, a_2 the deceleration.

Calculation of the travel time for case 1

It applies $s \leq s_{crit}$.

The travel time for case 1 is calculated as follows:

$$s = \frac{1}{2} {v_x}^2 \left(\frac{1}{a_1} + \frac{1}{a_2} \right)$$

$$v_x = \sqrt{\frac{2s}{\left(\frac{1}{a_1} + \frac{1}{a_2} \right)}}$$

$$t = \frac{v_x}{a_1} + \frac{v_x}{a_2}$$

$$t = \left(\frac{1}{a_1} + \frac{1}{a_2} \right) \sqrt{\frac{2s}{\left(\frac{1}{a_1} + \frac{1}{a_2} \right)}} \tag{2}$$

Calculation of the travel time for case 2

It applies $s > s_{crit}$.

The travel time for case 2 is calculated as follows:

$$t = \frac{v}{a_1} + \frac{v}{a_2} + \frac{s - s_{crit}}{v}$$

$$t = \frac{v}{a_1} + \frac{v}{a_2} + \frac{s - \frac{1}{2} v^2 \left(\frac{1}{a_1} + \frac{1}{a_2} \right)}{v}$$

$$t = \frac{v}{2} \left(\frac{1}{a_1} + \frac{1}{a_2} \right) + \frac{s}{v} \tag{3}$$

Calculation of the energy consumption of the vertical movement (elevator)

For the calculation of the energy consumption of the elevator movement, it is assumed that the elevator needs energy when moving upwards, and that it can maybe recover energy while moving downwards. If no energy can be recovered, since no recuperation is used, the recuperation efficiency is 0 percent, so no energy is consumed and none is recovered for moving downwards.

The energy consumption of an elevator movement upwards is calculated as follows:

$$E_{Elevator,up} = \frac{1}{\eta_{Elevator}} (m_{Elevator} + m_{Shuttle} + m_{Tote}) \cdot g \cdot s_{Elevator} \quad (4)$$

$\eta_{Elevator}$ is the efficiency factor of the elevator. $m_{Elevator}$ is the moving mass of the elevator. $m_{Shuttle}$ is the mass of the shuttle carrier. For tier-captive SBS/RS applies $m_{Shuttle} = 0$, since no shuttle carriers are transported by the elevator. m_{Tote} is the mass of the tote. If no tote is transported by the elevator, then $m_{Tote} = 0$. g is the gravitational acceleration. $s_{Elevator}$ is the vertical distance traveled.

The energy demand of an elevator movement downwards is calculated as follows:

$$E_{Elevator,down} = -\eta_{Elevator,Rekup} \cdot (m_{Elevator} + m_{Shuttle} + m_{Tote}) \cdot g \cdot s_{Elevator} \quad (5)$$

$-\eta_{Elevator,Rekup}$ is the recuperation efficiency factor of the elevator. If no recuperation is used, $-\eta_{Elevator,Rekup} = 0$. A negative value for the energy consumption corresponds to an energy recovery.

Calculation of the energy requirement of the horizontal movement (shuttle carrier or tote handling attachement) for case 1

Figure 4 shows the balance of forces of the horizontal movement during the acceleration process. The air resistance is neglected, due to low estimated impact.

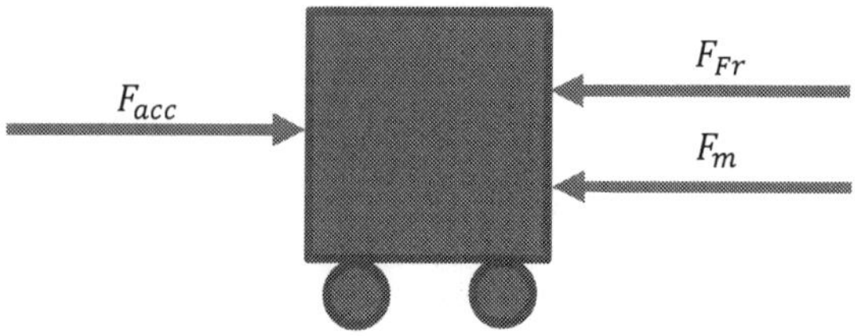

Figure 4: Balance of forces horizontal mass movement during acceleration

F_{acc} is the force needed to accelerate the mass. F_{Fr} is the friction force. F_m is the force of mass inertia.

The friction force is calculated as follows:

$$F_{Fr} = \mu_r \cdot F_N = \mu_r \cdot F_G = \mu_r \cdot (m_{Shuttle} + m_{Tote} + m_H) \cdot g$$

m_H denotes the mass of the tote handling attachment (only the moving mass of it). μ_r denotes the coefficient of friction. When calculating a movement with a shuttle carrier, μ_r uses the coefficient of friction for the shuttle carrier, and when there is movement of the tote handling attachment, the coefficient of friction is applied to the tote handling attachment. When a movement of the tote handling attachment is calculated, then $m_{Shuttle} = 0$. If there is no tote on the shuttle carrier unit during the move, then $m_{Tote} = 0$.

The force of mass inertia is calculated as follows:

$$F_m = (m_{Shuttle} + m_{Tote} + m_H) \cdot a_1$$

The balance of forces of horizontal acceleration gives:

$$F_{acc} = F_m + F_{Fr} = (m_{Shuttle} + m_{Tote} + m_H) \cdot a_1 + \mu_r \cdot (m_{Shuttle} + m_{Tote} + m_H) \cdot g$$

The energy requirement for horizontal acceleration thus results:

$$E_{acc} = F_{acc} \cdot s_1$$

$$E_{acc} = \mu_r \cdot (m_{Shuttle} + m_{Tote} + m_H) \cdot g \cdot s_1 + (m_{Shuttle} + m_{Tote} + m_H) \cdot a_1 \cdot s_1$$

$$E_{acc} = \mu_r \cdot (m_{Shuttle} + m_{Tote} + m_H) \cdot g \cdot s_1 + (m_{Shuttle} + m_{Tote} + m_H) \cdot a_1 \cdot \frac{1}{2} a_1 {t_1}^2$$

$$E_{acc} = \mu_r \cdot (m_{Shuttle} + m_{Tote} + m_H) \cdot g \cdot s_1 + (m_{Shuttle} + m_{Tote} + m_H) \cdot a_1 \cdot \frac{1}{2} a_1 \frac{{v_x}^2}{{a_1}^2}$$

$$E_{acc} = \mu_r \cdot (m_{Shuttle} + m_{Tote} + m_H) \cdot \mathrm{g} \cdot s_1 + \frac{1}{2} (m_{Shuttle} + m_{Tote} + m_H) {v_x}^2$$

Ancillary calculation to express s_1 as a function of acceleration and deceleration and total distance traveled:

$$v_x = \sqrt{\frac{2s}{\left(\frac{1}{a_1} + \frac{1}{a_2}\right)}}$$

$$s_1 = \frac{1}{2} a_1 {t_1}^2$$

$$t_1 = \frac{v_x}{a_1}$$

$$s_1 = \frac{1}{2} a_1 \frac{{v_x}^2}{{a_1}^2}$$

$${v_x}^2 = \frac{2s}{\left(\frac{1}{a_1} + \frac{1}{a_2}\right)} = 2s \cdot \frac{a_1 \cdot a_2}{a_1 + a_2}$$

$$s_1 = \frac{1}{2} \frac{\left(\sqrt{\frac{2s}{\left(\frac{1}{a_1} + \frac{1}{a_2}\right)}}\right)^2}{a_1}$$

$$s_1 = \frac{1}{2} \frac{\left(\frac{2s}{\left(\frac{1}{a_1} + \frac{1}{a_2}\right)}\right)}{a_1}$$

$$s_1 = s \frac{\left(\frac{1}{\left(\frac{1}{a_1} + \frac{1}{a_2}\right)}\right)}{a_1}$$

$$s_1 = \frac{s}{1 + \frac{a_1}{a_2}}$$

$$s_1 = s \frac{a_2}{a_2 + a_1}$$

Thus, the energy consumption of the acceleration can be formulated as follows:

$$E_{acc} = \mu_r \cdot (m_{Shuttle} + m_{Tote} + m_H) \cdot g \cdot s \frac{a_2}{a_1 + a_2} + (m_{Shuttle} + m_{Tote} + m_H) \cdot s \cdot \frac{a_1 \cdot a_2}{(a_1 + a_2)}$$

If we substitute $a = \frac{a_1 \cdot a_2}{(a_1 + a_2)}$, then follows:

$$E_{acc} = (m_{Shuttle} + m_{Tote} + m_H) \cdot s \cdot a \cdot \left(\mu_r \cdot g \cdot \frac{1}{a_1} + 1\right) \quad (6)$$

With consideration of an efficiency factor η_h, the energy requirement is:

$$E_{acc,h} = \frac{1}{\eta_h}(m_{Shuttle} + m_{Tote} + m_H) \cdot s \cdot a \cdot \left(\mu_r \cdot g \cdot \frac{1}{a_1} + 1\right) \tag{7}$$

When a movement is calculated for a shuttle carrier, the efficiency factor for the shuttle carrier is used, when a movement of the tote handling attachement is calculated, the efficiency factor of the tote handling attachement is used.

In the following, the energy recovery during the deceleration is calculated. Figure 5 shows the balance of forces of horizontal mass movement during braking. Air resistance is neglected.

Figure 5: Force balance horizontal mass movement during deceleration

The balance of forces of the deceleration gives:

$$F_{dec} = F_m - F_{Fr}$$

The energy recovery of the deceleration thus results in:

$$E_{dec} = (F_m - F_{Fr})s_2$$

$$E_{dec} = (m_{Shuttle} + m_{Tote} + m_H) \cdot s \cdot a - \mu_r \cdot (m_{Shuttle} + m_{Tote} + m_H) \cdot g \cdot s \frac{a_1}{a_1 + a_2}$$

$$E_{dec} = (m_{Shuttle} + m_{Tote} + m_H) \cdot s \cdot a \cdot \left(1 - \mu_r \cdot g \cdot \frac{1}{a_2}\right) \tag{8}$$

With consideration of the efficiency of the recuperation $\eta_{h,r}$, the energy recovery results in:

$$E_{dec,h} = \eta_{h,r}(m_{Shuttle} + m_{Tote} + m_H) \cdot s \cdot a \cdot (1 - \mu_r \cdot g \cdot \frac{1}{a_2}) \tag{9}$$

The total energy consumption for case 1 is calculated as follows:

$$E_{case1,h} = E_{acc,h} - E_{dec,h}$$

$$E_{case1,h} = \frac{1}{\eta_h}(m_{Shuttle} + m_{Tote} + m_H) \cdot s \cdot a \cdot \left(\mu_r \cdot g \cdot \frac{1}{a_1} + 1\right) -$$

$$-\eta_{h,r}(m_{Shuttle} + m_{Tote} + m_H) \cdot s \cdot a \cdot (1 - \mu_r \cdot g \cdot \frac{1}{a_2})$$

$$E_{case1,h} = (m_{Shuttle} + m_{Tote} + m_H)s \cdot a\left(\frac{1}{\eta_h}\left(\mu_r \cdot g \cdot \frac{1}{a_1} + 1\right) - \eta_{h,r}\left(1 - \mu_r \cdot g \cdot \frac{1}{a_2}\right)\right) \tag{10}$$

Calculation of the energy consumption of the horizontal movement (shuttle carrier or tote handling attachement) for case 2

In the case of the accelerated movement, in case 2, the achieved velocity is known (maximum velocity), therefore, the formulas for calculating the acceleration and the deceleration can be simplified as follows.

The energy consumption for the acceleration is calculated as follows:

$$E_{acc} = \mu_r \cdot (m_{Shuttle} + m_{Tote} + m_H) \cdot g \cdot s_1 + \frac{1}{2}(m_{Shuttle} + m_{Tote} + m_H)v^2$$

$$E_{acc} = \mu_r \cdot (m_{Shuttle} + m_{Tote} + m_H) \cdot g \cdot \frac{v^2}{2a_1} + \frac{1}{2}(m_{Shuttle} + m_{Tote} + m_H)v^2$$

$$E_{acc} = \frac{1}{2}(m_{Shuttle} + m_{Tote} + m_H)v^2 \cdot \left(\mu_r \cdot g \cdot \frac{1}{a_1} + 1\right) \tag{11}$$

With consideration of the efficiency factor η_h, the energy consumption of the acceleration results:

$$E_{acc,h} = \frac{1}{2\eta_h}(m_{Shuttle} + m_{Tote} + m_H)v^2 \cdot \left(\mu_r \cdot g \cdot \frac{1}{a_1} + 1\right) \tag{12}$$

The energy recovery of the deceleration is calculated as follows:

$$E_{dec} = \frac{1}{2}(m_{Shuttle} + m_{Tote} + m_H)v^2 \cdot (1 - \mu_r \cdot g \cdot \frac{1}{a_2}) \tag{13}$$

Including the efficiency factor of the energy recovery $\eta_{h,r}$ of the deceleration results in:

$$E_{dec,h} = \frac{\eta_{h,r}}{2}(m_{Shuttle} + m_{Tote} + m_H)v^2 \cdot (1 - \mu_r \cdot g \cdot \frac{1}{a_2}) \tag{14}$$

The energy consumption of the constant velocity movement is calculated below. The balance of forces is shown in Figure 6.

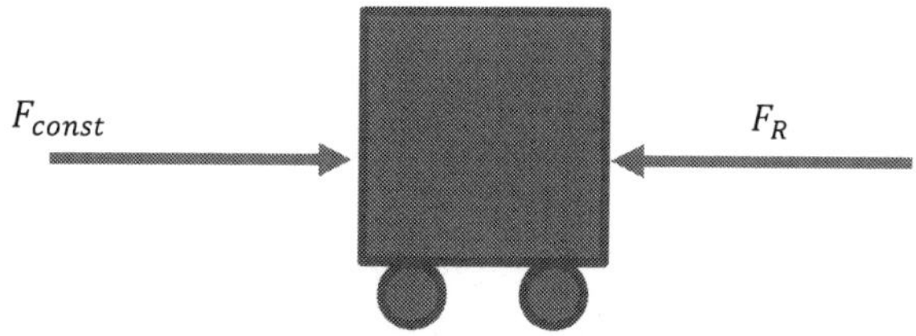

Figure 6: Balance of forces horizontal mass movement during constant velocity motion

The balance of forces results:

$$F_{const} = F_R$$

The energy consumption of the movement with constant velocity results in:

$$E_{const} = F_R \cdot s_{const}$$

$$E_{const} = \mu_r \cdot (m_{Shuttle} + m_{Tote} + m_H) \cdot g \cdot s_{const}$$

$$E_{const} = \mu_r \cdot (m_{Shuttle} + m_{Tote} + m_H) \cdot g \cdot (s - s_1 - s_2)$$

$$E_{const} = \mu_r \cdot (m_{Shuttle} + m_{Tote} + m_H) \cdot g \cdot \left(s - \frac{v^2}{2} \left(\frac{1}{a_1} + \frac{1}{a_2} \right) \right) \tag{15}$$

Taking into account the efficiency factor η_h results in the energy consumption:

$$E_{const,h} = \frac{1}{2\eta_h} \mu_r \cdot (m_{Shuttle} + m_{Tote} + m_H) \cdot g \cdot \left(s - \frac{v^2}{2} \left(\frac{1}{a_1} - \frac{1}{a_2} \right) \right) \tag{16}$$

The following summarizes the energy consumption of accelerated motion and constant motion:

$$E_{acc} + E_{const} = \mu_{roll} \cdot (m_{Shuttle} + m_{Tote} + m_H) \cdot g \cdot s_1 + \frac{1}{2}(m_{Shuttle} + m_{Tote} + m_H)v^2 + \mu_r \cdot (m_{Shuttle} + m_{Tote} + m_H) \cdot g \cdot s_{const}$$

$$E_{acc} + E_{const} = \mu_r \cdot (m_{Shuttle} + m_{Tote} + m_H) \cdot g \cdot (s_1 + s_{const}) + \frac{1}{2}(m_{Shuttle} + m_{Tote} + m_H)v^2$$

$$E_{acc} + E_{const} = \mu_r \cdot (m_{Shuttle} + m_{Tote} + m_H) \cdot g \cdot (s - s_2) + \frac{1}{2}(m_{Shuttle} + m_{Tote} + m_H)v^2$$

$$E_{acc} + E_{const} = \mu_r \cdot (m_{Shuttle} + m_{Tote} + m_H) \cdot g \cdot (s - \frac{v^2}{2a_2}) + \frac{1}{2}(m_{Shuttle} + m_{Tote} + m_H)v^2$$

$$E_{acc} + E_{const} = (m_{Shuttle} + m_{Tote} + m_H)\left(\mu_r \cdot g \cdot \left(s - \frac{v^2}{2a_2}\right) + \frac{1}{2}v^2\right) \quad (17)$$

With consideration of the efficiency factor η_h, the energy consumption is:

$$E_{acc,h} + E_{const,h} = \frac{1}{2\eta_h}(m_{Shuttle} + m_{Tote} + m_H)\left(\mu_r \cdot g \cdot \left(s - \frac{v^2}{2a_2}\right) + \frac{1}{2}v^2\right) \quad (18)$$

The total energy consumption for case 2 is:

$$E_{case2,h} = E_{acc,h} + E_{const,h} - E_{dec,h}$$

$$E_{case2,h} = \frac{1}{2\eta_h}(m_{Shuttle} + m_{Tote} + m_H)\left(\mu_r \cdot g \cdot \left(s - \frac{v^2}{2a_2}\right) + \frac{1}{2}v^2\right) - \frac{\eta_{h,r}}{2}(m_{Shuttle} + m_{Tote} + m_H)v^2 \cdot \left(1 - \mu_r \cdot g \cdot \frac{1}{a_2}\right)$$

$$E_{case2,h} = (m_{Shuttle} + m_{Tote} + m_H)\left(\frac{1}{2\eta_h}\left(\mu_r \cdot g \cdot \left(s - \frac{v^2}{2a_2}\right) + \frac{1}{2}v^2\right) - \frac{\eta_{h,r}}{2}v^2 \cdot \left(1 - \mu_r \cdot g \cdot \frac{1}{a_2}\right)\right) \quad (19)$$

2.3 Simulation Model

The simulation model was developed as part of the research project SmartShuttle. This paper focuses is on the description of the situation-dependent adaptation of velocity for SBS/RS. The simulation model, however, includes other storage management policies, such as class-based storage, sequencing of retrieval requests, dwell point strategies, and warehouse reorganisation. The model can be configured online (http://smartshuttle.hs-heilbronn.de), the simulation run is started after entering the configuration data (size of the warehouse, velocity of elevators and shuttle carriers, etc.) on the server. The user receives the results automatically after the simulation run is completed. This takes approximately five to thirty minutes.

The simulation model can be used for tier-captive and tier-to-tier SBS/RS, see Figure 1. In addition, a combined version can also be simulated by adding an elevetator in the middle of the aisle for transporting shuttle carriers to an originally tier-captive SBS/RS, see also Figure 1 (then not all tiers need to be occupied by shuttle carriers). In the combined version, the elevator in the middle is only responsible for transferring unloaded shuttle carriers to another tier on the basis of corresponding requests. The simulation is based on the software AutoMod. The visualization of the simulation model is three-dimensional. Figure 7 shows a visualization example of the simulation model with a tier-to-tier configuration.

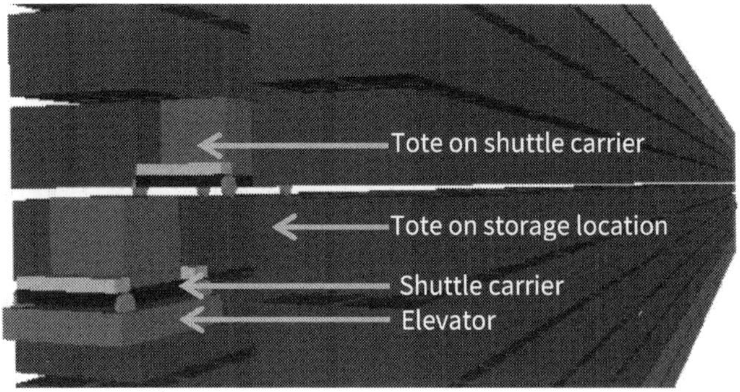

Figure 7: Visualization of the simulation model

For the simulations performed for this paper, the capacity of the elevator and shuttle carriers is one. This paper uses the simulation model for tier-to-tier SBS/RS. The simulation model can simulate single- and double-deep SBS/RS. Also retrieval requests can be given as input, over a period of twenty four hours. The retrieval requests entered per hour are equally distributed during the hour in the SBS/RS. As soon as a retrieval request has been completed by the SBS/RS, it waits for a configurable time before it is sent to the SBS/RS as a storage request. Elevator and shuttle carriers travel singe- or dual-command cycles, depending on the current request situation. An attempt will be made whenever possible to travel dual-command cycles. Each tier can accommodate a maximum of one shuttle carrier. Figure 8 shows the process of single- and dual-command cycles of elevator and shuttle carriers.

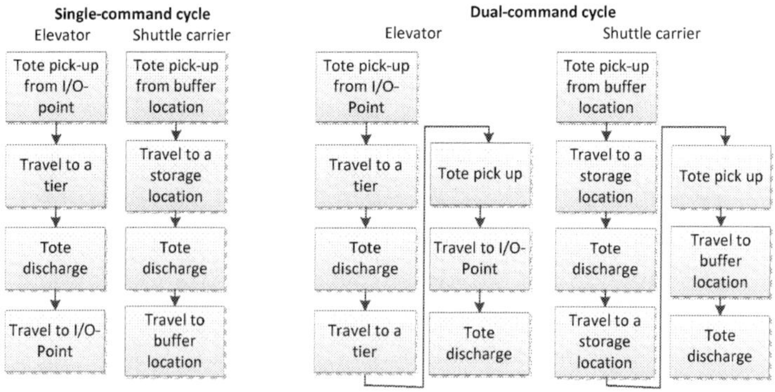

Figure 8: Single- and dual-command cycle of elevator and shuttle carrier

The following outputs of the simulation run were measured: energy consumption of the shuttle carriers, energy consumption of the entire SBS/RS, throughput of the SBS/RS (one aisle is considered) and retrieval request processing time. The energy consumption of all shuttle carriers is measured every hour over a simulation period of twenty four hours. The troughput of the SBS/RS and the retrieval request processing time is also measured every hour. The retrieval request processing time refers to a single position of an order. In result when a tote leaves the SBS/RS, the retrieval request processing time for it is measured. The retrieval requests are assigned to an order (positions of an order). The quantity of retrieval requests that belong to an order influences the retrieval request processing time, since the SBS/RS then simultaneously enters this quantity of retrieval requests at an equally distributed time within the hour. If there are many orders entering the system, then the retrieval request processing time increases. For this study, the amount of retrieval requests assigned to an order is equally distributed between one and nine. The storage management policy closest-location is used. Elevator and shuttle carriers travel to the closest possible location to store a tote. That leads to random storage assignment in the filled area of the aisle, because every tote in the SBS/RS is chosen with the same probability to be retrieved. In result totes that entering the SBS/RS could be stored in this (new) free storage locations. The dwell-point is point-of-service-completion (POSC). The retrieval request processing follows the first-in-first-out (FIFO) principle by the shuttle carriers within the tiers. The elevator also follows the FIFO-principle. However, in the case of many retrieval requests, some retrieval requests may have longer waiting times, since there is no shuttle carrier in the corresponding tier and the elevator travels to the closest location to transport a shuttle carrier to another tier.

2.4 Algorithm for adaption of velocity

The algorithm is based on a closed loop control with three-step controller with hysteresis, as described in the literature, e. g. in [1]. Figure 9 shows the controller output (velocity) depending on the controller input (waiting retrieval requests).

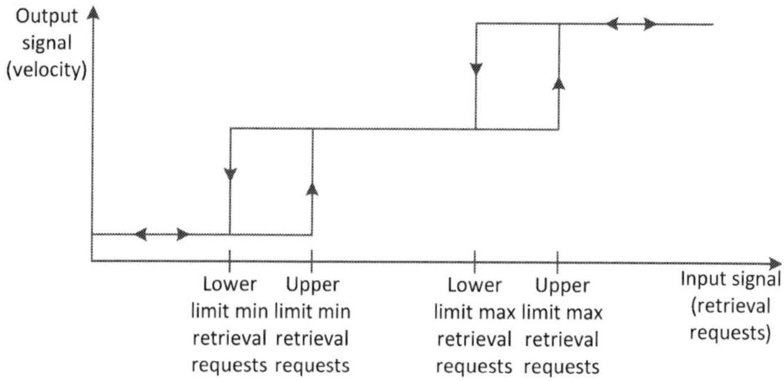

Figure 9: Output signal from the three-step controller, dependent from the input signal

The algorithm includes storage requests as another input signal. If the storage requests exceed a configurable limit, the highest velocity is issued, regardless of the number of waiting retrieval requests. This can prevent long queues forming at the entry point of the SBS/RS. The behavior of the three-step controller, as shown in Figure 9, only occurs as long as there are fewer waiting storage requests than the configured limit. In addition to the velocity adjustment, an adaptation of the acceleration and the deceleration takes place.

The algorithm is shown below:

if retrieval requests >= upper limit max retrieval requests or storage requests >= limit storage requests

set velocity, acceleration, deceleration to max value

else

if retrieval requests =< lower limit max retrieval requests and velocity = max) or (retrieval requests >= upper limit retrieval requests min and velocity = min)

set velocity, acceleration, deceleration to middle value

else

if retrieval requests <= lower limit retrieval requests

set velocity, acceleration, deceleration to min value

In the simulation model, the algorithm is executed after each completed cycle of the elevator and the variable values for velocity, acceleration and deceleration are set. The shuttle carriers check the variables before each travel and the values are updated.

2.5 Results

In the following, the influence of a situation-dependent adaption of the velocity on the energy consumption for different variants is shown.

Table 1 shows the constant parameter values of the simulated SBS/RS. Table 2 shows the variants.

Table 1: Constant parameter values

Parameter	**Value**
Distance between tiers [m]	0.4
Distance between first tier and I/O-point [m]	-1
Distance between aisle and first storage position in lane [m]	0.5
Distance between aisle and second storage position in lane [m]	1
Storage positions per tier	100
Distance between storage positions	0.5
Velocity tote handling attachment, unloaded	2
Velocity tote handling attachment, loaded	1
Acceleration and deceleration tote handling attachment, unloaded	2

Parameter	Value
Acceleration and deceleration tote handling attachment, loaded	1
Friction coefficient for shuttle carrier	0.06
Friction coefficient for tote handling attachment	0.5
Storage ratio [%]	95
Mass elevator	100
Mass shuttle carrier	100
Mass tote	50
Mass tote handling attachment	15
efficiency factor for energy consumption elevator [%]	70
efficiency factor for energy consumption shuttle carrier [%]	70

Parameter	Value
efficiency factor for energy consumption handling unit [%]	70
efficiency factor for energy recuperation elevator [%]	40
efficiency factor for energy recuperation shuttle carrier [%]	40
efficiency factor for energy recuperation tote handling attachment [%]	40
Time Gap until a retrieved tote enters the SBS/RS to be stored [s]	500 +-100
SBS/RS	Tier-to-tier, double-deep
Tiers	12
Velocity elevator	4
Acceleration elevator	4

Parameter	Value
Velocity shuttle carrier	7
Acceleration shuttle carrier	6
Pick-up and set-down time shuttle carrier	4
Velocity reduction middle [%]	60
Velocity reduction min [%]	90

Table 2: Variants

Variant	1	2	3	4	5	6	7	8
	100,	100,	100,	100,	100,	100,	100,	100,
	70,	70,	200,	200,	200,	200,	70,	70,
	90,	90,	300,	300,	300,	300,	90,	90,
Retrieval	150,	150,	150,	150,	150,	150,	150,	150,
requests	200,	200,	100,	100,	100,	100,	200,	200,
per hour(	300,	300,	50,	50,	50,	50,	300,	300,
	50,	50,	30,	30,	30,	30,	50,	50,
	30,	30,	10,	10,	10,	10,	30,	30,
	170,	170,	20,	20,	20,	20,	170,	170,

	120, 100, 50, 30, 140, 230, 120, 100, 90, 40, 30, 20, 10, 25, 10	120, 100, 50, 30, 140, 230, 120, 100, 90, 40, 30, 20, 10, 25, 10	30, 10, 20, 30, 100, 150, 120, 100, 90, 40, 30, 20, 10, 25, 10	30, 10, 20, 30, 100, 150, 120, 100, 90, 40, 30, 20, 10, 25, 10	30, 10, 20, 30, 100, 150, 120, 100, 90, 40, 30, 20, 10, 25, 10	30, 10, 20, 30, 100, 150, 120, 100, 90, 40, 30, 20, 10, 25, 10	120, 100, 50, 30, 140, 230, 120, 100, 90, 40, 30, 20, 10, 25, 10	120, 100, 50, 30, 140, 230, 120, 100, 90, 40, 30, 20, 10, 25, 10
Upper limit max retrieval requests	50	20	50	20	50	20	50	20
Lower limit max retrieval requests	40	10	40	10	40	10	40	10

Upper limit min retrieval requests	30	5	30	5	30	5	30	5
Lower limit min retrieval requests	20	2	20	2	20	2	20	2
Shuttle carriers	5	5	5	5	2	2	3	3
Pick-up and set-down time elevator	4	4	4	4	4	4	1	1

Figure 10 to 15 shows the energy consumption for the shuttle carriers (only for traveling, without tote handling), the throughput of the SBS/RS and the retrieval processing time for all variants. The retrieval request processing time is the average of all retrieval request processing times of one hour. It is calculated separately for each hour.

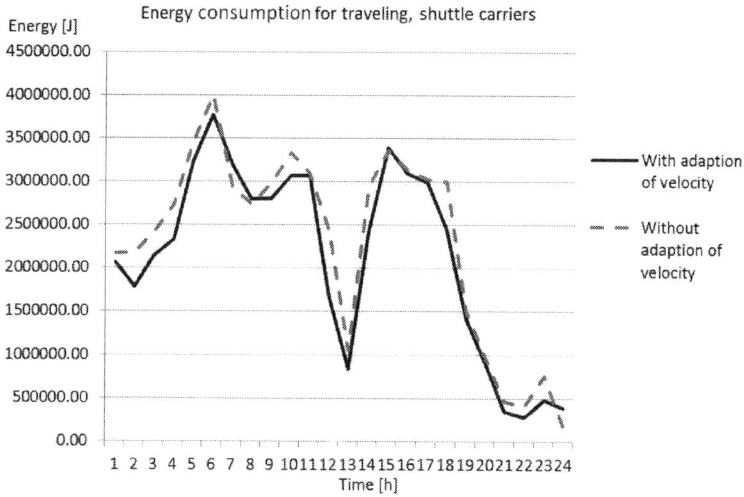

Figure 10: Energy consumption for traveling, shuttle carriers, variant 1

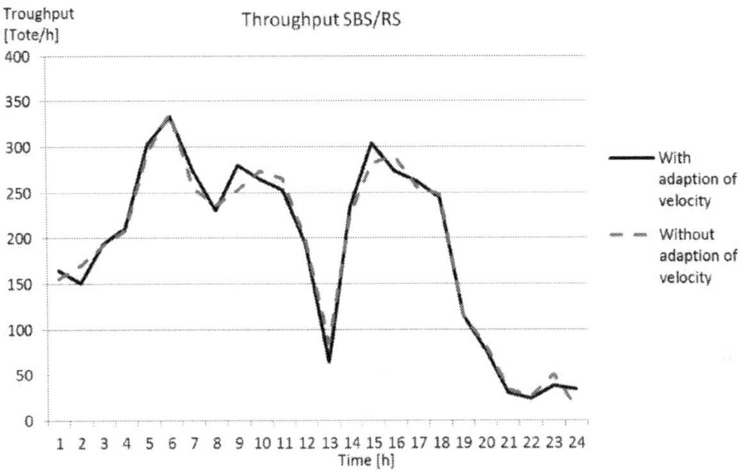

Figure 11: Throughput SBS/RS, variant 1

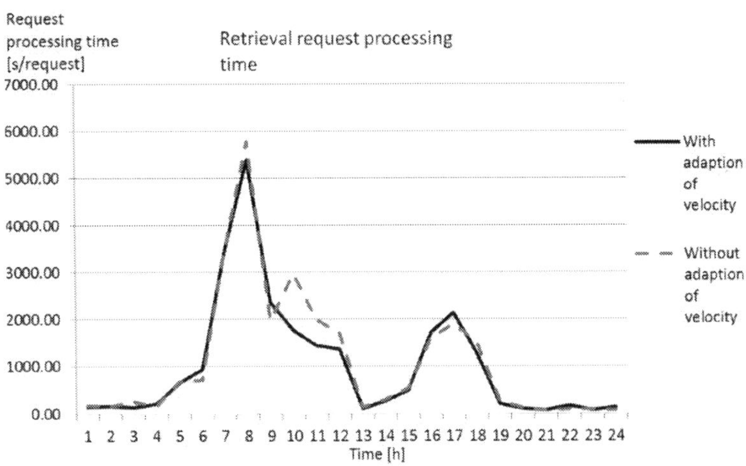

Figure 12: Retrieval request processing time, variant 1

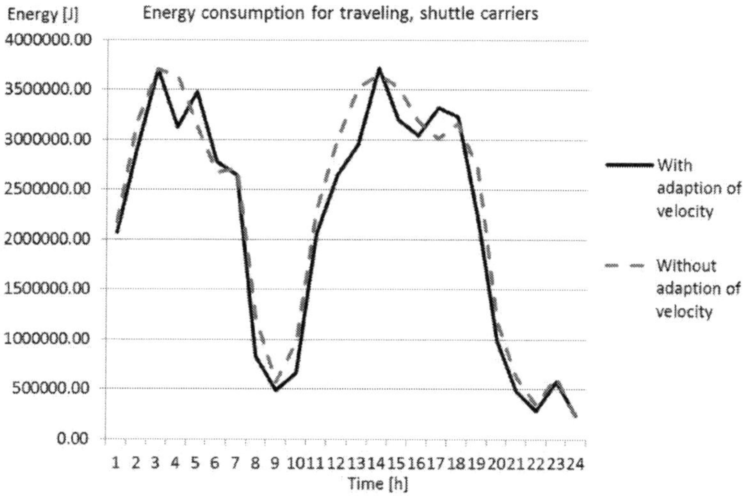

Figure 13: Energy consumption for traveling, shuttle carriers, variant 3

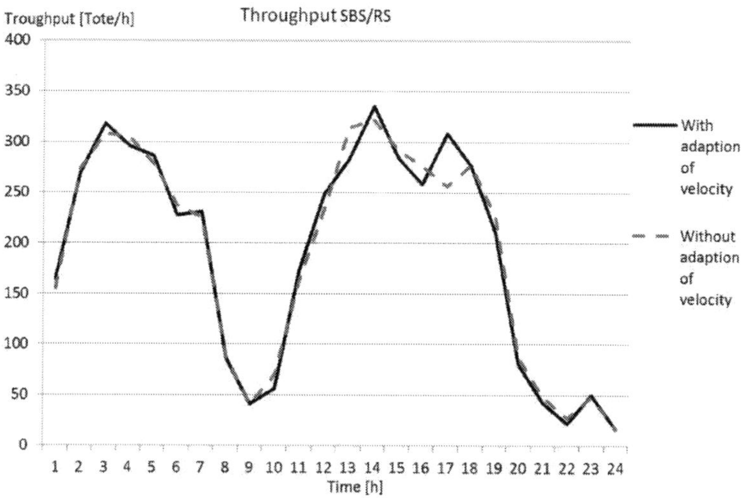

Figure 14: Throughput SBS/RS, variant 3

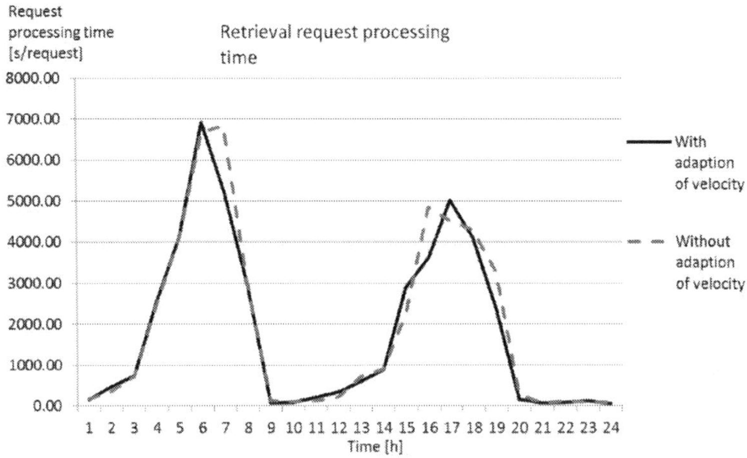

Figure 15: Retrieval request processing time, variant 3

Table 3 shows the energy consumption for the shuttle carriers and the SBS/RS (including tote handling, elevator and shuttle carrier traveling) without constant (or standby) energy consumption for elevator and shuttle carriers. Constant energy consumption is the consumption of energy that is independent from traveling or tote handling. Whether an elevator or shuttle carrier is active with traveling/tote handling or not, this value remains the same.

Table 3: Energy consumption of variants

	Energy consumption shuttle carriers [J/(24 h)]	**Energy consumption for SBS/RS [J/(24 h)]**	**Reduction of energy consumption shuttle carriers [%]**	**Reduction of energy consumption SBS/RS [%]**
Variant 1 with adaption of velocity	50918139	71377549	7.85	5.86
Variant 2 with adaption of velocity	51128641	71292362	7.47	5.97
Variant 1 and 2 without adaption of velocity	55258511	75817101		

	Energy consumption shuttle carriers [J/(24 h)]	**Energy consumption for SBS/RS [J/(24 h)]**	**Reduction of energy consumption shuttle carriers [%]**	**Reduction of energy consumption SBS/RS [%]**
Variant 3 with adaption of velocity	51647973	72075765	6.14	4.07
Variant 4 with adaption of velocity	52486542	72757432	4.62	3.17
Variant 3 & 4 without adaption of velocity	55028070	75137493		

	Energy consumption shuttle carriers [J/(24 h)]	**Energy consumption for SBS/RS [J/(24 h)]**	**Reduction of energy consumption shuttle carriers [%]**	**Reduction of energy consumption SBS/RS [%]**
Variant 5 with adaption of velocity	53132358	73301028	1.89	1.23
Variant 6 with adaption of velocity	53790160	74044842	0.67	0.23
Variant 5 & 6 without adaption of velocity	54153257	74217491		

	Energy consumption shuttle carriers [J/(24 h)]	Energy consumption for SBS/RS [J/(24 h)]	Reduction of energy consumption shuttle carriers [%]	Reduction of energy consumption SBS/RS [%]
Variant 7 with adaptation of velocity	50094455	71169149	12.00	8.76
Variant 8 with adaptation of velocity	52937432	74093209	7.00	5.01
Variant 7 & 8 without adaption of velocity	56925019	78000624		

As can be seen in the figures, throughput is only slightly changed by the adaptation of the velocity. In the first four variants, the lift is the major bottleneck, as the SBS/RS is equipped with five shuttle carriers. For variants five and six, the shuttle carriers are the bottleneck, since only two shuttle carriers are used. The retrieval request processing time increases slightly due to the velocity adaption during peak times for variants five and six. Variants five and six have an retrieval request entry that uses the SBS/RS with many retrieval request at peak times and rarely otherwise. Since only two shuttle carriers are available, the achievable maximum throughput is lower. This creates a long queue at the peak. The SBS/RS therefore runs at maximum velocity most of the time. Only when the queue is reduced, the velocity is reduced. Reducing of the queue takes longer than with variants three and four (which have the same hourly retrieval requests) due to fewer shuttle carriers in the SBS/RS. Since only very few retrieval requests are received between the two peaks, only a few retrieval requests are processed correspondingly at a reduced velocity.

Variant seven has the highest savings potential, here the retrieval request intake is rather evenly distributed, there are also two peaks, but in the meantime more retrieval requests are coming into the system. Accordingly, it is more often possible to travel at reduced velocity. Variation seven also has broader limits for the algorithm, so the velocity is reduced earlier respectively increased sooner. Nevertheless, the throughput and the retrieval request processing time is only slightly affected.

The influence of the broadened limits for the algorithm is most apparent when comparing variants five and six. Variant five has the broader limits and the retrieval request processing time is significantly higher at the second peak than at variant six, in which narrower limits are used.

Accordingly, the application of the algorithm depends on a suitable parameterization of the limits. The broader the limits, the more energy can be saved. At the same time, from a certain point on, a significant reduction in throughput and retrieval request processing time occurs. These should be avoided. Narrowed limits leads to less influence of the algorithm to save energy.

The retrieval request situation also has an impact on energy savings: if the SBS/RS receives more retrieval requests than it can handle, a long queue is formed. Then, the velocity cannot be reduced until the queuing is largely reduced. If there are subsequently no phases with request entries below the achievable throughput of the SBS/RS, there is little potential for saving energy. Velocity can only be reduced while the retrieval requests intake is less than maximum throughput.

The variants show that often energy savings of more than 5% are possible without having a significant effect on the throughput achieved and the retrieval request processing time. The energy consumption for the traveling of the shuttle carriers could be reduced in the variants up to 12 percent and the total energy consumption up to 8.76 percent.

3. Conclusion

This paper shows an analytical model for the energy calculation of SBS/RS and an algorithm for situation-dependent velocity adaptation. The analytical model is based on the balance of forces of the moving masses. For the horizontal movement the friction was considered. This model can be used to determine the energy consumption of the elevators, the shuttle carriers and the tote handling attachement. The algorithm is based on a three-step controller. The input signal is the number of retrieval requests. The output signal is one of the three velocity levels. The algorithm is configurable and can be adapted to a specific SBS/RS. If the limits for adjusting the velocity are chosen to be relatively high, a higher amount of energy can be saved. Excessive limits can lower throughput, however. The algorithm reduces the energy consumption, but at the same time the throughput is hardly influenced by proper parameterization. The algorithm and the energy model were applied within a simulation model of an SBS/RS. Results for energy saving and minimal impact on throughput and retrieval request processing time were shown.

Further interesting research topics for future work:

- The analytical energy model does not calculate energy savings by lowering the velocity of the elevator, so the velocity of the elevator was not lowered by the algorithm. By extending the model, it would be possible to calculate reductions of energy consumption by velocity regulation of the elevator. The algorithm allows shuttle carriers and elevators to reduce their velocity.
- The algorithm can be optimized in terms of energy savings, possibly by increasing the steps (multi-step controller) or another closed loop control.
- The input size for the algorithm can be changed. It does not necessarily have to be the quantity of retrieval requests that causes the velocity adaption, it may also be the waiting time of the shuttle carriers or the utilization of shuttle carriers. It is also conceivable to adjust the velocity to request-related priorities or specified deadlines.

- Methods of artificial intelligence, e.g. for deep reinforcement learning for adapting the velocity and optimize energy savings for SBS/RS.
- Furthermore, a velocity reduction also leads to less maintenance costs, and this relationship could be explored in the future.

References

Akpunar, A., Yetkin, E. and Lerher, T., 2017. Energy efficient design of autonomous vehicle based storage and retrieval system. Istrazivanja i projektovanja za privredu, [e-journal] 15(1), pp. 25–34. http://dx.doi.org/10.5937/jaes15-12132.

Borovinšek, M., Ekren, B. Y., Burinskienė, A. and Lerher, T., 2017. MULTI-OBJECTIVE OPTIMISATION MODEL OF SHUTTLE-BASED STORAGE AND RETRIEVAL SYSTEM. Transport, [e-journal] 32(2), pp. 120–137. http://dx.doi.org/10.3846/16484142.2016.1186732.

Eder, M. and Kartnig, G., 2017. Calculation method to determine the troughput and the energy consumption of S/R shuttle systems. In: N. Zrnic, S. Bosnjak, and G. Kartnig. XXII International Conference on "Material Handling, Constructions and Logistics" - MHCL 2017. 4th-6th October 2017, pp. 201–206.

Ekren, B. Y., Akpunar, A., Sari, Z. and Lerher, T., 2018. A tool for time, variance and energy related performance estimations in a shuttle-based storage and retrieval system. Applied Mathematical Modelling, [e-journal] 63, pp. 109–127. http://dx.doi.org/10.1016/j.apm.2018.06.037.

Kriehn, T., Schloz, F., Wehking, K.-H. and Fittinghoff, M., 2017. Storage management policies for throughput optimization of shuttle-based storage and retrieval systems. In: N. Zrnic, S. Bosnjak, and G. Kartnig. XXII International Conference on "Material Handling, Constructions and Logistics" - MHCL 2017. 4th-6th October 2017, pp. 177–184.

Lerher, T., Zrnić, N. Đ. and Jerman, B., op. 2016. Throughput and energy related performance calculations for shuttle based storage and retrieval systems. New York: Nova Science Publishers, Inc.

Schulz, G., 2010. Regelungstechnik 1. Lineare und Nichtlineare Regelung, Rechnergestützter Reglerentwurf. 4th ed. [e-book]. München: Oldenbourg. Available at: Schulz, Gerd (VerfasserIn). <http://www.oldenbourg-link.com/isbn/9783486707946.

VDI-2692, 2015. Automated vehicle storage and retrieval systems for small unit loads.

Disruptive Technologies - Integration in Existing Supply Chain Processes

Stephanie Niehues[1], *Tan Gürpinar*[1]

1 – TU Dortmund

Purpose: Based on technological progress, companies innovate and design a variety of new processes to be implemented in order to gain a competitive advantage. Nevertheless, many companies face issues during the initial integration of these emerging technologies with their existing business processes. This paper will collect and analyze existing procedures to leverage process innovations by means of emerging information- and production technologies with disruptive potential.

Methodology: An exploratory research of relevant procedures for the implementation of new processes is chosen to give a systematic overview of existing models. Furthermore, requirements for the technology integration of emerging information technologies and production technologies are collected within a systematic content analysis and based on these requirements, the existing models are compared and evaluated.

Findings: None of the existing models meet the requirements of a technology-based integration of process innovations. In order to integrate particularly emerging technology based innovations into business processes, existing models must be further developed, particularly with regard to the flexibility requirements of industry 4.0 process changes.

Originality: This paper collects and compares all important models for the implementation of process innovations, based on requirements specific to the integration of emerging information- and production technologies. Blockchain technology and additive manufacturing are used as exemplary current technologies with disruptive potential.

Keywords: Process Innovations, Blockchain, Additive Manufacturing, Supply Chain

First received: 19.May.2019 **Revised:** 28.May.2019 **Accepted:** 23.June.2019

1 Introduction

The development of emerging technologies, especially of those that have a disruptive potential, are increasingly changing existing business processes and affecting the entire supply chain of organizations. In recent years, more and more administrative processes have become obsolete and new processes based on given disruptive technologies have diffused into the processes of organizations. Many of these so called process innovations have been integrated into existing business processes without a strategic approach.

One important field to represent process innovations is the field of IT technologies, where we have process innovations like the blockchain technology that has the potential to radically change business processes. Organizations from various industries already started pilot projects to transfer the technology to their specific case of application. The blockchain, representing a ledger of data that gets distributed over several network nodes, builds a base not only for digital money transactions, but also for chaincode, the so called smart contracts, which represent the missing link for the realization of the vision of Industry 4.0 and the Internet of Things. (Xu, Weber and Staples, 2019, p. 3 ff.) Apart from IT technologies, we also have manufacturing technologies with disruptive potential. Building up on the digitalization of processes, new product manufacturing possibilities like additive manufacturing are out into place. The technology also possesses an enormous potential to affect the entire supply chain by creating the ability to save costs and pare down process steps in several stages (Moreira, Ferreira and Zimmermann, 2018, p. 262)

In this paper, the most important models for the integration of processes will be listed in order to evaluate their suitability for the integration of disruptive technologies like blockchain and additive manufacturing along the following research questions:

- What requirements can be seen for the integration of process innovations, with emphasis on new technologies?
- In which areas do existing integration models have to be adapted, in order to achieve a structured integration of new technologies?

2 Theoretical background

2.1 Innovation

Innovation, as defined by Porter, includes both technological improvements and better methods or ways of doing things. This can manifest itself in product changes, process changes, new marketing approaches, new sales forms and new concepts on scope and reach. (Porter, 2014)

In the result-oriented view, innovations are consequently understood as "technological, economic, legal and social innovations in the form of products, processes, contract forms and distribution channels". (Corsten and Gössinger, 2016, p. 6 ff.) The various innovations are differentiated in more detail by describing the subject of innovation (performance, process, market or social), the degree of innovation (incremental or radical) and the reference unit for determining the novelty characteristic (business-oriented, customer-oriented, competitive) (Schallmo and Brecht, 2017, pp. 23–24).

2.2 Process Innovations

Process innovations can be described as process improvements which include product creation process procedures and/or information distribution procedures. This type of innovation focuses on technologies that make processes more effective and consistent (Winkler and Kaluza, 2008). In combining the terms "process" and "innovation", process innovations can thus be understood as the renewal of an entire process or of individual process components (e.g. tasks and services). With regard to the degree of innovation, it should be noted that both radical and incremental further developments of processes occur. (Schallmo and Brecht, 2017, p. 25)

To build the link to disruptive technologies, blockchain technology can be seen as a radical process innovation by influencing the entire settlement of payment processes that can be secured and automatized (Holotiuk, Moormann and Pisani, 2019, p. 199 ff.). Furthermore, a process innovation can be represented by the adoption of technically new or significantly improved production methods (OECD, 2005, p. 53; Raymond and St-Pierre, 2010, p. 48). Thus, the introduction of additive manufacturing can be seen as a radical process innovation for organizations and supply chains. (Marzi, et al., 2018, p. 185).

3 Methodology

To evaluate existing models for the integration of new process, in a first step, existing models are collected within a systematic review following the approach of Fettke & Loos. After the systematic collection, the models are documented, in order to evaluate them on the basis of set criteria. (Fettke and Loos, 2004, p. 3)

In a second step, the requirements to evaluate the models are collected. Therefore we take into account requirements for the integration of innovations in general and requirements of IT- or rather production technologies for their integration in supply chain processes. Based on those requirements, the existing models are compared in regard to their suitability for the integration of disruptive technologies in existing supply chain processes.

3.1 Existing models for process management

3.1.1 Allweyer

The business-process-management-cycle of Allweyer (Figure 1) is primarily focused on IT-related processes, but can also be simultaneously applied to different fields (Allweyer, 2012, p. 89).

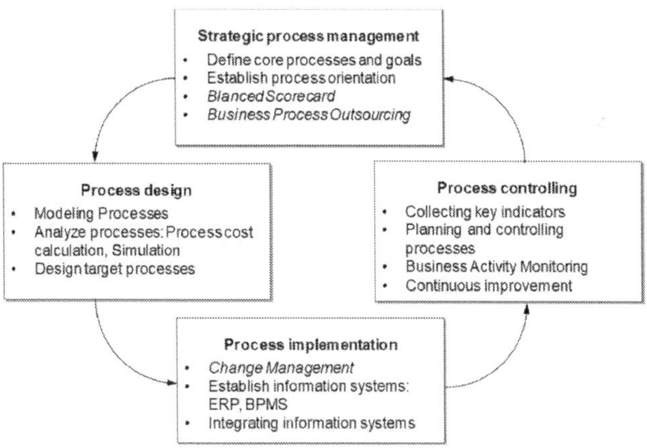

Figure 1: Business process management cycle (Allweyer, 2012, p. 91)

The first phase of the cycle, strategic process management, essentially refers to the design of the company and its relations with the environment. The next phase, "process design", has the task to identify, document and analyze the business processes of a company (Allweyer, 2012, p. 91 ff.). The third phase of the business process cycle deals with the question of how the designed business processes will be put into practice. (Allweyer, 2012, pp. 145, 314-315). The last phase, "process controlling", closes the business process management cycle. Essential tasks are the planning, monitoring and evaluation of the processes carried out during operation. (Allweyer, 2012, p. 385).

3.1.2 Brecht

Brecht's process model (Figure 2) contains five total phases, each of which is based on one another. These five phases are comprised of multiple subsections.

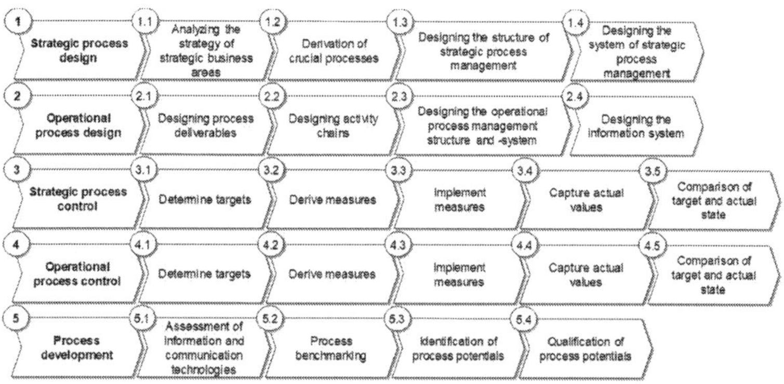

Figure 2: Procedural model by Brecht (Schallmo, Brecht and Ramosaj, 2018, p. 38)

In the first phase of strategic process design, the strategic business areas of the company are analyzed with the help of process architecture planning. In the second step, the strategic process design is refined until it is possible to collect the necessary reference variables needed to measure the new process performance. This is done so that not only will the processes of the developed process map be implemented as far as possible in the strategic process control phase, but they will also be regularly examined with regard to their output. In the phase of operational process control, the developed processes of process design are planned and implemented (Brecht, 2002, p. 207 f.). The process development phase represents the activities that evaluate the process potential gained while also initiating possible process optimizations. (Brecht, 2002, p. 202)

3.1.3 Davenport

The process model of Thomas H. Davenport can be divided into five phases, as shown in the figure below (Figure 3).

Figure 3: Procedural model by Davenport (Davenport, 1997, p. 25)

In the first phase, the processes that are to be improved or adapted must be identified. As such, all processes of the process map are examined with regard to this aspect. In the second phase the central question is where to start improving and changing processes in order to achieve the greatest possible effect. (Davenport, 1997, p. 36).

In the third phase, the goal is to determine the future characteristics of processes and then generate specific, measurable criteria for said processes. It thus links the strategic corporate goals with the operational procedures (Davenport, 1997, p. 119). Before the new process can be created, the existing process is first evaluated on the basis of the created process vision in the fourth phase. (Davenport, 1997, p. 137). The final phase, development and prototyping, elaborates the new processes and introduces them into the company. In addition, the organizational structure for the new process is determined (Davenport, 1997, p. 153).

3.1.4 Österle

The process model for the introduction of process innovations by Österle is divided into a total of eleven phases (Figure 4).

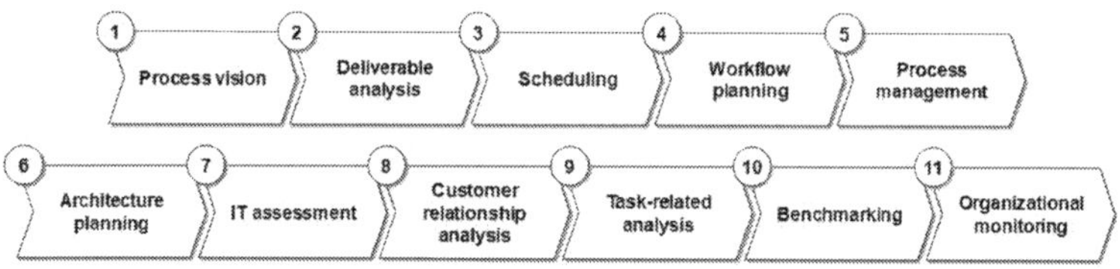

Figure 4: Procedure model by Österle (Schallmo and Brecht, 2017, p. 42)

The first phase, the creation of a process vision, aids the process design in searching for fundamentally new solutions and subsequently outlines the new process. (Österle, 1995, p. 63 ff.). A further performance analysis has the function of recording, evaluating and documenting services in detail (phase two). The next phase of the process planning aims at determining the tasks of a process and their sequence by assigning them to organizational units. (Österle, 1995, p. 85) In phases four and five, workflow management systems are presented to support the process control (Österle, 1995, pp. 100–101).

One level below, we have architecture planning as the focal point. According to the definition, this is a part of strategy development (Österle, 1995, p. 128). In phase seven, the IT Assessment focuses on the search for new potentials in the IT field. Phase eight, focuses on the analyzation of the co-operation between customer and supplier in order to increase customer benefits. Furthermore, the "task-related analysis" deals with the search for improvement potentials in the respective processes on the basis of the process characteristics or command variables. Prior to the last phase, standards are set for the process including all its components, until the procedure model finally deals with organizational monitoring (Österle, 1995, p 159ff.).

3.1.5 Schallmo & Brecht

The Schallmo and Brecht model is the most up-to-date procedural model of our selection, with respect to time. The authors often refer to the explanations of the Österle. However, in their own model, they limit themselves to a total of seven phases (as opposed to the eleven phases seen in the Österle model) as shown below (Figure 5).

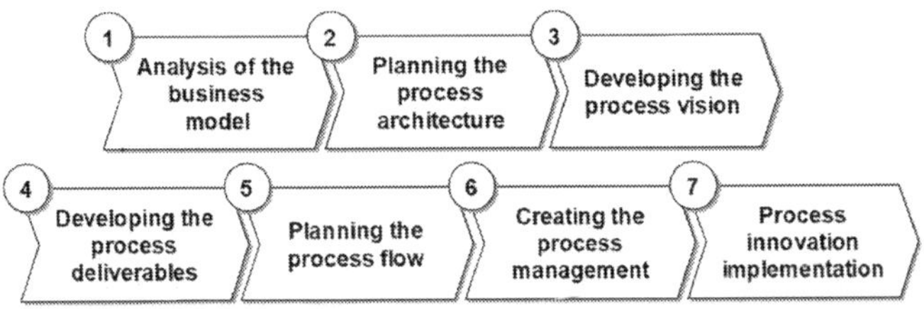

Figure 5: Procedure model by Schallmo and Brecht (Schallmo and Brecht, 2017, p. 67)

In the first phase of the "Business Model Analysis" model, the foundation is laid for future phases of the process model (Schallmo and Brecht, 2017, p. 85). The second phase of the procedural model deals with the planning of the process architecture. (Schallmo and Brecht, 2017, p. 103). The third phase of the procedural model focuses on the development of a process vision. (Schallmo and Brecht, 2017, p. 126). In the fourth phase, process services are to be developed. Important activities are the creation of a clear presentation of the process to be considered in interaction with other processes. (Schallmo and Brecht, 2017, p. 148). The fifth phase of the procedure model focuses on the creation of process flow diagrams and the necessary steps for the creation of such a flow chart (Schallmo and Brecht, 2017, p. 158). In the sixth phase of the procedure model the process control is created. (Schallmo and Brecht, 2017, p. 170). The last phase deals with techniques for the implementation of process design (Schallmo and Brecht, 2017, p. 180).

3.1.6 Becker

The procedure model according to Becker et al. is comprised of the seven phases shown below. (Figure 6)

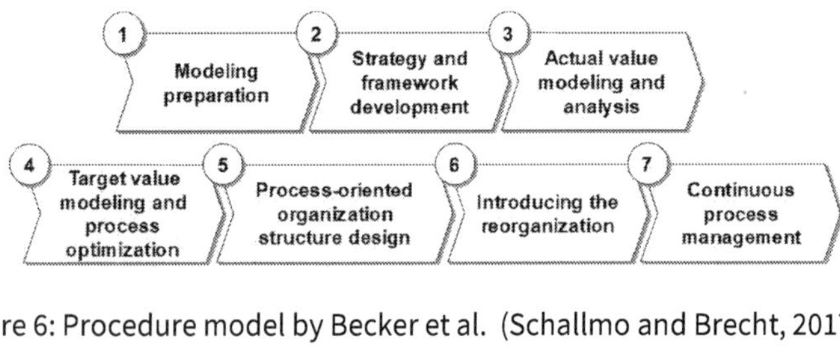

Figure 6: Procedure model by Becker et al. (Schallmo and Brecht, 2017, p. 37)

In the first phase of the procedure model all measures are taken to ensure the creation of high-quality process models (Becker, Kugeler and Rosemann, 2012, p. 51). The focus of the second phase deals with either the creation of a new process or the modification of an existing framework. (Becker, Kugeler and Rosemann, 2012, p. 113). In the third phase, the actual state of the company or process must be presented and modelled. (Becker, Kugeler and Rosemann, 2012, p. 166).

Based on the third step, the fourth phase can be used to initiate target modeling and process optimization. (Becker, Kugeler and Rosemann, 2012, p. 197). The fifth phase of the procedure model aims to efficiently optimize business processes with regard to the criteria time, quality, and costs (Becker, Kugeler and Rosemann, 2012, p. 229).

The penultimate stage deals with the implementation of the created processes in the company, while the seventh and final stage of the Becker process model deals with continuous process management (Becker, Kugeler and Rosemann, 2012, 8 ff.).

3.1.7 Schmelzer & Sesselmann

The business process model of Schmelzer und Sesselmann consists of four phases (Figure 7). A communication and training program runs parallel to phases one through four.

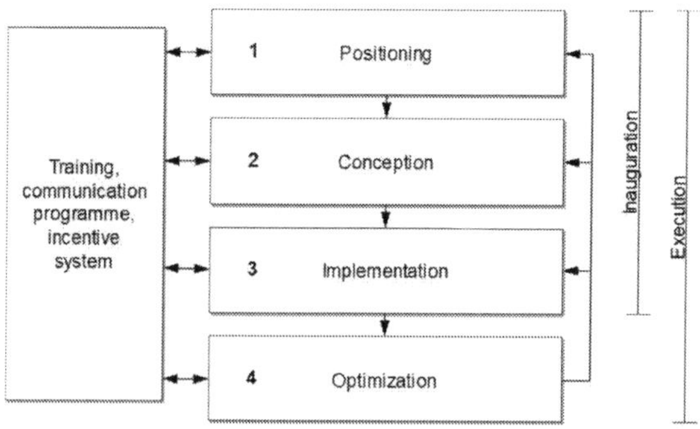

Figure 7: Procedure model by Schmelzer & Sesselmann (Schmelzer and Sesselmann, 2013, p. 545)

In the first project phase of the business process management, the initial situation is analyzed and the strategic orientation of the organization is reviewed. The focus of the conception phase is the identification and definition of business processes with their sub-processes as well as the framework decisions for the implementation phase (Schmelzer and Sesselmann, 2013, p. 551 ff.).

The third project phase essentially follows the milestones of the action plan adopted in a management workshop. These three phases all work together to reach the final project phase, which has the highest priority in the project. (Schmelzer and Sesselmann, 2013, p. 566 ff.).

3.2 Requirements of technology integration

Technology integration is a special challenge to most organizations, both in terms of activity volume and period of time. In particular the integration of innovations offers several requirements. These requirements will be summarized in the following chapters and categorized following the human-technology-organization-approach (HTO) (Karltun, et al., 2017).

3.2.1 Human-centered requirements

Motivation of the staff

As a first point to address human-centered aspects, it is imperative that the organization is able to motivate their staff to comply with the necessary technology. Following (Delany and D'Agostini, 2015, p. 10), it can be differentiated between promotion- or prevention-focused employees, that each need a different motivational approach.

Thus, the entire technological change process should be designed openly and transparently during its course. This allows employees to participate in the content of the project. (Ahsen, 2010, p. 10; Wellbrock and Göpfert, 2015, p. 285). In addition to transparency issues, knowledge management should also be taken into account for staff in innovation projects. (Hervas-Oliver, Sempere-Ripoll and Boronat-Moll, 2014, p. 882)

Strategy to adapt changes

The change process of the organization is of great importance in the context of technology integration. The corporate environment in which the integration is carried out, has to be adapted in such a way that the result of the integration can be used successfully. (Pleschak and Sabisch, 1996, pp. 128–129; Hervas-Oliver, Sempere-Ripoll and Boronat-Moll, 2014, p. 882)

Human factors play an especially important role when integrating emerging technologies into modern supply chain processes. Since employees of different supply chain partners will be affected by the integration, a clear overarching strategy needs to be developed and integrated in a flow-oriented way. (Wellbrock and Göpfert, 2015, p. 291)

3.2.2 Technology-centered requirements

Consideration of challenges and opportunities

Integration challenges and opportunities can be identified by the organization through an assessment of the existent technologies and their environment. For IT technologies this would also include business applications and supporting infrastructure. Required for a consideration in terms of challenges, it is necessary to raise one-time and time-phased costs, as well as risk related information(Nahass, Smith and Curragh, p. 4)

Technical decisions

As a further requirement category, the organization needs to take professional decisions regarding the actual technical integration. Therefore, it is necessary to consider adaptions to the current technological infrastructure and possible upgrades to be applied in order to fulfill future needs. (Delany and D'Agostini, 2015, pp. 22–25) Another aspect of interest would be the degree of autonomy that plays a particularly important role for IT technologies. Therefore, (Hasselbring, 2000, p. 35) it is a frequent requirement to have technologies that are about to be integrated, remain autonomous. (Nahass, Smith and Curragh, p. 4).

Process and product innovation

In many areas where innovations take place, they do not take place independently. Various innovations are mutually dependent. This applies in particular to the implementation of process innovations. Thus, the introduction of new process technologies, such as additive manufacturing, is a component of process innovations. (Pleschak and Sabisch, 1996, pp. 20–23). "Process and product technologies often are equal partners in innovation because process technology has to be in place to support new product innovations and without this capability new product developments will fail." (Brown, 2001, p. 467) Under this assumption it follows that a profound coordination between product and process innovation is necessary in a reciprocal relationship (Lischka, 2011, pp. 18–23).

Strategy to integrate the technology

In order to finally introduce and manage new technologies, it is important to develop a vision that is easily transferable to a technological strategy. This strategy will also strongly support the aforementioned aspect of staff involvement, because only those who know the purpose and the goal of process change will fully support the integration process (Harrison, 1990, p. 165). Furthermore, a technological edge over competitors can only happen if the supply chain is formulated with a broad production strategy and is implemented together with all its partners (Birasnav and Bienstock, 2019, pp. 143–144).

3.2.3 Organizational-centered requirements

Planning and structured approach

First and foremost for the organizational aspects, early planning activities and a clear understanding of the technology's attributes and functions are essential requirements for a successful integration. This stage is strongly connected to the organization's business model and builds the basis for several existing integration approaches to be chosen from. (Hasselbring, 2000, pp. 33–36)

To get a clear understanding of the technology, the organization should examine key attributes, considering all areas from which various requirements may possibly originate. Following Gollapudi, these ten attributes are key for an examination of the proposed technology: usability, interoperability, common business views, agility, scalability, reliability, openness, manageability, infrastructure and security (Gollapudi, Jangeti and Kotapati, 2012).

Holistic approach

Moreover, an important aspect is that the integration approach supports all aspects of the process innovation: from planning to introduction phase. A holistic approach to the integration of disruptive technologies is necessary.

Through a holistic approach and thus given transparency throughout all phases of the innovation project, necessary corporate structures that are affected by the innovation are taken into account and can be included in the transformation process (Wilfling, 2013, p. 44).

Complexity reduction

For most companies, innovation processes are unique because they cannot be structurally located in the normal organizational structure. They run parallel to most processes and usually affect several departments. As a result, projects and especially innovation projects are conspicuously complex due to their structure (Ahsen, 2010, p. 10). Finding the right measure of complexity is therefore a major challenge. An important aspect of complexity control is therefore to break down the core task and divide it into several sub-tasks. (Pleschak and Sabisch, 1996, p. 29)

Control mechanism

Control of the process is an important instrument in process management and in every management task. Controls are used to check the effectiveness of actions and ensure the quality of results. Often, the control of an activity takes place after its completion. Due to the long period of strategic processes, such as innovation processes, a final check alone is not sufficient. Further controls during the course of the project are necessary (Hungenberg, 2014, p. 369; Wellbrock and Göpfert, 2015, p. 291). So-called milestones in the course of the project are planned as control points. These can be seen as points at which sub-sections are completed. Hence, these milestone checks are necessary to determine if there is satisfactory progression within the project (Hungenberg, 2014, p. 306).

Methodological instruments

Methodological competence is particularly important in project management. Project planning and controlling instruments must be understood and applied. These competencies are a particularly important means of achieving project goals and adhering to time and resource constraints. (Hölzle, 2007, pp. 15–17). A well planned, methodological approach to problems aids in both reducing and structuring the workload (Ahsen, 2010, p. 72). The importance of sufficient planning and targeted control methods in project management can be statistically proven in most cases (Albers and Gassmann, 2011, p. 494).

Flexibility Enhancement

Another evaluation criterion is the freedom associated with a flexible innovation process design. In general, many different solutions can be applicable in this regard. However, in order to achieve the best possible solution, subject-specific decisions are necessary (Pleschak and Sabisch, 1996, p. 57). Thus, an ongoing adaptation of the project to the developing conditions is essential in order to prevent disruptions over the course of the project (Pleschak and Sabisch, 1996, p. 168). It is often a challenge to find the balance between the necessary freedom and the necessary room for maneuvering. Additionally, a sufficient structure with measurable and tangible goals helps to achieve the best possible project result.

As a final step of the chapter, the requirements can be summarized and assigned to the higher level HTO-categories, as can be seen in Figure 8.

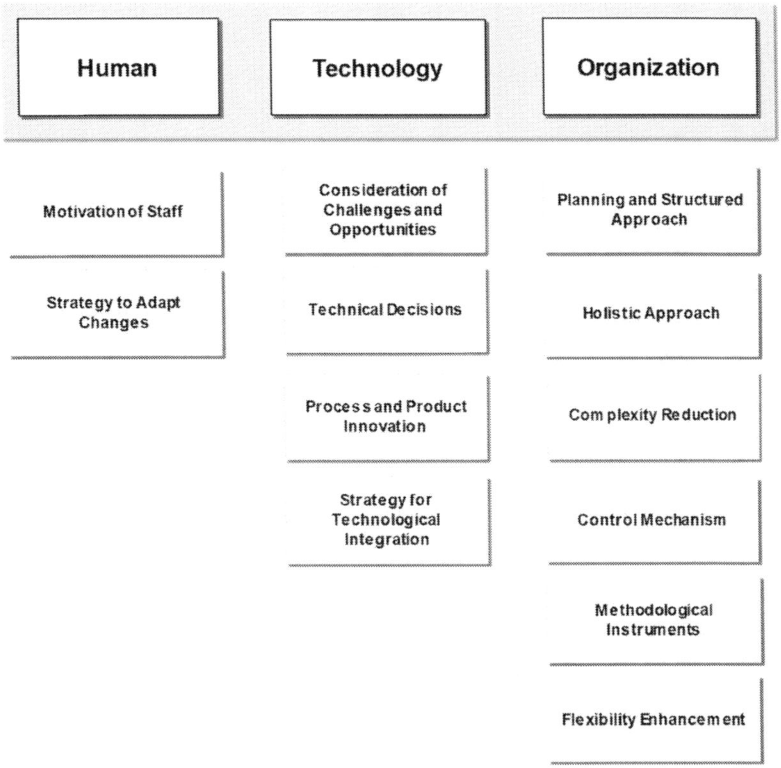

Figure 8: Summary of requirements

The integration of technological process innovations requires employees within the entire supply chain to develop a precise understanding of the technological components and their characteristics. Furthermore, it is important to pursue a clear integration strategy to consider challenges and opportunities. On the organizational side, a clear determination of existing processes and responsibilities has to be achieved. How these aspects are addressed by current models will be analyzed in the next chapter.

4 Model Evaluation

In order to examine the different models and make them comparable, the following table compares them with regard to the requirements for the integration of technologies as process innovations in corporate structures. The evaluation is based on the above-mentioned literature sources, which were also used for their brief introduction.

Three categories are used in the evaluation in order to ensure the clarity of the valuation and to obtain unambiguous evaluations without much dispersion (Ahsen, 2010, p. 49f). The three evaluation categories for the models are "fully satisfied" (+), "partially satisfied" (0) and "not satisfied" (-).

Table 1: Model Evaluation

	Allweyer, 2012	**Brecht, 2002**	**Becker, 2012**	**Davenport, 1997**	**Österle, 1995**	**Schallmo, Brecht, 2017**	**Schmelzer, Sessel-mann, 2013**
Motivation of the Staff	-	-	+	0	-	0	+
Strategy to Adapt Changes	-	0	+	-	0	0	0
Considera-tion of Challenges and Oppor-tunities	-	-	0	+	+	+	-
Technical Decisions	-	0	+	0	+	+	0
Process and Prod-uct Innova-tion	-	0	+	-	0	0	0
Strategy for Technologi-cal Integra-tion	0	-	-	-	+	+	-

	Allweyer, 2012	Brecht, 2002	Becker, 2012	Davenport, 1997	Österle, 1995	Schallmo, Brecht, 2017	Schmelzer, Sessel-mann, 2013
Planning and Structured Approach	+	0	+	+	+	+	0
Holistic Approach	+	0	+	0	0	+	+
Complexity Reduction	-	-	-	-	0	0	0
Control Mechanism	+	+	0	+	+	+	+
Methodological Instruments	+	+	+	+	+	+	+
Flexibility Enhancement	0	0	-	-	0	-	0

5 Findings

The evaluation table shows that a model, holistically oriented towards technological process innovations, is not yet available.

It becomes obvious that especially aspects from the technology cluster are underrepresented in current models. Especially in terms of challenges and opportunities, as well as technological decisions to be made, organizations have to consider much more far-reaching aspects. In particular blockchain technology contrasts with traditional technologies that used to aim at the optimization on operational level. In contrast, technical decisions related to the blockchain architecture, rights and obligations of the network nodes and the related consensus algorithm affect the supply chain on a much more strategical level. (Vasilievich Babkin, et al., 2018; Queiroz, Telles and Bonilla, 2019)

Also when considering the integration of additive manufacturing, it is particularly important to formulate technical decisions precisely in advance and to select a suitable processing method on that basis. (Dwivedi, Srivastava and Srivastava, 2017, pp. 975-977) In addition to the processing method, the procurement structure of materials in the supply chain, as well as the after-sales services, the logistics and expenses are further aspects integration models would need to consider. (Lindemann, et al., 2012, p. 179) Above all, the integration challenges of additive manufacturing lie in the production process itself. It is important to note that production speed and quality cannot be compared with other processing methods. (Dwivedi, Srivastava and Srivastava, 2017, p. 975)

Another aspect that should be highlighted in the technical cluster is "product and process innovation" and "strategy for technological integration". In this context, it is necessary that processes and their assignment to specific areas of the organization are clearly determined and recognized in the integration procedure. In particular when it comes to blockchain technology, its integration has to be accompanied by a necessary restructuring and reorganization of processes. This is because the technology affects multiple different processes in the company and beyond. Its impact hence, has the ability to automatize and change payment-, procurement- and data sharing processes radically. The involvement of supply chain partners therefore, plays a significant role for integration models that still lack a systematic concept in this context. (Toma et. al, 2019, pp. 288-299; Kamble, Gunasekaran and Arha, 2019)

Additive manufacturing has an impact on both product and process innovations like the spare parts management of after-sales services. This relationship becomes particularly visible in the design process where the new manufacturing process and its design liberties but also limits have to be established. (Dwivedi, Srivastava and Srivastava, 2017, p. 976)

Further organizational aspects that are noticeable in the evaluation of the models, are the lack in complexity reduction mechanisms and flexibility. However, the development of new and disruptive technologies makes it necessary to bring the more complex and interconnected change processes into manageable modules capable of processing. Innovations also lead to new requirements for the flexibility of change processes. A simultaneous change and adaptation process should be implemented into newly developed models. Flexibility is particularly necessary in the area of blockchain and additive manufacturing, because the technologies not only offer the possibility to use one specific application, but offers multiple application fields and possibilities throughout the entire product life cycle. It is therefore particularly important to not only remain on one path of change, but also to include changes in related fields and use scaling effects. (Lindemann, et al., 2012, pp. 177-180)

The involvement of employees and their needs, as the last cluster, already works very well for the most models. Above all, the Sesselmann & Schmelzer model best represents the integration of the employee into the change process with a continuous communication and information module. Also a practical control mechanism is available in most of the models. Only in this way investments can be tracked and justified, in order to strengthen management support for further process innovation activities.

6 Conclusion

To return to the initial research questions, chapter 5 already presented requirements that are relevant for the integration of technological process innovations. Through the exemplary investigation of blockchain technology and additive manufacturing, also special requirements for IT- and production technologies were highlighted.

To answer the second research question on how models for the integration of technological process innovations have to be adapted, it can be stated that the systematic review focused generic models for process integration, which did reflect some, but not all the requirements needed for the integration of technological process innovations. Further steps in conducting research therefore should include the development of a technology-centered model for the integration of technological process innovations. Following the principle of flexibility, the model should have possibilities to revise a transformation path and offer different modules for different types of innovations, as in our example IT- or production related technologies.

Above all, the evaluation also showed that the aspect of network integration is not yet considered in most models. However, the increasingly interconnected cooperation of companies in different value-added stages along the supply chain makes it necessary to integrate new technologies across companies.

In particular, considering the integration of blockchain technology, it has become clear that existing models do not currently meet the necessary requirements needed in the overall integration planning. In particular, the far reaching strategical consequences for the organization have to be addressed correctly, while existing models still aim at technologies that can be controlled on an operational level. Moreover, the increased technical complexity leads to a greater importance of initial technical decisions to be made, like the blockchain architecture type, the consensus algorithm and scalability approaches. Apart from that perspective, institutional and legal norms also emerge as highly applicable integration problems and do not find enough recognition in existing models. Likewise, we a have similar outcome for the integration of additive manufacturing. Here, it should be particularly important to consider both product and process innovations. This is because an appropriate utilization of 3D printers can usually only occur if the design liberties are taken into account.

Financial Disclosure

The work on this paper was supported by grants from the Federal State of North Rhine-Westphalia using funds from the European Regional Development Fund (EFRE) 2014-2020, "Investments in Growth and Employment", as well as from the Center of Excellence for Logistics & IT, an initiative of the Fraunhofer Society and the state of North Rhine-Westphalia, Germany.

References

Ahsen, A., 2010. Bewertung von Innovationen im Mittelstand. [e-book]. Berlin, Heidelberg: Springer-Verlag Berlin Heidelberg. <http://gbv.eblib.com/patron/FullRecord.aspx?p=510417>.

Albers, S. and Gassmann, O., eds., 2011. Handbuch Technologie- und Innovationsmanagement. 2nd ed. Wiesbaden: Gabler.

Allweyer, T., 2012. Geschäftsprozessmanagement. Strategie, Entwurf, Implementierung, Controlling. 5th ed. Herdecke, Bochum: W3L-Verl.

Becker, J., Kugeler, M. and Rosemann, M., eds., 2012. Prozessmanagement. Ein Leitfaden zur prozessorientierten Organisationsgestaltung. Berlin, Heidelberg: Springer Gabler.

Birasnav, M. and Bienstock, J., 2019. Supply chain integration, advanced manufacturing technology, and strategic leadership: An empirical study. Computers & Industrial Engineering, [e-journal] 130, pp. 142–157. http://dx.doi.org/10.1016/j.cie.2019.01.021.

Brecht, L., 2002. Process leadership. Methode des informationssystemgestützten Prozessmanagement. Zugl.: St. Gallen, Univ., Habil.-Schr., 2000. Hamburg: Kovač.

Brown, S., 2001. Managing process technology — further empirical evidence from manufacturing plants. Technovation, [e-journal] 21(8), pp. 467–478. http://dx.doi.org/10.1016/S0166-4972(00)00074-2.

Corsten, H. and Gössinger, R., 2016. Produktionswirtschaft. Einführung in das industrielle Produktionsmanagement. 14th ed. [e-book]. Berlin, Boston: De Gruyter Oldenbourg. <http://www.degruyter.com/search?f_0=isbnissn&q_0=9783110452778&searchTitles=true>.

Davenport, T. H., 1997. Process innovation. Reengineering work through information technology. Boston, Mass.: Harvard Business School Press.

Delany, R. and D'Agostini, R., 2015. The Challenges of Integrating New Technology into an Organization. La Salle.

Dwivedi, Gourav; Srivastava, Samir K.; Srivastava, Rajiv K. (2017): Analysis of barriers to implement additive manufacturing technology in the Indian automotive sector. In: Int Jnl Phys Dist & Log Manage 47 (10), S. 972–991. DOI: 10.1108/IJPDLM-07-2017-0222.

Fettke, P. and Loos, P., 2004. Systematische Erhebung von Referenzmodellen - Ergebnisse der Voruntersuchung. Working Paper of the Research Group Information Systems & Management. Mainz.

Gollapudi, K., Jangeti, S. K. and Kotapati, A. R., 2012. ANALYSIS OF USING A BUSINESS INTELLIGENCE TOOL (COGNOS) IN A COMPANY TO RESULT IN MORE EFFICIENT AND INTUITIVE COMPANY IN THE CURRENT ERA. Business Intelligence Journal, 5(2), pp. 405–409.

Harrison, M., 1990. Advanced manufacturing technology management. London: Pitman.

Hasselbring, Wilhelm (2000): Information System Integration // Information system integration. In: Commun. ACM June 2000/Vol.43, No. 6 // 43 (6), S. 33–38. DOI: 10.1145/336460.336472.

Hauschildt, J., 2016. Innovationsmanagement. 6th ed. [e-book]. München: Franz Vahlen. <https://ebookcentral.proquest.com/lib/gbv/detail.action?docID=4688806>.

Hervas-Oliver, J.-L., Sempere-Ripoll, F. and Boronat-Moll, C., 2014. Process innovation strategy in SMEs, organizational innovation and performance: a misleading debate? Small Business Economics, [e-journal] 43(4), pp. 873–886. http://dx.doi.org/10.1007/s11187-014-9567-3.

Holotiuk, F., Moormann, J. and Pisani, F., 2019. Blockchain in the Payments Industry: Developing a Discussion Agenda Based on Pain Points and Opportunities. In: H. Treiblmaier, and R. Beck, eds. 2019. Business Transformation through Blockchain. Volume I. Cham: Springer International Publishing; Imprint: Palgrave Macmillan, pp. 197–231.

Hölzle, P., 2007. Projekt-Management. Kompetent führen, Erfolge präsentieren. 2nd ed. [e-book]. München: Rudolf Haufe Verlag GmbH & Co. KG. <https://www.wiso-net.de/document/HAUF,AHAU,VHAU__9783448075021191>.

Hungenberg, H., 2014. Strategisches Management in Unternehmen. Ziele - Prozesse - Verfahren. 8th ed. [e-book]. Wiesbaden: Springer Gabler. <http://dx.doi.org/10.1007/978-3-658-06681-9>.

Kamble, Sachin; Gunasekaran, Angappa; Arha, Himanshu (2019): Understanding the Blockchain technology adoption in supply chains-Indian context. In: International Journal of Production Research 57 (7), S. 2009–2033. DOI: 10.1080/00207543.2018.1518610.

Karltun, Anette; Karltun, Johan; Berglund, Martina; Eklund, Jörgen (2017): HTO - A complementary ergonomics approach. In: Applied ergonomics 59 (Pt A), S. 182–190. DOI: 10.1016/j.apergo.2016.08.024.

Kianian, Babak; Tavassoli, Sam; Larsson, Tobias C. (2015): The Role of Additive Manufacturing Technology in Job Creation: An Exploratory Case Study of Suppliers of Additive Manufacturing in Sweden. In: Procedia CIRP 26, S. 93–98. DOI: 10.1016/j.procir.2014.07.109.

Lindemann, Christian; Jahnke, Ulrich; Habdank, Mattthias; Koch, Rainer (2012): Analyzing Product Lifecycle Costs for a Better Understanding of Cost Drivers in Additive Manufacturing. In: Conference: Solid Freeform Fabrication Symposium - An Additive Manufacturing ConferenceAt: Austin, TX, USA 23, S. 177–188. Online verfügbar unter https://sffsymposium.engr.utexas.edu/Manuscripts/2012/2012-12-Lindemann.pdf.

Lischka, J.-M., 2011. Management von Prozessinnovationen. Ein empirisch gestütztes Erfolgsfaktorenmodell. Zugl.: Berlin, Techn. Univ., Diss., 2010. [e-book]. Wiesbaden: Gabler Verlag / Springer Fachmedien Wiesbaden GmbH Wiesbaden. <http://dx.doi.org/10.1007/978-3-8349-6207-2>.

Marzi, G., Zollo, L., Boccardi, A. and Ciappei, C., 2018. Additive Manufacturing in SMEs: Empirical Evidences from Italy. International Journal of Innovation and Technology Management, [e-journal] 15(01), p. 1850007–1850007. http://dx.doi.org/10.1142/S0219877018500074.

Moreira, A. C., Ferreira, L. M. D. F. and Zimmermann, R. A., 2018. Innovation and Supply Chain Management. Cham: Springer International Publishing.

Nahass, G., Smith, J. and Curragh, M. Information Technology Integration. Putting IT to work in driving deal success. In: PwC's Deals M&A Integration practice.

Organisation for Economic Co-operation and Development and Statistical Office of the European Communities, Luxembourg, 2005. Oslo Manual. Guidelines for Collecting and Interpreting Innovation Data, 3rd Edition. [online]. Paris. Available at: <http://dx.doi.org/10.1787/9789264013100-en>.

Österle, H., 1995. Business Engineering Prozeß- und Systementwicklung. Band 1: Entwurfstechniken. [e-book]. Berlin, Heidelberg, s.l.: Springer Berlin Heidelberg. <http://dx.doi.org/10.1007/978-3-662-06188-6>.

Pleschak, F. and Sabisch, H., 1996. Innovationsmanagement. Stuttgart: Schäffer-Poeschel.

Porter, M. E., 2014. Wettbewerbsvorteile. Spitzenleistungen erreichen und behaupten. 8th ed. [e-book]. Frankfurt, New York: Campus. <http://www.content-select.com/index.php?id=bib_view&ean=9783593423289>.

Queiroz, Maciel M.; Telles, Renato; Bonilla, Silvia H. (2019): Blockchain and supply chain management integration: a systematic review of the literature. In: Supp Chain Mnagmnt 12 (1), S. 671. DOI: 10.1108/SCM-03-2018-0143.

Raymond, L. and St-Pierre, J., 2010. R&D as a determinant of innovation in manufacturing SMEs: An attempt at empirical clarification. Technovation, [e-journal] 30(1), pp. 48–56. http://dx.doi.org/10.1016/j.technovation.2009.05.005.

Schallmo, D. and Brecht, L., 2017. Prozessinnovation erfolgreich anwenden. Grundlagen und methodisches Vorgehen: Ein Management- und Lehrbuch mit Aufgaben und Fragen. 2nd ed. [e-book]. Berlin, Heidelberg: Springer Gabler. <http://dx.doi.org/10.1007/978-3-662-54504-1>.

Schallmo, D. R. A., Brecht, L. and Ramosaj, B., 2018. Process Innovation: Enabling Change by Technology. Basic Principles and Methodology: A Management Manual and Textbook with Exercises and Review Questions. [e-book]. Berlin, Heidelberg: Springer Berlin Heidelberg. <http://dx.doi.org/10.1007/978-3-662-56555-1>.

Schmelzer, H. J. and Sesselmann, W., 2013. Geschäftsprozessmanagement in der Praxis. Kunden zufriedenstellen, Produktivität steigern, Wert erhöhen: [das Standardwerk. 8th ed. München: Hanser.

Sealy, W., 2012. Additive manufacturing as a disruptive Technology: How to avoid the pitfall. American Journal of Engineering and Technology Research, 12(1), pp. 86–93.

Vasilievich Babkin, Aleksandr; Dmitrievna Burkaltseva, Diana; Viktorovich Betskov, Aleksandr; Shapievich Kilyaskhanov, Hizri; Sergeevich Tyulin, Andrey; Vladimirovna Kurianova, Irina (2018): Automation Digitalization Blockchain: Trends and Implementation Problems. In: IJET 7 (3.14), S. 254. DOI: 10.14419/ijet.v7i3.14.16903.

Wellbrock, W. and Göpfert, I., 2015. Innovative Supply-Chain-Management-Konzepte. Branchenübergreifende Bedarfsanalyse sowie Konzipierung eines Entwicklungsprozessmodells. [e-book]. Wiesbaden: Springer Gabler. <http://search.ebscohost.com/login.aspx?direct=true&scope=site&db=nlebk&AN=969012>.

Wilfling, S., 2013. Management organisationaler Anpassungsprozesse. Transformationsvorhaben durch geeignete Informationsbasis erfolgreich planen und umsetzen. Zugl.: Hagen, FernUniv., Diss., 2011 u.d.T.: Wilfling, Sabine: Entwicklung eines Informationsmodells zur Gestaltung organisationaler Anpassungsfähigkeit. [e-book]. Wiesbaden: Springer. <http://dx.doi.org/10.1007/978-3-658-00676-1>.

Winkler, H. and Kaluza, B., 2008. Einsatz einer Wertschöpfungsrechnung zur Erfassung und Bewertung von Produkt- und Prozessinnovationen in Wertschöpfungsnetzwerken. In: D. Specht, ed. 2008. Produkt- und Prozessinnovationen in Wertschöpfungsketten. Wiesbaden: Gabler, pp. 1–31.

Xu, W., Weber, I. and Staples, M., 2019. Architecture for block chain applications. Cham, Switzerland: Springer.

Identifying Research Gaps in Supply Chain Innovation

Fatemeh Seidiaghilabadi[1], *Zahra Seidiaghilabadi*[1], *Aida Miralmasi*[1]

1 – Islamic Azad University

Purpose: In recent years, supply chain innovation (SCI) has become a central characteristic of firms in creating dynamic competitive advantages. The present study aims to examine the SCI literature in the last five years.

Originality: Although there is much research on supply chain management (SCM), innovations, sustainable supply chain (SSC), and green supply chain (GSC), limited research has identified gaps in the SCI.

Findings: The study offers useful implications for researchers, managers, marketers, and other stakeholders involved in the SCI. The study also motivates the researchers to conduct additional studies in the area of SCI. Furthermore, the results of this study contribute to building new research agendas concerning SCI.

Methodology: The study adopts a qualitative content analysis approach to examine the literature on SCI (from 2014 to 2019). Qualitative software MAXQDA12 was applied for literature content' classification and coding.

Keywords: Supply Chain Innovation, Qualitative Content Analysis, MAXQDA

First received: 19.May.2019 **Revised:** 16.June.2019 **Accepted:** 24.June.2019

1 Introduction

A large amount of wealth and creating values occurs in the supply chain (SC) and supply chain management (SCM). SCM has an indispensable role in fostering organizational performance and improving supply chain innovation (SCI). SCM is synonymous with combined logistics systems, focusing upon reducing the storage of goods through the supply chain (Chong et al., 2011).

The concept of supply chain(SC) refers to the procedures from the early raw resources to final and completed product connecting across supplier-user businesses, or as the roles inside and outside a corporation that allow the value chain to produce goods and offer facilities to the user (Chitale & Gupta, 2014; Cox, 1999).

Researchers shows that 'innovation' is the skill of producing economic values from new ideas that represent progress in the economic development of organizations. Innovation is a significant factor in generating success and competitive advantages within organizations while at the same time, it has also brought significant challenges for organizations (Gao, Xu, Ruan, & Lu, 2017; Schumpeter, 1934; Wu & Tsai, 2018).

Supply chain systems are continually varying as an outcome of social as well as technological developments (Zijmet et al.,2019). The efficient functioning of SC is particularly essential to those companies which attempted to improve their SC systems' effectiveness (Azevedo, 2013).

Based on resource-based view (RBV), SCI integrates innovative activities with logistic approach (logistics-related services), innovative marketing-focused activities (customer needs), and other related activities (e.g., developing technologies, creating new knowledge and technical skills) to augment joint profit (Bello et al., 2004; Jajja, Kannan, Brah, & Hassan, 2017; Paulraj & Chen, 2004; Wong & Ngai, 2019). SCI is a change in processes, technology, and network, meaning that SCI improves competitive advantage and organizational performance.

By analyzing the literature of the supply chain innovation, we found that although numerous articles have been published regarding the SCI, none of them have thoroughly addressed the gaps in the SCI. Wong and Ngai (2019), systematically reviewed 18 years of the SCI literature in order to find the research gaps. However, their study focused only on a general overview of SCI employing Gregor's (2006) theory classification. It is still necessary to examine existing research gaps. In this study, we reviewed the SCI literature from 2014 to 2019 to identify research gaps and address some future research directions on this critical topic.

Hence, in this study we review and analyze the available literature, looking for an answer to the question of "RQ: how to make the supply chain more competitive?". Therefore, to reach the aim of the study, in the next section, an overview of the SCI literature will be given. Then we explain research methodology and present result of preliminary analyses. Content analysis by MAXQDA 2018 was applied. Finally, we reflect findings, and highlight some possible future research perspectives.

2 Literature Review

SCI has been received many scholars' attention (Abdelkafi & Pero, 2018; Arlbjørn et al.,2011; Gao et al., 2017; Iddris, 2016; Munksgaard et al.,2014; Sabri et al.,2018; Tan et al., 2015; Yoon et al., 2016; Zimmermann et al., 2016) . In the 1980s, organizations focused on a range of diverse systems to achieve sustainable competitive advantages. However, since competitors mimicked these competitive advantages, they did not have the necessary stability. Indeed, attempts to optimize organizational processes without taking into account of external companies, especially suppliers and customers, seemed ineffective, and meanwhile, organizations in cooperation with each other had better performance. It was here that the supply chain concept was born. (Stavrulaki & Davis, 2010).

Porter (1990) and Drucker (1985) pointed out that innovation is a characteristic feature of organizational sustainability, and organizations with high levels of knowledge in technology are required to pay more attention to innovation strategies. Innovation is indeed the process of transforming opportunities into new ideas, as well as the application of such new ideas to foster specific improvements (West & Farr, 1989).

SCI is a procedure that can improve organizational processes to manage effective SCM through integrated interactions with suppliers, producers, distributors, and customers. (Lin, 2007). Thus, SCI causes time and cost reduction, developing novel operational techniques and reliable delivery system for coping with growing changes in the business (Lee et al.,2011). Chapman et al. (2003), propose that service industries should focus on SCI for effective delivery services. Researchers agree that supply chain innovation helps companies to maintain their competitive position and improve supply chain performance (Flint, Larsson, Gammelgaard, & Mentzer, 2005; Franks, 2000; Krabbe, 2007; Lee et al., 2011).

Flint and Larsson (2007), examined innovation management process and SC Learning as an SCI' antecedents and considered as part as creating a customer. Herzlinger (2006), Highlighted three types of innovation in service sectors (i.e., customer-focused, technology-based, integrated innovation) in the SCI process. Hui et al. (2015) found that participatory management influences the innovative performance of the supply chain.

Dubey, Singh, and Tiwari (2012), found that innovation in supply chain processes (e.g., sales management and the number of orders) had a significant effect on the SC performance. Thus, Suppliers and manufacturers required to transform their business through innovation in SC to gain success among competitive industries (Wong & Ngai, 2019). According to Shapiro & Wagner (2009), modeling technologies allow managers to manage data more effectively in their companies, thus creating rationality in a supply chain network.

The supply chains are responsible for including any fast transfer and distribution of technological innovations (Sabri et al., 2018). Cai et al.,(2009), found that SC-affiliated companies, which produce innovative products for the general market, might utilize a supply chain model to meet their daily innovative needs in the supply chain.

Arlbjørn et al., (2011), considered SCI as a gradual or a fundamental change in the supply chain network, technology or business processes in the SC framework in their systematic review.

Moreover, recent discussions of sustainability and sustainable have been developed in supply chain. Sustainable supply chain (SSC) means managing of material, information, and coordination considering economic, social, and environmental dimensions. Researchers have claimed that emerging companies perform better in the sustainable supply chain (e.g., Nidumolu et al., 2015; PAGELL & WU, 2009). The sustainable supply chain is rooted in the traditional concept of SC. Therefore, evaluation of the SSC performance can be useful in creating transparency and fostering innovation in the supply chain (Schaltegger & Burritt, 2014). It has been confirmed that supply chain management can encourage innovation in organizations (Chong et al., 2011). Given the sustainability of current competitive business environments, researchers consider innovation as central to the supply chain (Jellali & Benaissa, 2015).

The effect of SCI on companies depends on their size. In large-sized firms, when innovations are developed to improve supply chain performance, it may result in new businesses. Accordingly, SCI leads to the expansion of the portfolio for a large-sized company. This phenomenon happens when a company uses logistics innovation to optimize its core processes and launch new values in the marketplace (Abdelkafi & Pero, 2018).

Reviewing the supply chain, Wong & Ngai (2019), found that the literature has tended to focus mainly on the concept of supply chain structures, ignoring the development of measures for operationalizing the concept. Therefore, by conducting in-depth analysis and analyzing the content of the studies, research gaps in SCI can be identified. As noted earlier, the present study examines the SCI literature in the last five years (2014-2019). Table 1 summarizes the articles, publication year, and field of study.

Table 1: The SCI Literature from 2014 to 2019

Studies	Journals/Proceedings	Area of Study
Munksgaard et al., 2014	Operations Management Research	Proposed an SCI model with three interactive components: network structure, technology, and business processes.
Artsiomchyk & Zhivitskaya, 2015	IFAC-PapersOnLine	Proposes an integrated approach to help companies manage innovation and design sustainable supply chains.
Jangga, Ali, Ismail, & Sahari, 2015	Procedia Economics and Finance	Uncertainty and flexibility of the supply chain
Tan et al., 2015	International Journal of Production Economics	Using big data to gain competitive advantage through SCI capabilities.

Studies	Journals/Proceedings	Area of Study
Nasr, Kilgour, & Noori, 2015	European Journal of Operational Research	Examine innovation distribution versus protection between SC partners.
Yoon et al., 2016	Technological Forecasting and Social Change	Illustrated that SCI plays an essential role in fostering operational procedures for SC effectiveness.
Iddris, 2016	International Journal of Innovation Science	Developing SCI'capability constructs
Rajabian Tabesh, Batt, & Butler, 2016	Journal of Food Products Marketing	Relationship between different theoretical constructs and how they affect SCI and performance.
Stentoft, Mikkelsen, & Jensen, 2016	Supply Chain Forum: An International Journal	Comparison of external and internal products from SCI' perspective

Studies	Journals/Proceedings	Area of Study
Gao et al., 2017	Journal of Cleaner Production	Reviewing 107 studies published from 1996 to 2014. Proposing a theoretical framework encompassing the meaning of sustainable supply chain innovation.
Shah & Naghi Ganji, 2017	British Food Journal	Lean production and SCI in food industry
Stentoft & Rajkumar, 2018	Innovation and Supply Chain Management	Connection between SCI and organizational and market performance.
Sabri et al., 2018	Journal of Engineering and Technology Management	Implementation of process and product innovation in SC framework.

Studies	Journals/Proceedings	Area of Study
Shou, Che, Dai, & Jia, 2018	International Journal of Operations & Production Management	Examining complementarity and similarity in SCs
Kwak, Seo, & Mason, 2018	International Journal of Operations & Production Management	the linkage between SCI, risk management capabilities, and competitive advantage in international supply chains.
Ted et al.,2018	The International Journal of Logistics Management	SCI and sustainable transport
Wu & Tsai, 2018	Transportation Research	setting new business models based on new SC and logistics methods.
Abdelkafi & Pero, 2018	Business Process Management Journal	Exploratory analysis of SCI' models

Studies	Journals/Proceedings	Area of Study
Chen, Dimitrov, & Pun (2019)	Omega	Diverse types of subsidy, collaboration and effort, and companies' incomes in SCI
Ardito, Messeni Petruzzelli, Dezi, & Castellano (2018)	Journal of Business Research	Exploring the influence of SC' knowledge sourcing actions on innovation initiatives.
Wong & Ngai (2019)	Industrial Marketing Management	Systematically reviewing the construct of SCI.
Reimann, Xiong, & Zhou (2019)	European Journal of Operational Research	Examining the relationship between remanufacturing and the opportunity to lower the variable remanufacturing cost via innovation process

Studies	Journals/Proceedings	Area of Study
Lv & Qi(2019)	Computers & Industrial Engineering	Choosing the interactive partner of SC based on innovation resources
Hsin Chang, Hong Wong, & Sheng Chiu, (2019)	Information & Management	Development of the conceptual framework for understanding factors of business systems leveraging (BSL)
Russell & Swanson (2019)	The International Journal of Logistics Management	Probing gaps between information processing and SC dexterity.

3 Research Methodology

There are three approaches to explore the status of knowledge in a particular field of study: Delphi-based method, meta-analysis, and content analysis (Li & Cavusgil, 1995). The Delphi method draws on a panel of experts familiar with the field. In the meta-analysis, the results of multiple empirical studies are combined and analyzed based on effect size. Finally, content analysis is a scientific approach used to summarize and analyze qualitative data (Neuendorf, 2016). Content analysis provides researchers with a tool to systematically extract and analyze multiple data. In qualitative research, there are different approaches to data analysis depends on the nature of the questions and research problems. In this study, the researchers used content analysis to discover underlying aspects of SCI. Thus, content analysis is an appropriate tool to extract, analyze, and interpret the gathered data. Content analysis is an appropriate strategy, especially when the data are in the form of documents and texts. By analyzing and describing the previous theoretical research concerning with supply chain innovation, the authors first review, analyze and summarize the SCI research, then present a framework for studying supply chain research, and finally, suggest several ideas for future research. The stages of content analysis used in this study are shown in Figure 1.

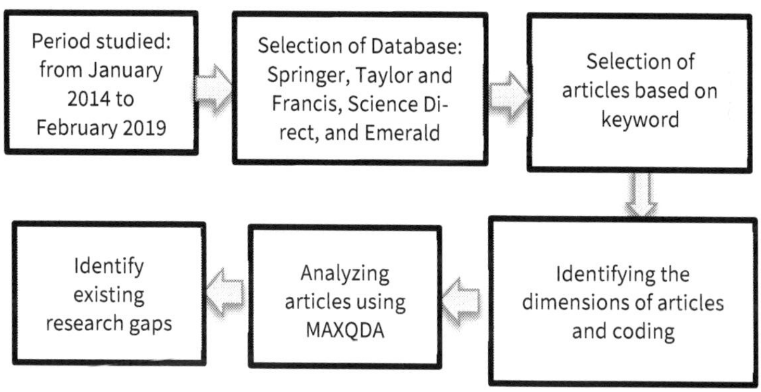

Figure 1: Stages Involved in Content Analysis

In the first phase, this study collected articles related to SCI. The authors focused on articles published from January 2014 to February 2019, searching for keywords such as 'supply chain innovation' 'innovation in the supply chain' The full text of each article was carefully analyzed in order to remove articles unrelated to SCI.

In the next phase, articles were selected and collected from accredited databases such as Science Direct, Taylor & Francis, Emerald, and Springer. The reason was that these databases publish relevant and appropriate articles related to SCI. The study did not include prefaces, editing notes, book reviews, general reports, BA, MA, and Ph.D. theses, and textbooks. Although we cannot guarantee that this research is comprehensive, we believe that the selected journals and articles contain a holistic structure of research done on SCI. Finally, extensive literature analysis generated 25 articles related to SCI.

4 Analysis and Results

Since the primary purpose of this study was to investigate and identify Identifying research gaps in supply chain innovation, our search resulted in 25 articles related to the concept. In addition, this study employed the content analysis, several sections of the literature were analyzed (e.g., Introduction, literature reviews, methodology, and research limitations). Such classification highlights the growth of research on supply chain innovation.

Table 2: publication types and numbers

Database type	Count	%
Article	21	84%
Proceedings	2	8%
Review	2	8%

Table 2 shows the structure of SCI literature database ranging from 2014-2019. As shown in table 2, the majority of publications are articles (84%).

The growth of research in SCI literature was determined by conducting a

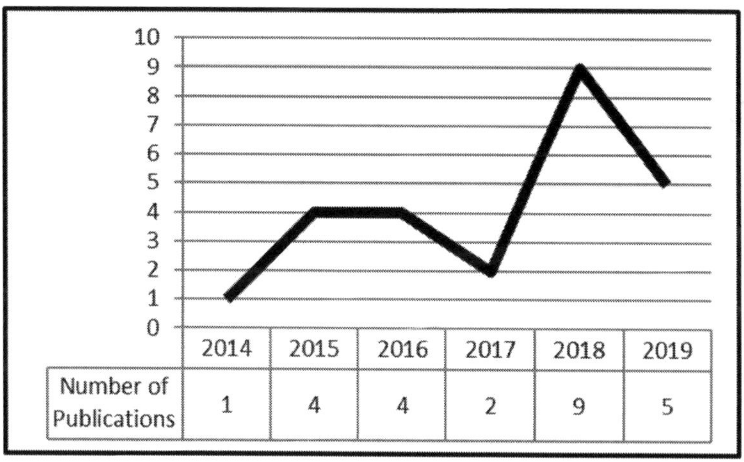

frequency analysis of articles published annually. (See Figure 2)

Figure 2: related to SCI' publications (2014-2019)

As revealed in Figure 2, since 2015, the number of articles related to supply chain articles has increased steadily. In 2015 and 2016, proper attention was paid to SCI. Although In 2017, SCI studies were declined substantially, there is a continued increase in SCI studies in 2018. Since the present study included two months of 2019, it is hard to predict any increase or decrease in SCI studies. However, it can be argued that the attention to SCI has been increasing since 2015, as there were not many relevant articles in the previous year.

According to Table 1, SCI studies were published in journals, including Operations Management Research, European Journal of Operational Research, etc. Our literature analysis showed that several topics related to supply chain innovation are of rising interest to marketers in various industries and businesses.

In this study, we conducted a content analysis approach to detect research gaps in supply chain innovation using MAXQDA. We hope that future research will draw more on content analysis methods and provide significant opportunities for research in the area of SCI.

4.1 Summary of the results of content analysis using MAXQDA

In this section, a summary of the key steps in conducting content analysis is presented. Content analysis can be used to analyze textual, audio and video data. First, data are analyzed using open, axial and selective coding stages. Then MAXQDA is used to analyze the qualitative data effectively. Finally, initial codes, emerging categories and related dimensions are determined in the open coding stage. In the axial coding, the main categories are identified and elaborated.

The first step in conducting content analysis is to determine the variables and factors. When the conceptualization and usability of variables are completed, the researcher can continue the coding process(Neuendorf, 2016). If the utility of the variables does not match with the definition of variables, the coded data will be wrong in the next stage, and inaccurate results will be presented. In this study, the open coding stage began considering articles and literature. In this stage, the relevant text was carefully analyzed and related SCI information was extracted. The result of this process was some notes considered as initial data. Concepts were extracted from these initial data. In the first stage, 970 initial codes (e.g., increased market share and operational performance of supply chain innovation, formulation of strategic objectives, human capital and skilled force in technical projects, interdependence among companies) were extracted. After removing and adjusting the initial codes, 34 principal codes were obtained (see Table 3 and Figure 3). When concepts created, it is necessary to group them under categories having higher explanatory power. When a category is identified, clarification of its features and dimensions requires careful attention. Therefore, at this stage, the constant comparative analysis was employed to find similarities and differences between the concepts and similar concepts were placed in a category. Table 3 introduces categories related to the concepts.

After specifying the categories, the axial coding step begins. The axis coding is the process of connecting categories at the level of features and dimensions. It is called "axial" because the researcher attempts to make a connection among categories. As Corbin & Strauss (2008) observed, categories that have a logical connection with each other are all subsumed under the central construct. i.e., supply chain innovation.

Table 3: Extracted Codes from literature

Selective Code	**Open Codes (Categories)**	**Initial/Primary Codes**
Supply chain innovation	SCI' moderators	The rapid growth of the industry, Firm size, Technological intensity, Equity
	SCI' driving forces	Product characteristics
		Aggregation of ideas, Knowledge, and skills
		Internal coordination
		Research and development

Selective Code	Open Codes (Categories)	Initial/Primary Codes
		Workforce training
		Focus on knowledge
		Interdependence among companies
		Flexibility towards the environment
		Product timing
		Aligning business strategy with supply chain
		Functional supply chain management
		Leadership and senior management
		Effective logistics processes

Selective Code	Open Codes (Categories)	Initial/Primary Codes
		Intra-organizational processes
		Innovation in logistics
		Innovation in business processes
		Innovation in network structure
		Technological innovation
		Human capital and skilled force in technical projects
		Formulating strategic goals (general to particular)
	Consequences of SCI	Customer value

Selective Code	Open Codes (Categories)	Initial/Primary Codes
		Reduced cost
		Competitive advantage
		Risk management
		Improving the supply of goods
		Operational performance of innovation in supply chain
		Creating values
		Delivering value
		Gaining values
		Supply chain development

According to Table 3 and Figure 3, in this study, three categories of open coding including SCI' moderators (4 items), SCI' driving forces (20 items), and the Consequences of SCI (10 items) were generated. Figure 3 shows the frequency of each category of items in terms of its importance.

Figure 3 shows the importance of categories. Since innovation plays a crucial role in creating competitive advantages, the ability to generate innovation is becoming an essential area of attention in innovation research. In this study, we considered this capability as "SCI' driving forces" – a critical factor in the survival of companies. Based on the output matrix of MAXQDA, and Figure 3, 20 codes were identified as a driving force in the supply chain. From these codes, the most essentials are the interdependence among companies, flexibility towards the environment, aggregation of ideas, knowledge, and skills, innovation in business processes and technology, product characteristics, and human capital (see Figure 3). It was found that the nature of products and product demand are associated with supply chain innovation. Characteristics such as product life-cycle management, product diversification, product sales forecast, and product distribution intensity stimulate innovation in the supply chain. Product characteristics exert an essential effect on the design and application of SCI. When delivering any product to the market, the product itself affects the supply chain. Thus, the supply chain requires redesigning in order to deliver the product to the marketplace successfully.

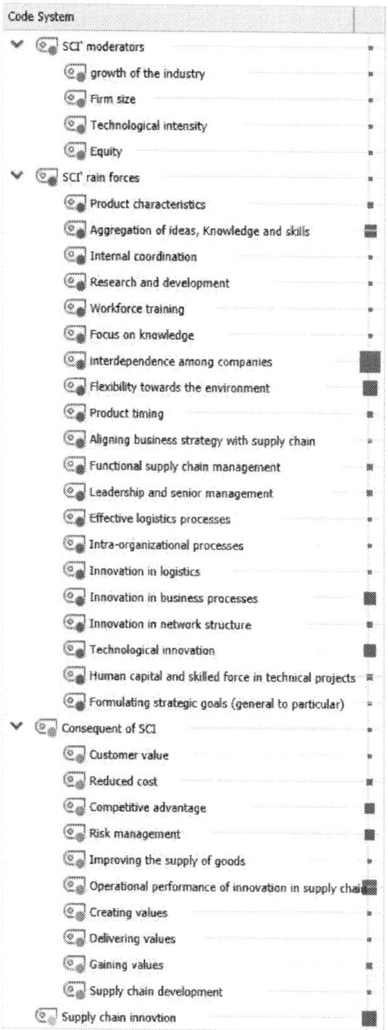

Figure3: MAXQDA Code Matrix Browser

Flexibility to the environment helps organizations meet their current needs, without losing the ability to produce. Since some organizations no longer compete as an independent organization, they prefer to compete as a supply chain. Therefore, the concept of flexibility has been expanded from an organizational focus to the supply chain. Moreover, an expert with technical skills can influence innovation in the supply chain. Human capital has always been emphasized as an important source of innovation (McGuirk et al., 2015; van Uden, Knoben, & Vermeulen, 2017). Human capital is a major driving force behind an organization's economic growth and a crucial factor in generating success and innovation. Aggregation of ideas, knowledge, and skills is also a vital part of organizational strategy in the process of supply chain innovation. Companies seek opportunities to foster their corporate knowledge. When the knowledge is shared, performance, as a result, will be increased because the primary source of innovation in the supply chain is the knowledge that the organization has access to it. Therefore, successful organizations are those who can acquire the most useful and reliable knowledge in their business and use them effectively. Such valuable knowledge must be managed skillfully. Innovative organizations actively promote their members' knowledge, provide a high level of job security, and help their employees change.

Information and Communication Technology (ICT) is another area rapidly affecting business processes. Since technological innovation is the result of product innovation (services) and innovation in business processes, as a capability and stimulus, it can expand supply chain activities and processes. The development, diffusion, and use of innovation in practice can be act as a stimulus for the process of SCI.

The role of SCI in generating competition and growth is a precise fact, as observed by managers, policy makers, and researchers. Ideally, SCI can improve efficiency and operational performance, manage risks, reduce costs, and create a competitive advantage for the organization. Supply chain innovation can both increase and reduce business risks (Kwak et al., 2018). Any risk associated with the flow materials and information can disrupt the operation. Therefore, it can be argued that SCI not only advances the supply chain' performance but also develops the ability to manage related risks. In the traditional supply chain, opportunities or risks cannot be controlled by companies, and it is supply chain innovation that provides an effective solution to such problems. Depending on the organization's priorities, supply chain innovation can improve products and services, reduce costs, prevent waste of resources, and increase efficiency. Reducing costs is, in fact, one of the key outputs of innovation in the supply chain, leading to increased efficiency, competitiveness, and profitability.

On the other hand, the rapid growth of the industry, firm size, technological intensity, and equity are the moderators of SCI. Firm size plays a leading role in SCI. SCI' functions differently in large sized companies as it generates portfolios and improves business performance. Technological intensity is effective in improving and adjusting supply chain innovation. Although the application of advanced technologies increases costs in the initial stages, it ultimately leads to improved performance and efficiency.

5 Discussion

This study explored previous research on supply chain innovation using qualitative content analysis. The employment of content analysis is vital in

SCI because dominant quantitative methods might fail to meet the expectations of individuals and organizations. Therefore, the present study contributes to supply chain and SCI, and provides opportunities for the advancement of SCI through the employment of content analysis methods. Although content analysis methods have been used in SCM literature, SCI studies have not deployed a content analysis methodology.

Additionally, among the 25 articles in this study, only two systematic review articles dealt with SCI (Gao et al., 2017; Wong & Ngai, 2019). This indicates a lack of qualitative studies employing content analysis methods. Thus, our study demonstrates that research gaps in supply chain innovation can be more effectively studied through qualitative content analysis methods.

According to the results of the present study, research opportunities were discussed in three areas (moderators ,driving forces and consequences). Therefore, considering advances in technology and knowledge, managers should realize the significance of innovation and the management of innovation in all parts of the organization. There are many opportunities for future research. For example, in order to promote supply chain innovations, it is essential to engage all members of the organization (e.g., employees, managers) in research and development, and to take advantage of customer ideas information of activities in the organization. Considering issues such as high demand as well as competition at the international and supranational level, the importance of the final customer should be prioritized. Besides, it is essential to pay attention to innovation facilitators/moderators when evaluating supply chain innovation.

The present study has three limitations. First, we focused on those articles published in the English language, ignoring other articles in non-English languages. This bias might result in a lack of useful information in the area of supply chain innovation. Second, we restricted our study to four databases. Local-scientific journals and theses can offer a rich picture of the supply chain. Third, we did not analyze the articles separately based on the research setting and industry under investigation. Addressing these limitations, further research can provide rich insights into supply chain innovation

Acknowledgements

We would like to show our appreciation to Iran' Analysis Academy Institute for providing the opportunity for our research. We thank Dr. Mohsen Moradi, head of Institute and two anonymous reviewers for their comments that improved the earlier manuscript.

References

Abdelkafi, N., & Pero, M. 2018. Supply chain innovation-driven business models: Exploratory analysis and implications for management. Business Process Management Journal, 24(2), 589–608. https://doi.org/10.1108/BPMJ-05-2016-0109

Ardito, L., Messeni Petruzzelli, A., Dezi, L., & Castellano, S. 2018. The influence of inbound open innovation on ambidexterity performance: Does it pay to source knowledge from supply chain stakeholders? Journal of Business Research, (December), 0–1. https://doi.org/10.1016/j.jbusres.2018.12.043

Arlbjørn, J. S., de Haas, H., & Munksgaard, K. B. 2011. Exploring supply chain innovation. Logistics Research, 3(1), 3–18. https://doi.org/10.1007/s12159-010-0044-3

Artsiomchyk, Y., & Zhivitskaya, H. 2015. Designing sustainable supply chain under innovation influence. IFAC-PapersOnLine, 28(3), 1695–1699. https://doi.org/10.1016/j.ifacol.2015.06.330

Azevedo, A. 2013. Advances in Sustainable and Competitive Manufacturing Systems. Springer Science & Business Media. https://doi.org/10.1007/978-3-319-00557-7

Bello, D. C., Lohtia, R., & Sangtani, V. 2004. An institutional analysis of supply chain innovations in global marketing channels. Industrial Marketing Management, 33(1), 57–64.

Cai, J., Liu, X., Xiao, Z., & Liu, J. 2009. Improving supply chain performance management: A systematic approach to analyzing iterative KPI accomplishment. Decision Support Systems, 46(2), 512–521. https://doi.org/10.1016/J.DSS.2008.09.004

Chapman, R. L., Soosay, C., & Kandampully, J. 2003. Innovation in logistic services and the new business model. International Journal of Physical Distribution & Logistics Management, 33(7), 630–650. https://doi.org/10.1108/09600030310499295

Chen, J. Y., Dimitrov, S., & Pun, H. 2019. The impact of government subsidy on supply Chains' sustainability innovation. Omega (United Kingdom), 86, 42–58. https://doi.org/10.1016/j.omega.2018.06.012

Chitale, K., & Gupta, C. 2014. Materials management a supply chain perspective: text and cases. PHI Learning Pvt. Ltd.

Chong, A. Y. L., Chan, F. T. S., Ooi, K.-B., & Sim, J.-J. 2011. Can Malaysian firms improve organizational/innovation performance via SCM? - ABI/INFORM Complete - ProQuest. Industrial Management & Data Systems, 111(3), 410–431. Retrieved from http://search.proquest.com/abicomplete/docview/859629828/fulltextPDF/AB2808DB2F9E4346PQ/36?accountid=15297

Corbin, J., & Strauss, A. 2008. Basics of Qualitative Research (3rd ed.): Techniques and Procedures for Developing Grounded Theory. Thousand Oaks, California: SAGE Publications, Inc. https://doi.org/10.4135/9781452230153

Cox, A. 1999. A research agenda for supply chain and business management thinking. Supply Chain Management: An International Journal, 4(4), 209–212. https://doi.org/10.1108/13598549910284534

Dubey, R., Singh, T., & Tiwari, S. 2012. Supply Chain Innovation is a Key to Superior Firm Performance an Insight from Indian Cement Manufacturing. International Journal of Innovation Science, 4(4), 217–230. https://doi.org/10.1260/1757-2223.4.4.217

Flint, D. J., & Larsson, E. 2007. Supply chain innovation. Handbook of Global Supply Chain Management, 49(4), 475–487. https://doi.org/10.4135/9781412976169.n28

Flint, D. J., Larsson, E., Gammelgaard, B., & Mentzer, J. T. 2005. LOGISTICS INNOVATION: A CUSTOMER VALUE-ORIENTED SOCIAL PROCESS. Journal of Business Logistics, 26(1), 113–147. https://doi.org/10.1002/j.2158-1592.2005.tb00196.x

Franks, J. 2000. Supply chain innovation. Work Study, 49(4), 152–155.

Gao, D., Xu, Z., Ruan, Y. Z., & Lu, H. 2017. From a systematic literature review to integrated definition for sustainable supply chain innovation (SSCI. Journal of Cleaner Production, 142, 1518–1538. https://doi.org/10.1016/j.jclepro.2016.11.153

Gregor, S. 2006. The nature of theory in information systems. MIS Quarterly, 611–642.

Herzlinger, R. E. 2006. Why innovation in health care is so hard. Harvard Business Review, 84(5), 58—66, 156. Retrieved from http://europepmc.org/abstract/MED/16649698

Hsin Chang, H., Hong Wong, K., & Sheng Chiu, W. 2019. The effects of business systems leveraging on supply chain performance: Process innovation and uncertainty as moderators. Information and Management, (January), 1–17. https://doi.org/10.1016/j.im.2019.01.002

Hui, Z., He-Cheng, W., & Min-Fei, Z. 2015. Partnership Management, Supply Chain Collaboration, and Firm Innovation Performance: An Empirical Examination. International Journal of Innovation Science, 7(2), 127–138. https://doi.org/10.1260/1757-2223.7.2.127

Iddris, F. 2016. Measurement of innovation capability in supply chain: An exploratory study. International Journal of Innovation Science, 8(4), 331–349. https://doi.org/10.1108/IJIS-07-2016-0015

Jajja, M. S. S., Kannan, V. R., Brah, S. A., & Hassan, S. Z. 2017. Linkages between firm innovation strategy, suppliers, product innovation, and business performance. International Journal of Operations & Production Management, 37(8), 1054–1075. https://doi.org/10.1108/ijopm-09-2014-0424

Jangga, R., Ali, N. M., Ismail, M., & Sahari, N. 2015. Effect of Environmental Uncertainty and Supply Chain Flexibility Towards Supply Chain Innovation: An exploratory Study. Procedia Economics and Finance, 31(15), 262–268. https://doi.org/10.1016/s2212-5671(15)01228-9

Jellali, A., & Benaissa, M. 2015. Sustainable performance evaluation of the supply chain. In 2015 4th International Conference on Advanced Logistics and Transport (ICALT) (pp. 151–156. IEEE. https://doi.org/10.1109/ICAdLT.2015.7136612

Krabbe, M. 2007. Leverage supply chain innovation: distribution center innovation puts you ahead of the competition. Industrial Engineer, 39(12), 26–31.

Kwak, D. W., Seo, Y. J., & Mason, R. 2018. Investigating the relationship between supply chain innovation, risk management capabilities and competitive advantage in global supply chains. International Journal of Operations and Production Management, 38(1), 2–21. https://doi.org/10.1108/IJOPM-06-2015-0390

Lee, S. M., Lee, D., & Schniederjans, M. J. 2011. Supply chain innovation and organizational performance in the healthcare industry. International Journal of Operations & Production Management, 31(11), 1193–1214.

Li, T., & Cavusgil, S. T. 1995. A classification and assessment of research streams in International Marketing. International Business Review, 4(3), 251–277. https://doi.org/10.1016/0969-5931(95)00009-O

Lin, C.-Y. 2007. Supply chain performance and the adoption of new logistics technologies for logistics service providers in Taiwan. Journal of Statistics and Management Systems, 10(4), 519–543. https://doi.org/10.1080/09720510.2007.10701270

Lv, B., & Qi, X. 2019. Research on partner combination selection of the supply chain collaborative product innovation based on product innovative resources. Computers and Industrial Engineering, 128(February 2018), 245–253. https://doi.org/10.1016/j.cie.2018.12.041

McGuirk, H., Lenihan, H., & Hart, M. 2015. Measuring the impact of innovative human capital on small firms' propensity to innovate. Research Policy, 44(4), 965–976. https://doi.org/10.1016/j.respol.2014.11.008

Munksgaard, K. B., Stentoft, J., & Paulraj, A. 2014. Value-based supply chain innovation. Operations Management Research, 7(3–4), 50–62. https://doi.org/10.1007/s12063-014-0092-y

Nasr, E. S., Kilgour, M. D., & Noori, H. 2015. Strategizing niceness in co-opetition: The case of knowledge exchange in supply chain innovation projects. European Journal of Operational Research, 244(3), 845–854. https://doi.org/10.1016/j.ejor.2015.02.011

Neuendorf, K. A. 2016. The Content Analysis Guidebook. Sage Publications.

Nidumolu, R., Prahalad, C. K., & Rangaswami, M. R. 2015. Why sustainability is now the key driver of innovation. IEEE Engineering Management Review, 43(2), 85–91. https://doi.org/10.1109/EMR.2015.7123233

PAGELL, M., & WU, Z. 2009. BUILDING A MORE COMPLETE THEORY OF SUSTAINABLE SUPPLY CHAIN MANAGEMENT USING CASE STUDIES OF 10 EXEMPLARS. Journal of Supply Chain Management, 45(2), 37–56. https://doi.org/10.1111/j.1745-493X.2009.03162.x

Paulraj, A., & Chen, I. J. 2004. Towards a theory of supply chain management: the constructs and measurements. Journal of Operations Management, 22(2), 119–150. https://doi.org/10.1016/j.jom.2003.12.007

Rajabian Tabesh, A., Batt, P. J., & Butler, B. 2016. Modelling the Impact of Environmental and Organizational Determinants on Green Supply Chain Innovation and Performance. Journal of Food Products Marketing, 22(4), 436–454. https://doi.org/10.1080/10454446.2014.949987

Reimann, M., Xiong, Y., & Zhou, Y. 2019. Managing a closed-loop supply chain with process innovation for remanufacturing. European Journal of Operational Research, 276(2), 510–518. https://doi.org/10.1016/j.ejor.2019.01.028

Russell, D. M., & Swanson, D. 2019. Transforming information into supply chain agility: an agility adaptation typology. International Journal of Logistics Management, 30(1), 329–355. https://doi.org/10.1108/IJLM-09-2017-0237

Sabri, Y., Micheli, G. J. L., & Nuur, C. 2018. Exploring the impact of innovation implementation on supply chain configuration. Journal of Engineering and Technology Management - JET-M, 49(December 2016), 60–75. https://doi.org/10.1016/j.jengtecman.2018.06.001

Schaltegger, S., & Burritt, R. 2014. Measuring and managing sustainability performance of supply chains. Supply Chain Management: An International Journal, 19(3), 232–241. https://doi.org/10.1108/SCM-02-2014-0061

Schumpeter, J. A. 1934. Change and the Entrepreneur. Essays of JA Schumpeter.

Shah, S. R., & Naghi Ganji, E. 2017. Lean production and supply chain innovation in baked foods supplier to improve performance. British Food Journal, 119(11), 2421–2447. https://doi.org/10.1108/BFJ-03-2017-0122

Shapiro, J. F., & Wagner, S. N. 2009. STRATEGIC INVENTORY OPTIMIZATION. Journal of Business Logistics, 30(2), 161–173. https://doi.org/10.1002/j.2158-1592.2009.tb00117.x

Shou, Y., Che, W., Dai, J., & Jia, F. 2018. Inter-organizational fit and environmental innovation in supply chains: A configuration approach. International Journal of Operations and Production Management, 38(8), 1683–1704. https://doi.org/10.1108/IJOPM-08-2017-0470

Stavrulaki, E., & Davis, M. 2010. Aligning products with supply chain processes and strategy. The International Journal of Logistics Management, 21(1), 127–151.

Stentoft, J., Mikkelsen, O. S., & Jensen, J. K. 2016. Offshoring and backshoring manufacturing from a supply chain innovation perspective. Supply Chain Forum, 17(4), 190–204. https://doi.org/10.1080/16258312.2016.1239465

Stentoft, J., & Rajkumar, C. 2018. Innovation and Supply Chain Management. https://doi.org/10.1007/978-3-319-74304-2

Tan, K. H., Zhan, Y. Z., Ji, G., Ye, F., & Chang, C. 2015. Harvesting big data to enhance supply chain innovation capabilities: An analytic infrastructure based on deduction graph. International Journal of Production Economics, 165, 223–233. https://doi.org/10.1016/j.ijpe.2014.12.034

Ted Lirn, T. C., Wu, Y. J., & Coronado Mondragon, A. E. 2018. ISL 2016: Sustainable transport and supply chain innovation. The International Journal of Logistics Management, 29(1), 2–4. https://doi.org/10.1108/ijlm-12-2017-0336

van Uden, A., Knoben, J., & Vermeulen, P. 2017. Human capital and innovation in Sub-Saharan countries: a firm-level study. Innovation, 19(2), 103–124. https://doi.org/10.1080/14479338.2016.1237303

West, M. A., & Farr, J. L. 1989. Innovation at work: Psychological perspectives. Social Behaviour, 4(1), 15–30.

Wong, D. T. W., & Ngai, E. W. T. 2019. Critical review of supply chain innovation research (1999–2016. Industrial Marketing Management, (January), 1–30. https://doi.org/10.1016/j.indmarman.2019.01.017

Wu, Y. J., & Tsai, K. 2018. Making connections: Supply chain innovation research collaboration. Transportation Research Part E: Logistics and Transportation Review, 113(February), 222–224. https://doi.org/10.1016/j.tre.2018.02.004

Yoon, S. N., Lee, D. H., & Schniederjans, M. 2016. Effects of innovation leadership and supply chain innovation on supply chain efficiency: Focusing on hospital size. Technological Forecasting and Social Change, 113, 412–421. https://doi.org/10.1016/j.techfore.2016.07.015

Zijm, H., Klumpp, M., Heragu, S., & Regattieri, A. 2019. Operations, Logistics andSupply Chain Management: Definitions and Objectives. In H. Zijm, M. Klumpp, A. Regattieri, & S. Heragu (Eds.), Operations, Logistics and Supply Chain Management (pp. 27–42. Cham: Springer International Publishing. https://doi.org/10.1007/978-3-319-92447-2_3

Zimmermann, R., DF Ferreira, L. M., & Carrizo Moreira, A. 2016. The influence of supply chain on the innovation process: a systematic literature review. Supply Chain Management: An International Journal, 21(3), 289–304.

Can Gamification Reduce the Shortage of Skilled Logistics Personnel?

Florian Hofbauer[1] *and Lisa-Maria Putz*[1]

1 – University of Applied Sciences Upper Austria

Purpose: The logistics sector faces a worldwide shortage of skilled personnel. Gamification represents a new approach to attract people's interest by applying game elements in a non-game context. In our study, we develop a theoretical framework to improve the image of the logistics sector using gamification for future studies.

Methodology: Based on the findings of the literature review, we discuss gamification and its potential to improve the image of the logistics sector. Moreover, the core of this paper is to develop a theoretical framework to apply career choice theories for gamification to attract logistics personnel.

Findings: In total, six career choice theories were identified as appropriate for a theoretical framework. We suggest using these six theories for future empirical studies to measure how gamification influences people's career choices.

Originality: This paper introduces gamification to attract people to the logistics sector as a new approach with substantial potential. It provides an initial valuation of the potential of gamification to improve the image of logistics jobs. Due the novelty of the topic, the theoretical framework provides a starting point for future empirical studies.

Keywords: Logistics, Career choice, Theoretical framework, Gamification

First received: 17.May.2019 **Revised:** 29.May.2019 **Accepted:** 20.June.2019

1 Introduction

Logistics represents a worldwide growth sector in terms of activity and expenditure. The logistics sector plays a vital role in the European economy, with approximately 7% of the total GDP generated in the EU-28. The transport sector, which is part of the logistics sector, accounts for more than 11 million jobs in Europe (Ecorys, et al., 2015). Despite other sector facing a jobs reduction due to automation and digitalization, the vast majority of logistics jobs are still done by humans. In fact, logistics activities are labor-intensive on both operational and managerial levels (McKinnon, et al., 2017). The logistics performance of companies is highly dependent on the availability and quality of skilled people (Gravier and Farris, 2008; Ecorys, et al., 2015). However, current studies suggest that there is a global shortage of people with interest and skills within all areas of the logistics sector (Fawcett and Rutner, 2014; McKinnon, et al., 2017). For example, Dubey and Gunasekaran (2015) and Maloni, et al. (2016) report a lack of junior, middle and senior level supply managers. Staats, et al. (2017) emphasize the shortage of truck drivers, who are urgently needed worldwide. The situation is similar for order pickers, who play a pivotal role in logistics (Grosse, Glock and Neumann, 2017).

Previous studies claim that one reason for the shortage of skilled people is the poor image of the logistics sector. To ensure the worldwide competiveness needed in the logistics sector, the attraction of people into the logistics industry represents a major objective of the European Union (Ecorys, et al., 2015). One new approach to inform people and attract them to the sector could be gamification. Gamification is defined as "the use of game design elements in non-game contexts" (Deterding, et al., 2011). It is a new approach that aims at engaging people using game elements such as competitions, feedback functions or social elements within a non-game context, instead of full-fledged games (Huotari and Hamari, 2012; Treiblmaier, Putz and Lowry, 2018). Previous studies found that gamification can be efficiently used to motivate people into adopting a certain behavior and influence their attitude regarding a certain topic (such as e.g. sustainability) (Corner, Kane and Owen, 2017; Kasurinen and Knutas, 2018). Up to now, gamification has been repeatedly applied in operational logistics processes such as picking. In fact, game elements are used to make order picking more fun and thus increase employees' engagement and retention to the company (Warmelink, et al., 2018; Putz, Hofbauer and Mates, 2019).

It has been proven that gamification can increase educational performance and information gain (Kapp, 2012), that behavior and attitude can be positively changed (Marcucci, Gatta and Le Pira, 2018). Moreover, gamification has been successfully implemented in the logistics sector (Warmelink, et al., 2018; Putz, Hofbauer and Mates, 2019). In this paper, we follow a new approach: We suggest that gamification can influence people's career choice, help to attract people to the logistics sector and support the reduction in the shortage of skilled workers.

Previous empirical studies on gamification and its impact on the attitude towards career choice show that gamification has the potential to increase interest in specific job sectors. Pérez-Manzano and Almela-Baeza (2018) used gamification-based applications to increase participants' interest in science and to promote scientific and technological careers. McGuire, et al. (2017) created gamified workplace simulations to increase student motivation and awareness of career opportunities. Their mixed results show that a well-designed tutorial is required to make sure that the purpose of the gamified information system is clearly communicated to the participants. Furthermore, the empirical studies show that appropriate measurement tools are required to measure the impact of gamified information systems on the attitude towards career choice as compared to non-gamified systems.

The goal of this paper is to develop a theoretical framework for future studies into the investigation of gamification and career choices in the logistics sector. This paper is organized as follows: In the methodology section, we describe how we proceeded to develop the theoretical framework. Subsequently, we present the theories and a comprehensive summary of our findings. We conclude the paper by highlighting several recommendations that might serve as a starting point for future research.

2 Method

The investigation of the impact of gamification on the attitude towards career choice requires a theoretical framework as a basis for further research. As expected, no such framework has been found in our literature search. Therefore, we decided to focus on theories which have been used in the context of the attitude towards career choice. The literature search was carried out in April 2019 using the SCOPUS database. We used the search term "theor*" in combination with several synonyms of "career choice" (Datta, 2017). The search was limited to title and resulted in 103 hits which were further inspected for inclusion or exclusion as shown in Fig.1. As a first step, we limited our search results to papers written in English that comprise the term "attitude" in their full-text. The remaining 25 papers were further examined regarding their content relevance. Eight studies are not concerned with the attitude towards career choice and were therefore excluded. Four studies used theories which seemed to lack sufficient scientific support to have an impact on career choice. Further literature search and analysis of these theories confirmed this assumption, therefore the four studies were excluded. 13 papers were identified for full analysis and for the creation of a theoretical framework.

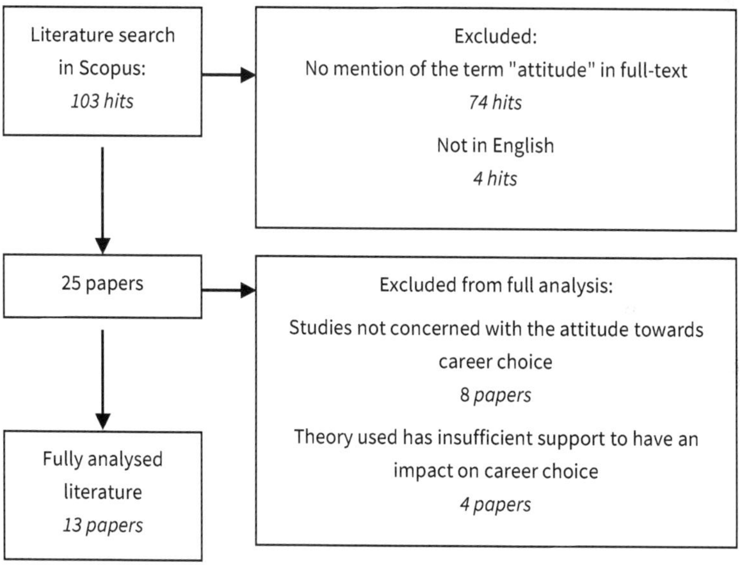

Figure 1: Flow chart of the literature search procedure following the approach of Warmelink, et al. (2018).

3 Results and Discussion

13 papers were identified as relevant for further investigation, resulting in six career choice theories. These six career choice theories are: Social cognitive career theory (SCCT), technology acceptance model (TAM), social cognitive theory (SCT), theory of reasoned action (TRA), self-determination theory (SDT) and Hollands' theory of career choice (RAISEC). Table 1 represents a list of the 13 papers identified for full analysis and shows their respective research focus and theories used. The papers were analyzed regarding the applied research area and the theories which were used to identify the career choices. 11 of the 13 papers were applied in the research area of education. The most used theories are the SCCT and the RAISEC (both applied in 4 out of 13 papers), followed by the TRA theory (with 3 out of 13 papers).

Table 1: Theoretical framework

Source	**Research area**	**SCCT**	**TAM**	**SCT**	**TRA**	**SDT**	**RIASEC**
Ray, Bala and Dasgupta, 2019	Education (technical studies)	X	X				
Gewinner, 2017	Education			X			
Zhang, Rashid and Mohammed, 2017	Education (tourism)				X		
Rosenkranz, Wang and Hu, 2017	Education (medical)					X	
Kaminsky and Behrend, 2015	Education (STEM)	X					
Rajabi, Papzan and Zahedi, 2012	Education (agriculture)	X					
Monson, 2012	Dental hygienists						X
Law, 2010	Education (accounting)				X		

Source	**Research area**	**SCCT**	**TAM**	**SCT**	**TRA**	**SDT**	**RIASEC**
Perdue, Reardon and Peterson, 2007	Telecommunications			X			X
Cunningham, et al. , 2005	Education (sports and leisure)	X					
Felton, Dimnik and Northey, 1995	Education (accounting)				X		
Guthrie and Herman, 1982	Vocational maturity						X
Laudeman and Griffeth, 1978	Education						X

The six theories are briefly described in the following section. The theories are analyzed regarding their practicability for gamification to reduce the shortage of skilled personnel in logistics.

3.1 Social cognitive career theory

Social cognitive career theory developed by Lent, Brown and Hackett (1994; 2000) is based on Bandura's (1986) social cognitive theory and aims to explain "three interrelated aspects of career development: (1) how basic academic and career interests develop, (2) how educational and career choices are made, and (3) how academic and career success is obtained" (Lent, Brown and Hackett, 2006, p. 750). The focus lies on the cognitive-person variables self-efficacy, outcome expectations and goals, and how they interact with other personal and environmental factors (e.g. gender, ethnicity, social supports, barriers) to make educational and career choices (Lent, Brown and Hackett, 2000).

In the studies from the literature review, SCCT has been mainly used as a starting point to test the impact of additional specific factors on career choice. Ray, Bala and Dasgupta (2019) used SCCT in combination with the technology acceptance model (TAM) to test the role of authenticity and perceived benefits of online courses on career choice. The conclusions were that personal input strongly influences the learning experiences and authenticity and perceived benefits from the course are crucial factors for course choice and the associated career choice. Kaminsky and Behrend (2015) expanded the SCCT model with the factor "calling", as a predictor of career outcome expectations, interests and goals. The conclusions were that calling provides more predictive power than self-efficacy in terms of career outcome expectations and interests but is a weaker predictor of goals. Rajabi, Papzan and Zahedi (2012) used SCCT to test the cognitive and personal factors for career choice intentions of agricultural students. The conclusions were that the personality factors extraversion, openness, conscientiousness and agreeableness play a significant role for career choice intention along self-efficacy belief and career outcome expectations, while personal factors such as gender, major, average grade and neuroticism had no significant impact. Cunningham, et al. (2005) used SCCT to investigate student intentions to enter the sport and leisure industry. The study generally supports the SCCT model and showed that a multifaceted approach to measure outcome expectations is beneficial because various types of outcome expectations, such as expected power and satisfaction with the job environment differ significantly regarding their impact on choice goals.

The results of these studies and the frequent use of the theory indicate that SCCT is well-suited for the investigation of specific occupational fields and can be expanded with additional factors. With regard to the logistics industry, this model offers the opportunity to investigate the reasons for skilled labor shortage in detail. Self-efficacy expectations could give insight into whether logistics jobs are perceived as too difficult. The model could reveal existing barriers and clarify if outcome expectations (e.g. salary, decisive power, and satisfaction with the job environment) from respondents differ from what the logistics industry currently offers. SCCT is also perceived as a suitable theory for measuring the impact of gamification on the attitude towards career choice because gamification could be easily implemented as an influencing factor on learning experiences.

3.2 Technology acceptance model

The technology acceptance model from Davis (1989) „theorizes that an individual's behavioral intention to use a system is determined by two beliefs: perceived usefulness, defined as the extent to which a person believes that using the system will enhance his or her job performance, and perceived ease of use, defined as the extent to which a person believes that using the system will be free of effort" (Venkatesh and Davis, 2000, p. 187).

Ray, Bala and Dasgupta (2019) used TAM in combination with SCCT to consider the technological and educational factor of online courses towards career choice. They combined intention and interest from the respective model in one variable. Perceived usefulness and perceived use from TAM were included in the perceived benefits varia-ble. They conclude that individuals' intention to learn a course for their career choice is strongly influenced by perceived benefits (Ray, Bala and Dasgupta, 2019).

In the logistics industry, TAM is frequently used to measure the acceptance of newly implemented technologies (Qi, et al., 2009; Chen and Chao, 2011). As for gamification, TAM could be a suitable approach to measure the impact of gamification on perceived ease of use, when gamification is combined with new technologies such as augmented reality (Bräuer and Mazarakis, 2019; Logistikum Steyr, 2019).

3.3 Social cognitive theory

Social Cognitive Theory from (Bandura, 1986) theorizes that learning occurs in a social context with a dynamic and reciprocal interaction of the person, environment, and behavior. Individuals observe the behavior of others and the related behavioral consequences, remember the sequence of events and use this information as guidance for their own behavior. Depending on whether the behavior is rewarded or punished, the observer will choose to replicate the behavior or not (Bandura, 1986).

Gewinner (2017) used SCT as a basis for development of a theoretical approach for gendered occupational choices and stereotypes. Perdue, Reardon and Peterson (2007) explored the relationship between person–environment congruence, self-efficacy, and environmental identity and job satisfaction in a multinational telecommunications corporation. They concluded that self-efficacy was significantly related to job satisfaction. Self-efficacy is strongly linked with social cognitive theory because it is developed from external experiences and self-perception and has a strong impact on human thought, motivation and action (Bandura, 1991).

Similar to SCCT, SCT could provide information for the logistics industry about which environmental factors turn out to be supportive or barriers for a career choice in logistics. In addition, the behavioral factors could give an insight into which behavior or public appearance of the logistics industry would lead to an improvement in the image and, subsequently, to an increasing interest in working in the logistics field. In terms of gamification, SCT could be used to verify whether observation of gamified actions are perceived as more likely to be replicated than non-gamified actions.

3.4 Theory of reasoned action

The theory of reasoned action by Fishbein and Ajzen (1975) describes the behavioral process in which attitudes and subjective norms determine an individual's intention to demonstrate a certain behavior. The attitude towards a behavior is determined by the individual's belief on the likelihood of the behavior to generate the desired outcomes. Subjective norms are a product of the consumer's perceived social desirability of the observed behavior and the motivation to meet the social requirements (Fishbein and Ajzen, 1975).

Analysis of the respective studies from our theoretical framework showed that the TRA model is an appropriate choice to measure the influence of the attitude towards a profession and subjective norms on intention of career choice. Law (2010) used the TRA model to examine factors influencing the accounting student's career choice in public accounting practices. He concluded that his results support the TRA model and that intrinsic factors, flexibility of career options, gender and parental influence had significant influence on the decision to choose a career in public accounting, while financial rewards and high school accounting had no influence. Felton, Dimnik and Northey (1995) used the TRA model to examine the factors of student's choice of a career in chartered accountancy. The study supports the TRA model and concludes that the students' choice of a chartered accountancy career depends on the attitude towards this profession, which in turn is determined by the expected outcomes. Most relevant outcomes of a career as a chartered accountant are long-term earning, advancement opportunities, variety in the work, chance to make a contribution and flexibility of career options, which should be promoted by accounting recruiters. Zhang, Rashid and Mohammed (2017) used the social norms variable from the TRA model as orientation for creating a simplified research framework to measure parental factors on student's career decision to enter the hospitality industry in China.

The positive results of the studies indicate that the TRA model would be well suited to measure the influence of attitude towards logistics professions and social norms on intention of a career choice in the logistics field. An analysis of these variables could identify the causes of logistics' image and could give insight into whether the reasons lie more in the expected outcomes when choosing a logistics job or the social environment of individuals. Results from this analysis could be used as guidance as to what the logistics industry should focus on in the future to attract skilled workers. In addition, the TRA model can be used to measure the impact of gamification on the attitude towards career choice. Gamification could be implemented as an influencing factor on the determinants of the attitude towards a logistics profession and subjective norms.

3.5 Self-determination theory

The self-determination theory (SDT) by (Ryan and Deci, 2000a; 2000b) focuses on intrinsic motivation and the associated factors that cause a person to make decisions without external influence. SDT identifies autonomy, competence and relatedness as main factors for intrinsic motivation. Autonomy is defined as an individual's estimate of the extent to which the outcome of actions can be determined and controlled. Competence means that the individual has goals and the skills to achieve them. Relatedness refers to the feeling of being respected and cared for (Ryan and Deci, 2000a; 2000b).

Rosenkranz, Wang and Hu (2015) use SDT to identify the factors that motivate and demotivate medical students to do research. They concluded that students with practical experience and research experience in group work have a more positive attitude towards research. Furthermore, the motivation to do research is especially associated with the following intrinsic motivators: confidence, clinical relevance and research as a social activity. SDT could also be used to identify and measure the factors of intrinsic motivation for logistics professions. Based on the findings, gamification could be used to positively influence these factors making the occupation more fun and interesting and to create relatedness with the job.

3.6 Theory of vocational choice

The Holland Codes (RIASEC) refer to Holland's (1997) theory of vocational choice. The term RIASEC refers to the initials of the six personality types: Realistic, investigative, artistic, social, enterprising and conventional. Holland sees interests as basic personality orientations. Accordingly, the career choice is influenced by the individual's characteristics. The best fit between a person and a job is when the individual orientation is congruent with the orientation of the job. A good fit leads to higher job satisfaction, more career success and greater career development stability. (Holland, 1997)

Analysis from our body of literature revealed that RAISEC is frequently used to deter-mine the personality types for specific professions. Monson (2012) used RIASEC to identify the characteristics of dental hygienists. He concluded that Holland's six personality types can help students as a guide for their career choice. Perdue, Reardon and Peterson (2007) used RAISEC to evaluate the congruence between interests and job (person-environment congruence) of employees in a multinational telecommunications corporation. Against their expectations, person-environment congruence showed no significant relation to job satisfaction. They concluded that missed moderating factors might be the reason for this result. Guthrie and Herman (1982) determined the relationship of vocational maturity with Holland's theory of vocational choice. Therefore, they used RAISEC to identify the personality type of students enrolled in high- and mid-level programs in two post-secondary institutions. Results showed that congruency was significantly related to vocational maturity. Laudeman and Griffeth (1978) tested the validity of the Holland codes by identifying the personality type of college students with the RAISEC model. Their results generally supported Holland's postulated personality typologies and value dimensions.

The application of Holland's six personality types for the respective logistics professions could give potential candidates information on whether their own interests are congruent with the orientation of the job. In addition, the information could be used for the targeted search and recruitment of individuals with the desired characteristics. Gamification can be used to facilitate the identification of individual's personality types and to compare them with the orientation of logistics jobs.

4 Conclusion and further research

In this paper, we analyzed the existing empirical literature about career choice theories. We found six appropriate theories that are suitable for investigating the effects of gamification on the image of the logistics sector. The six theories are (1) the social cognitive career theory (SCCT), (2) the technology acceptance model (TAM), (3) the social cognitive theory (SCT), (4) the theory of reasoned action (TRA), (5) the self-determination theory (SDT) and (6) Hollands' theory of career choice (RAISEC). The vast majority of the theories were applied in an educational context.

SCCT is well-suited to investigating the reasons for skilled labor shortage in the logistics industry in detail. Furthermore, the model can easily be expanded with additional factors, such as gamification, which could be implemented as an influencing factor in learning experiences. TAM could be used to find out whether newly applied technologies are perceived as a support or a barrier, in terms of career decision making and to measure the impact of gamification on the perceived ease of use of new technologies. As gamification has no direct impact on career choice in this theory, we recommend to combine TAM with other theories to measure the direct impact of gamification on the attitude towards career choice. SCT could provide information on which behavior or public appearance of the logistics industry would improve the individual's interest in working in the logistics field. Furthermore, SCT could measure whether gamified actions are more likely to be replicated than non-gamified actions. TRA is also a suitable option to identify the causes of logistics' image and could give insight into whether the reasons lie more in the expected outcomes when choosing a logistics job or the social environment of individuals. Additionally, the TRA model can be used to measure the impact of gamification on the determinants of the attitude towards a profession and subjective norms. SDT could be used to identify and measure the intrinsic motivators of logistics professions and the impact of gamification on these intrinsic factors. Holland's six personality types are well-suited to checking whether individuals' interests are congruent with the orientation of the job. Gamification can help to facilitate the identification of individual's personality types and to convey the work environment type of a profession. Furthermore, it can be tested if specific personality types show a more positive reaction on the use of gamification. Job sectors with these personality types should consider the implementation of game elements in the promotion of their specific profession.

As with any research, ours has limitations. The literature review was limited to the Scopus database. While we are confident of the comprehensiveness of our literature search, it is possible that publications have been missed due to the lack of a listing in the database or due to indexing errors.

Based on the identification of appropriate theories, the next step would be to develop metrics for measuring the respective constructs. We would suggest beginning with multi-item scales that researchers can derive from existing research. Collecting empirical data in longitudinal studies, as claimed by Nacke and Deterding (2017), to investigate how gamification influences people's career choices would be the next step. With this paper, we hope to inspire the academic community to investigate gamification, particularly in a logistics career-oriented context. We would encourage this work to be considered as a starting point for conversations that re-think, re-contextualize, and challenge new (career) theories for logistics. At the same time, this paper represents a deepening of the research within the gamification field in such a way that the topic of career choice is theoretically prepared for future research.

We hope that the theoretical framework provided will serve as a basis for investigation of whether gamification can help to moderate existing barriers before and during career choice, or if it can help to improve self-efficacy expectations. Furthermore, we would like to obtain further insights into whether gamification can add additional value to technologies which support career choice and if gamer types could be linked with Holland's six personality types.

Financial Disclosure

This paper is part of the research project REWWay which was funded by viadonau and is part of the research field 'sustainable transport systems,' which was funded by the State of Upper Austria as part of the research program 'FTI Struktur Land Oberösterreich.'

References

Bandura, A., 1986. Social Foundations of Thought and Action. a Social Cognitive Theory. Upper Saddle River, NJ: Prentice Hall.

Bandura, A., 1991. Social cognitive theory of self-regulation. Organizational Behavior and Human Decision Processes, [e-journal] 50(2), pp. 248–287. http://dx.doi.org/10.1016/0749-5978(91)90022-L.

Bräuer, P. and Mazarakis, A., 2019. Badges or learderboard? How to gamify an augmented reality warehouse setting. In: J. Koivisto, and J. Hamari, eds. 2019. Proceedings of the 3rd International GamiFIN Conference: CEUR-WS, pp. 229–240.

Chen, C.-F. and Chao, W.-H., 2011. Habitual or reasoned? Using the theory of planned behavior, technology acceptance model, and habit to examine switching intentions toward public transit. Transportation Research Part F: Traffic Psychology and Behaviour, 14(2), pp. 128–137.

Corner, A., Kane, G. and Owen, P., 2017. Behaviour change for sustainability. London: Routledge.

Cunningham, G. B., Bruening, J., Sartore, M. L., Sagas, M. and Fink, J. S., 2005. The application of social cognitive career theory to sport and leisure career choices. Journal of Career Development, [e-journal] 32(2), pp. 122–138. http://dx.doi.org/10.1177/0894845305279164.

Datta, P., 2017. Supply network resilience. A systematic literature review and future research. The International Journal of Logistics Management, [e-journal] 28(4), pp. 1387–1424. http://dx.doi.org/10.1108/IJLM-03-2016-0064.

Davis, F. D., 1989. Perceived usefulness, perceived ease of use, and user acceptance of information technology. MIS Quarterly, 13(3), pp. 319–340.

Deterding, S., Dixon, D., Khaled, R. and Nacke, L., 2011. From game design elements to gamefulness. In: A. Lugmayr, H. Franssila, C. Safran, and I. Hammouda, eds. 2011. Proceedings of the 15th International Academic MindTrek Conference Envisioning Future Media Environments. New York, NY: ACM, p. 9–15.

Dubey, R. and Gunasekaran, A., 2015. Shortage of sustainable supply chain talent: an industrial training framework. Industrial & Commercial Training, [e-journal] 47(2), pp. 86–94. http://dx.doi.org/10.1108/ICT-08-2014-0052.

Ecorys, Fraunhofer, TCI, Prognos and AUEB-RC/TRANSLOG, 2015. Fact-finding studies in support of the development of an EU strategy for freight transport logistics. Lot 1: Analysis of the EU logistics sector. <https://ec.europa.eu/transport/sites/transport/files/themes/strategies/studies/doc/2015-01-freight-logistics-lot1-logistics-sector.pdf> [Accessed 19 September 2018].

Fawcett, S. and Rutner, S., 2014. A longitudinal view of supply chain education. The International Journal of Logistics Management, [e-journal] 25(1), pp. 180–201. http://dx.doi.org/10.1108/IJLM-07-2012-0061.

Felton, S., Dimnik, T. and Northey, M., 1995. A Theory of Reasoned Action model of the Chartered Accountant career choice. Journal of Accounting Education, [e-journal] 13(1), pp. 1–19. http://dx.doi.org/10.1016/0748-5751(94)00027-1.

Fishbein, M. and Ajzen, I., 1975. Belief, Attitude, Intention and Behavior. An Introduction to Theory and Research. Reading, Mass. Don Mills, Ontario, ON: Addison-Wesley Pub. Co.

Gewinner, I., 2017. Gendered career choices and stereotypes: A theoretical approach: Emerald Group Publishing Ltd.

Gravier, M. J. and Farris, M. T., 2008. An analysis of logistics pedagogical literature: past and future trends in curriculum, content, and pedagogy. The International Journal of Logistics Management, [e-journal] 19(2), pp. 233–253. http://dx.doi.org/10.1108/09574090810895979.

Grosse, E. H., Glock, C. H. and Neumann, W. P., 2017. Human factors in order picking: a content analysis of the literature. International Journal of Production Research, [e-journal] 55(5), pp. 1260–1276. http://dx.doi.org/10.1080/00207543.2016.1186296.

Guthrie, W. R. and Herman, A., 1982. Vocational maturity and its relationship to Holland's theory of vocational choice. Journal of Vocational Behavior, [e-journal] 21(2), pp. 196–205. http://dx.doi.org/10.1016/0001-8791(82)90029-X.

Holland, J. L., 1997. Making vocational choices. A theory of vocational personalities and work environments. 3rd ed. Lutz, Fla.: Psychological Assessment Resources.

Huotari, K. and Hamari, J., 2012. Defining gamification - a service marketing perspective. In: . 2012. 16th International Academic MindTrek Conference, pp. 17–22.

Kaminsky, S. E. and Behrend, T. S., 2015. Career Choice and Calling: Integrating Calling and Social Cognitive Career Theory. Journal of Career Assessment, [e-journal] 23(3), pp. 383–398. http://dx.doi.org/10.1177/1069072714547167.

Kapp, K. M., 2012. The gamification of learning and instruction. Game-based methods and strategies for training and education. San Francisco, CA: Pfeiffer.

Kasurinen, J. and Knutas, A., 2018. Publication trends in gamification. a systematic mapping study. Computer Science Review, [e-journal] 27, pp. 33–44. http://dx.doi.org/10.1016/j.cosrev.2017.10.003.

Laudeman, K. A. and Griffeth, P., 1978. Holland's theory of vocational choice and postulated value dimensions. Educational and Psychological Measurement, [e-journal] 38(4), pp. 1165–1175. http://dx.doi.org/10.1177/001316447803800438.

Law, P. K., 2010. A theory of reasoned action model of accounting students' career choice in public accounting practices in the post-Enron. Journal of Applied Accounting Research, [e-journal] 11(1), pp. 58–73. http://dx.doi.org/10.1108/09675421011050036.

Lent, R. W., Brown, S. D. and Hackett, G., 1994. Toward a Unifying Social Cognitive Theory of Career and Academic Interest, Choice, and Performance. Journal of Vocational Behavior, [e-journal] 45(1), pp. 79–122. http://dx.doi.org/10.1006/jvbe.1994.1027.

Lent, R. W., Brown, S. D. and Hackett, G., 2000. Contextual supports and barriers to career choice: A social cognitive analysis. Journal of Counseling Psychology, [e-journal] 47(1), pp. 36–49. http://dx.doi.org/10.1037/0022-0167.47.1.36.

Lent, R. W., Brown, S. D. and Hackett, G., 2006. Social cognitive career theory. In: J. H. Greenhaus, and G. A. Callanan, eds. 2006. Encyclopedia of career development. Thousand Oaks, Calif.: SAGE Publications, pp. 750–754.

Logistikum Steyr, 2019. LOGISTIFY – understand the world of logistics! - REWWay. [online]. Available at: <https://www.rewway.at/en/logistify-learning-app/> [Accessed 14 May 2019].

Maloni, Scherrer, Campbell and Boyd, 2016. Attracting students to the field of logistics, part 1. Transportation Journal, [e-journal] 55(4), p. 420–420. http://dx.doi.org/10.5325/transportationj.55.4.0420.

Marcucci, E., Gatta, V. and Le Pira, M., 2018. Gamification design to foster stakeholder engagement and behavior change: an application to urban freight transport. Transportation Research Part A: Policy and Practice, 118, pp. 119–132.

McGuire, A., Broin, D. O., Power, K., White, P. J. and Deevy, C., 2017. Increasing student motivation and awareness of career opportunities through gamification. In: M. Pivec, and J. Gründler, eds. 2017. Proceedings of the 11th European Conference on Game-Based Learning. ECGBL 2017. Reading, UK: Academic Conferences and Publishing International Limited, pp. 898–906.

McKinnon, A., Floethmann, C., Hoberg, K. and Busch, C., 2017. Logistics competencies, skills, and training: a global overview: The World Bank.

Monson, A. L., 2012. Characteristics of dental hygienists based on Holland's career choice theory. Journal of dental hygiene : JDH / American Dental Hygienists' Association, 86(2), pp. 141–149.

Nacke, L. E. and Deterding, S., 2017. The maturing of gamification research. Advanced Human-Computer Interaction, [e-journal] 71, pp. 450–454. http://dx.doi.org/10.1016/j.chb.2016.11.062.

Perdue, S. V., Reardon, R. C. and Peterson, G. W., 2007. Person - Environment congruence, self-efficacy, and environmental identity in relation to job satisfaction: A career decision theory perspective. Journal of Employment Counseling, [e-journal] 44(1), pp. 29–39. http://dx.doi.org/10.1002/j.2161-1920.2007.tb00022.x.

Pérez-Manzano, A. and Almela-Baeza, J., 2018. Gamification and trans-media for scientific promotion and for encouraging scientific careers in adolescents. Comunicar, [e-journal] 26(55), pp. 93–103. http://dx.doi.org/10.3916/C55-2018-09.

Putz, L.-M., Hofbauer, F. and Mates, M., 2019. A vignette study among order pickers about the acceptance of gamification. In: J. Koivisto, and J. Hamari, eds. 2019. Proceedings of the 3rd International GamiFIN Conference: CEUR-WS, pp. 154–166.

Qi, J., Li, L., Li, Y. and Shu, H., 2009. An extension of technology acceptance model: Analysis of the adoption of mobile data services in China. Systems Research and Behavioral Science, [e-journal] 26(3), pp. 391–407. http://dx.doi.org/10.1002/sres.964.

Rajabi, S., Papzan, A. and Zahedi, G., 2012. Application of Social Cognitive Career Theory to investigate the effective factors of the career decision-making intention in iranian agriculture students by using ANN. SAGE Open, [e-journal] 2(4), pp. 1–13. http://dx.doi.org/10.1177/2158244012467024.

Ray, A., Bala, P. K. and Dasgupta, S. A., 2019. Role of authenticity and perceived benefits of online courses on technology based career choice in India: A modified technology adoption model based on career theory. International Journal of Information Management, [e-journal] 47, pp. 140–151. http://dx.doi.org/10.1016/j.ijinfomgt.2019.01.015.

Rosenkranz, S. K., Wang, S. and Hu, W., 2015. Motivating medical students to do research: A mixed methods study using Self-Determination Theory Career choice, professional education and development. BMC Medical Education, [e-journal] 15(1). http://dx.doi.org/10.1186/s12909-015-0379-1.

Ryan, R. M. and Deci, E. L., 2000a. Intrinsic and Extrinsic Motivations: Classic Definitions and New Directions. Contemporary Educational Psychology, 25(1), pp. 54–67.

Ryan, R. M. and Deci, E. L., 2000b. Self-determination theory and the facilitation of intrinsic motivation, social development, and well-being. American psychologist, 55(1), pp. 68–78.

Staats, U., Lohaus, D., Christmann, A. and Woitschek, M., 2017. Fighting against a shortage of truck drivers in logistics: measures that employers can take to promote drivers' work ability and health. Work (Reading, Mass.), [e-journal] 58(3), pp. 383–397. http://dx.doi.org/10.3233/WOR-172626.

Treiblmaier, H., Putz, L.-M. and Lowry, P. B., 2018. Research commentary. Setting a definition, context, and theory-based research agenda for the gamification of non-naming applications. AIS Transactions on Human-Computer Interaction, [e-journal] 10(3), pp. 129–163. http://dx.doi.org/10.17705/1thci.00107.

Venkatesh, V. and Davis, F. D., 2000. A theoretical extension of the technology acceptance model. four longitudinal field studies. Management Science, 46(2), pp. 186–204.

Warmelink, H., Koivisto, J., Mayer, I., Vesa, M. and Hamari, J., 2018. Gamification of production and logistics operations: status quo and future directions. Journal of Business Research, [e-journal] in press. http://dx.doi.org/10.1016/j.jbusres.2018.09.011.

Zhang, H., Rashid, B. and Mohammed, A. A., 2017. A theoretical model of parental factors and students' career decisions to join the hospitality industry in China. Geojournal of Tourism and Geosites, 19(1), pp. 133–141.

Machine Learning in Demand Planning: Cross-industry Overview

Nikolas Ulrich Moroff[1] *and Saskia Sardesai*[1]

1 – Fraunhofer Institute for Material Flow and Logistics IML

Purpose: This paper aims to give an overview about the current state of research in the field of machine learning methods in demand planning. A cross-industry analysis for current machine learning approaches within the field of demand planning provides a decision-making support for the manufacturing industry.

Methodology: Based on a literature research, the applied machine learning methods in the field of demand planning are identified. The literature research focuses on machine learning applications across industries wherein demand planning plays a major role.

Findings: This comparative analysis of machine learning approaches provides/creates a decision support for the selection of algorithms and linked databases. Furthermore, the paper shows the industrial applicability of the presented methods in different use cases from various industries and formulates research needs to enable an integration of machine learning algorithms into the manufacturing industry.

Originality: The article provides a systematic and cross-industry overview of the use of machine learning methods in demand planning. It shows the link between established planning processes and new technologies to identify future areas of research

Keywords: Machine Learning, Demand Planning, Artificial Intelligence, Digitalization

First received: 14.May.2019 **Revised:** 29.May.2019 **Accepted:** 20.Jun.2019

1 Introduction

In recent years, digitalization has become more and more important, and hence it has a high significance for the manufacturing and planning processes of industrial companies. According to a study by the VDI, up to 25% of companies intend to deal with the topic of artificial intelligence (AI) and machine learning (ML), which is one of the main tasks of digitalization in the following years (Verein Deutscher Ingenieure e.V., 2018). Artificial intelligence describes a field of research that becomes increasingly relevant in the digitalization process for the automation of planning processes. Machine learning is a sub-area of artificial intelligence and describes a method for the implementation of self-learning systems. The study shows, that these topics already have a high priority in the economy today and their importance for future economic success is very present. Table 1 supports the estimated potential of the new technologies (Reder, 2018).

Table 1: Studies for AI and ML (Verein Deutscher Ingenieure e.V., 2018)

Study	**Date**	**Extrapolation (increase by)**	**Market**	**Until Year**
Accenture	July 2017	gross value added (12 trillion Euros)	12 industrial countries (incl. Germany)	2035
McKinsey	April 2018	gross domestic product (16 billion Euros)	Germany	2030
PwC	June 2018	gross domestic product (430 billion Euros)	Germany	2030
BMWi, iit Berlin	July 2018	gross value added (31,8 billion Euros)	Industry Germany	2023

The process of digitalization and its analysis began in the 1990s with the increasing creation of data records, digital transactions and networked systems (Zuboff, 2010; Bryanjolfsson and Mcafee, 2014). Especially in logistics and supply chain management, a high amount of transaction data is generated and collected, displaying the high potential for the use of this information (e.g. for forecast improvement) (Röniger, 2018).

Along with the increasing computing power, it is now possible to analyze the generated data sets of a company and render them usefully (Papenfort, Frank and Strughold, 2015). Methods and algorithms have to be developed for the usage of the data sets that can analyse large amounts of information in order to recognize patterns and rules that can be used as decision support for future processes and planning (Haasis, et al., 2015). One of these methods for analysing large amounts of datasets is machine learning, an algorithm that evolves using feedback information and historical data (Michie, 1968). Especially demand forecasting in the area of supply chain planning (e.g. in manufacturing industry) is an important application area for machine learning methods, as many different factors affect the sales market and conventional statistical methods reach their limits (Deutschländer, 2003).

2 State of the Art

2.1 Demand Planning

Demand planning describes the process of determining future material requirements according to quantity and time and represents a major task within the area of Supply Chain Management (SCM) (Tempelmeier, 2018). The SCM task model from Kuhn and Hellingrath describes the interfaces between the individual SCM tasks and the planning horizon of each individual module (Hellingrath and Kuhn, 2013). Originally, the SCM task model was designed to structure requirements for SCM software, to define all areas and respective interfaces in detail (Koch, 2012). Within the SCM task model, demand planning covers the interface between tactical and strategic planning. It serves as an input for further underlying planning processes (e.g. procurement, production and distribution planning). A large number of planning parameters are therefore dependent on the results of demand planning. Consequently, it is essential for the profitability of a company to achieve a high forecast quality and to minimize uncertainties (Ashayeri and Lemmes, 2006). There are several different forecasting methods, which form the basis for the strategic and operative planning of a company (Hammer, 1998). Current forecasting methods, though, analyze the past in order to draw conclusions about future events. It is assumed, that events always have the same implications and follow the same pattern. This approach is known as the time stability hypothesis. Based on this assumption, there are deviations between the forecast result and reality, which are evaluated as forecast errors (Hansmann, 2013; Stickel, Groffmann and Rau, 2013). Methodically, the forecasting procedures can be divided into two categories: quantitative and qualitative. The characteristics of the different methods are briefly summarized in Table 2.

In summary, it can be said that demand planning is based on the forecast results of the procedures briefly outlined before. The forecast thus determines the demand of goods for a fixed period (e.g. monthly). However, in order to achieve a high forecasting quality, an exponentially increasing computing effort and sufficient database is necessary, so that a structured and case-dependent selection of the forecasting method is essential (Feindt and Kerzel, 2014; Kühnapfel, 2014).

Due to the advancing development of information technologies, classical forecasting methods are slowly being replaced by more complex methods. The trend is tending towards the use of e.g. machine learning or neural networks to enhance the forecast with further internal and external data (Mertens and Rässler, 2012). Machine learning can combine quantitative and qualitative methods so that it is possible to integrate both, historical data and expert knowledge (semantic data) in the decision-making process (Shalev-Shwartz and Ben-David, 2014). Furthermore, it is possible to integrate additional internal sources (e. g. intranet, social media) of information that provide extra knowledge for the planning process. The following section of the article gives a brief overview of the challenges in forecasting demand in the manufacturing industry.

Table 2: Properties of forecast methods (Hansmann, 2013; Stickel, Groffmann and Rau, 2013)

Quantitative forecasting methods	Qualitative forecasting methods
Mathematical model which is applied to a company's historical data	Use of expert knowledge to generate recommendations for action
One-dimensional or multidimensional	Often no database available/required
Linear and non-linear models	

2.2 Challenges in the demand planning of manufacturing companies

Demand planning in the manufacturing industry is affected by a high degree of forecasting uncertainty. This is due to the many different factors which can influence the demand of a company and to the length of the forecasting horizon. These uncertainty factors can be divided into the following types: aleatoric and epistemic uncertainties (Helton, et al., 2010). Aleatoric uncertainties describe the stochastic influences on the forecast quality. In addition to statistical uncertainties, there are further factors (epistemic uncertainties) influencing the forecast quality, which are briefly described in the following list (Grömling, 2005; Egri, et al., 2008).

- Data revision: Data in the model is incorrect, resulting in a wrong basis for the forecast model.
- Unpredictable events: Statistical methods do not consider new events and hence do not forecast the resulting effect.
- Human factor: Humans tend to override the output and adhere to familiar principles (status-quo preservation).

To overcome some of the issues mentioned above, Habla analyzed machine learning methods in demand planning of a manufacturing company in order to improve the forecast quality. The author concluded that many areas in the modelling and application of the algorithm were still insufficiently illuminated. For example, the integration of trends and seasonal effects is not mentioned in his study and he also argues that data mining methods can be used to generate different aggregation levels to enable a robust forecast (Habla, et al., 2007). The main reasons lie in the high complexity of the data structure and the changing customer behavior. Because of the ever shorter product life cycles, thus, it is not possible to collect enough historical data over a long period of time or for a specific product phase (start-up, peak or phase-out phase) to make statistically sound statements regarding the demand forecast (Johnson, 2018; Mayr and Moser, 2018). The type of product is also decisive here: Innovative products have a significantly shorter product life cycle than standard products, hence their sales behavior can be predicted much less accurately (Lee, 2002). Even in case a good data basis is available, there are further challenges in the analysis and interpretation of the data. It is crucial to filter the essential data and ignore unimportant data. This can only be done in relation to the use case to be examined, since the importance of a data set highly depends on the target size and data structure (Alansari, et al., 2018). In 2017, Scheidler developed a process model to extract knowledge from the database of a supply chain in order to create a usable data basis. The process model focuses on the definition of the desired target quantity, the selection and cleansing as well as the preparation of the data stock and the application of a data mining procedure (Scheidler, 2017). Each of these tasks involves its own challenges. The following list presents the greatest challenges specifically for the use of machine learning methods in demand planning in the manufacturing industry. For the handling of the mentioned challenges artificial intelligence methods are increasingly applied.

- Creation of a database
- Data acquisition/recording; Data preparation; Data analysis
- Creation of data model as basis of the algorithm
- Consideration of influencing parameters (environmental factors, regional factors,
- Political factors) depending on the branch
- Definition of the Output Parameter (Modelling)
- Changing products and shortened product life cycles

The following section of the paper gives a brief overview of current machine learning methods and provides an initial overview of the most common methods used in demand planning.

2.3 Machine Learning Methods in Demand Planning

Machine Learning is a field of research that deals with the modelling and implementation of learning phenomena and is part of artificial intelligence

(Engemann & Sudmann, 2018). The first research work in this field began in the 1950s, but only through technological advancements and the use of graphics processing units (GPUs) is it possible to handle data sets quickly and cost-effectively since 2015 (Wesseler, 2018). To implement machine learning methods, no static programming rules are used, instead the program extracts regularities from sample data in order to subsequently develop a model. After a learning period, the model is then applied to new data sets.

The main difference between traditional programs and machine learning algorithms is the feedback loop within the model. This allows the model to get a feedback on the quality of the output parameter and then adjust its own decision rules in order to improve the result. A self-learning system is created through the feedback loop as the machine learning algorithms continue to improve along with the duration of use and the amount of data (Gottschlich, et al., 2018; Wittpahl, 2019). Due to these characteristics machine learning methods are suitable for the following problems (Géron, 2017):

- Complex problems without an existing good solution
- Problems with changing environmental influences
- Large amounts of data as a basis
- Complex lists with existing solutions

In summary, machine learning algorithms can handle large amounts of data and flexibly react to changing conditions and environmental influences. According to Géron, there are three different strategies for implementing a machine learning algorithms:

- With and without human supervision
- Incremental learning and non-incremental learning system
- Pattern recognition

In demand forecasting, supervised learning algorithms are preferably used, as they can handle large amounts of data very efficiently and precisely. The algorithm's approach here is to identify patterns based on training data and to derive rules that can then be transferred to new data sets (Marsland, 2011; Mello and Ponti, 2018). Example algorithms are Logistic Regression or Random Forest, but also artificial neural networks (ANN) which are explained more in detail in the following section, as they play an important role in the context of artificial intelligence and machine learning. The ANN is a replica of the organic neuronal structure of humans. It is a mathematical-statistical method, which processes information on the basis of the human neuronal nerve structure and is very well suited for prognosis procedures and prediction models. Structurally, ANN do not represent a classically programmed algorithm, but a structure that can be flexibly adapted to data or system changes (Crone and Preßmar, 2006). The data points of the artificial neural network are linked and trained to react to unknown situations. The knowledge of the neural network lies in the weighted connection between the individual data points, which influence the transmission of the data and the result of the output function. A key capability of the ANN is its generalization capability. Therewith, good results can also be obtained for data sets whose behavior has not previously been trained. Especially in connection with prognosis problems, neural networks have the advantage of being able to model linear, non-linear or all interrelationships between data sets, which allows a very large scope for design (Hornik,

Stinchcombe and White, 1989; Zell, 2000; Haykin, 2009). After the presentation of the individual implementation strategies and the ANN, the best known algorithms of machine learning will be discussed briefly. It is not our intention to present a complete list of all machine learning algorithms, as there is no clear definition or demarcation between the algorithms. The number of machine learning algorithms varies between 1,000 and 10,000 (Engemann and Sudmann, 2018). The best known machine learning algorithms can be subordinated into the tasks of classification, regression and clustering (Géron, 2017). In the following, these tasks are shortly explained.

2.3.1 Classification

Classification procedures are used to classify objects or data points according to their properties. Based on the underlying set of data, the classes are calculated. New data points are sorted into the calculated grid based on their attributes (e.g. K-Means-Algorithm). It is important to note that the classification problem is discrete, hence the assignment is always unambiguous (Marsland, 2011). After the assortment of the data point, the classes are recalculated on the basis of the new dataset to achieve a continuous improvement of the classification result. These methods are often used in the field of pattern recognition or artificial intelligence (Wittpahl, 2019).

2.3.2 Regression

Regression analysis is already used in many areas for forecasting demand because it predicts the trends of the future on the basis of historical values and a statistical model. For this purpose, the dependency between two variables (dependent and independent) is examined in order to define their relationship to each other. This can be a simple regression (an independent variable) or a multiple regression (several independent variables). For this reason, they are particularly useful in forecasting and scenario analysis because they can consider several different data sources (internal data, weather data, regional data, etc.) (Stoetzer, 2017; Welc and Esquerdo, 2018). For further development of regression analysis, it is linked to a machine learning architecture, to use new and actual data sets to constantly regenerate and optimize the model that has been developed on the basis of the historical data.

2.3.3 Clustering

Clustering procedures are used to group and classify data points according to their properties. For this purpose, it is essential that the similarity between data points is described in a measurable manner as to enable comparative analysis. Clustering algorithms work through two procedures one after the other. First, the similarity measure between the new data point and the existing data basis is determined. This is followed by a division of similar data into groups thus creating cluster. At this point, a distinction is made between hard and soft clustering algorithms, i.e. the unique assignment of a data point to a cluster. If it is a soft clustering algorithm, data points can also belong to several clusters. If a data point is always uniquely assigned to a cluster, it is a hard clustering algorithm (Ester and Sander, 2013; Backhaus, et al., 2016). This method is also often used to prepare data sets. For example, to summarize customers based on their specific properties so that they can be handled equally in the demand forecast (Murray, Agard and Barajas, 2015). In addition to the different methods and learning strategies of machine learning approaches, the data analyst has to consider the phases of learning, which segregate into a training phase and an application phase. During the training phase, the algorithm develops the knowledge from example data in order to use it in the application phase or, depending on the algorithm and learning strategy, to extend it afterwards (Neef, 2017). In the next section of the article, the requirements of demand planning processes in the classical manufacturing industry are first presented, followed by demand planning processes in other industries using machine learning.

3 Cross-industry application of ML in Demand Planning

3.1 Methodology comparative cross-industry analysis

In order to gain new findings for the implementation of machine learning methods, the methodology of a cross-industry comparative analysis is used. The findings are processed in a CRISP (Cross-industry standard process) model for standardized collection and clustering of the results (Wirth and Hipp, 2000).

Different industries are analyzed with regard to one problem in order to transfer solution approaches between the industries. Cross-industry innovations and analyses are increasingly implemented in product development to prevent conventional thinking (Schulthess, 2013). The aim of this analysis is to identify commonalities between different industries in order to transfer already established machine learning approaches to demand planning in the manufacturing industry. As the concept and the basic schema in demand planning is similar throughout the industries this allows a possible transfer to demand planning in SCM. Each industry has a certain demand variable to plan follow-up processes, e.g.:

- Energy: Energy consumption
- Finances: Market and shares development (Arroyo, Espinola and Maté, 2011)
- Agriculture: Crop development

In this paper, machine learning approaches in the sectors of energy, finance and agriculture are examined, since the financial and energy sectors in particular, are pioneers in digitalization according to a survey conducted by the Institute of the German Economy in Cologne (Demary, et al., 2016). The reason to choose these three industries are the different influencing factors which have a direct influence on supply chain processes. The energy industry is strongly influenced by climatic conditions, which impact the supply, and direct contact with the end consumer who has a high influence on the development of demand. In agriculture, regional and environmental factors are decisive for the expected crop yield. In contrast, in the financial sector these are usually global factors (e.g. market dependencies, political influences) that affect market development and performance on the financial market. The following Table lists the primary influencing factors and dependencies of the individual industries.

Table 3: Influencing variables in different industries

Industry	**Influence variables/dependencies**	**Source**
Energy	Climatic conditions Direct contact with the end consumer	(Lüdeke-Freund and Opel, 2014)
Finances	Market dependencies Political influences	(Onischka and Orbach, 2008)
Agriculture	Environmental/ regional influences	(Döring, et al., 2011)

3.2 Machine Learning Methods Forecasting Energy

The energy sector is highly dependent on current climatic conditions and has direct contact with end consumers (households and industry). For this reason, consumption data are the main input for machine learning algorithms and thus provide a large and detailed data basis. To prepare the data basis, the following steps are performed in order as named: specification (variable declaration and selection of secondary data sources (e.g. weather data or socio-economic data sets)), inspection (quality analysis) and preprocessing (filtering and data curation). These steps are necessary to provide a suitable input for the machine learning algorithm (Al-Alawi and Islam, 1996). Table 4 presents three research papers concerning predicting power demand using machine learning techniques.

Table 4: Machine Learning in Forecasting Electricity consumption

Database	**Machine Learning Algorithm**	**Results**	**Source**
Electricity consumption: 70,128 data points in the period 01.01.2008-31.12.2015 and weather data	ANN; Random forest Algorithm; Support Vector Machine Algorithm	Average percentage error 3.10% 96.3% of peak loads are covered	(Gajowniczek and Ząbkowski, 2017)
Electricity consumption Years: 1980-2000 (training data); 2001-2008 (test data)	ANN	Average percentage error 0.64	(Küçükdeniz, 2010)
Electricity consumption Years: 2006-2007 (training data); 2008 (test data); Influencing variables: Time; Seasons	Support Vector Regression	Average percentage error 0.41-0.53 depending on combination of influencing variables	(Setiawan, Koprinska and Agelidis, 2009)

These three papers were selected because they use different approaches and data bases, and all provide good results. In contrast to Küçükdeniz, Gajowniczek and Ząbkowsk focus on a short time horizon, which allows to evaluate the advantages of the different algorithms in relation to the considered application case. The last paper analyzes different time windows, but focuses only on support vector machines, therefore a comparison of the three publications gives a good overview of the state of the art in this industry.

The first column describes the used database, afterwards the used machine learning algorithm is described. The results column displays the quality of the results and the calculated error size. The results of the research work show that artificial neural networks and supervised machine learning methods provide the qualitatively best prognosis results. In addition, the quality of the results increases with the number of training data of the primary data set (see Küçükdeniz) or with the consideration of further secondary data sets (see Setiawan).

3.3 Machine Learning Methods Forecasting Finance

Intelligent forecasting methods are already widely used and established within the financial sector. Krollner has prepared a compilation of the techniques (time horizon, algorithm) and data bases (data evaluation, data selection, input parameters) used in this area in 2010 (Krollner, Vanstone and Finnie, 2010). The following figures are based on his literature research and are intended to give a brief overview of the use of machine learning methods in forecasting financial time series. As shown in Figure 1, the artificial neural networks (ANN) based algorithm is most widely used in the financial sector followed by evolutionary and optimization techniques. In addition several multiple and hybrid applications are employed. The trend moves towards the use of existing methods of artificial neural networks, which are upgraded with innovative training algorithms or combined with new technologies to hybrid systems. This development leads to the integration of machine learning methods to improve the training algorithms (Krollner, Vanstone and Finnie, 2010).

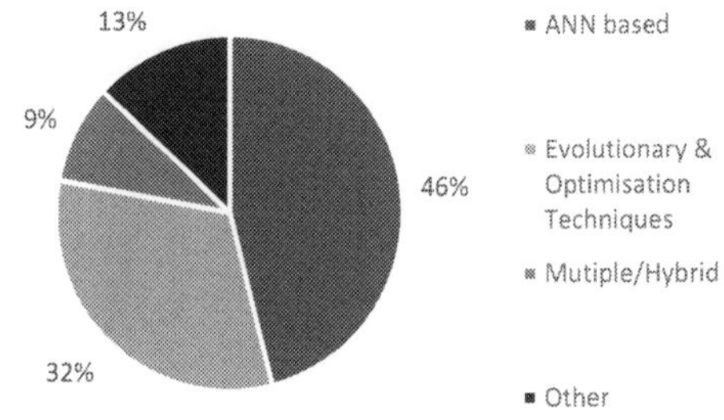

Figure 1: Machine Learning Methods for Forecasting in Finance sector (Krollner, Vanstone and Finnie, 2010)

Figure 2 displays secondary data used to improve the forecast quality in the financial sector. As the second graph shows, a wide variety of secondary data is referred to. The data can be separated into three categories: Firstly, finance-specific data such as the trading volume and exchange rate, secondly specific comparative rates such as oil price, gold price and thirdly, macro-economic data as the unemployment rate.

Before the respective data can be used, they must be processed. Krollner does not deal with the topic of data preparation, so that further research is necessary to give a detailed overview about concepts of data preparation.

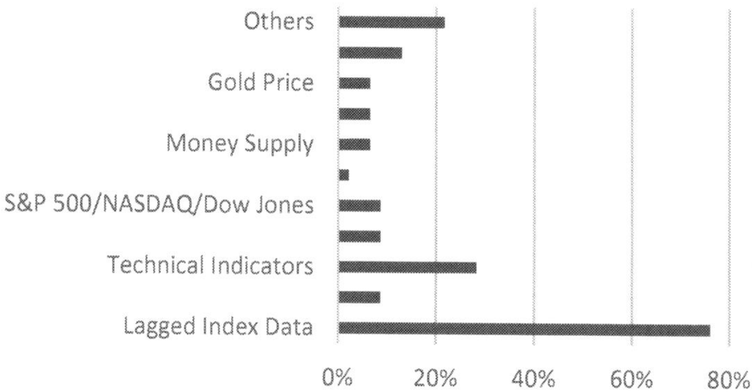

Figure 2: Percentage of used secondary data bases (Krollner, Vanstone and Finnie, 2010)

Tsang (Tsang, Yung and Li, 2004) has developed a method to process data sets specifically for the financial sector and to prepare them for use in machine learning algorithms. In his research work, the author described strategies and procedures for solving data problems, which are briefly listed below:

- Too few data: Generation of missing data's (Data mining based methods)
- Too many data: Classification and discretization
- Noise in the data: Regression-based data smoothing
- Season/Trend: Log difference
- Differently scaled data sets: Line scaling method

These approaches are already used in conventional historical data analysis and can be adopted to prepare the database for the machine learning algorithm.

3.4 Machine Learning Methods Forecasting Agriculture

Table 5: Used Machine Learning methods in Forecasting Agriculture (Liakos, et al., 2018)

Machine Learning Method	Quantity (of 46)
Bayesian Model	2
Ensemble Learnings	2
ANN	22
Regression and Clustering	4
Instance based models	1
Decision trees	1
Support Vector Machines	14

In contrast to the financial or energy sector, qualitative forecasting areas are also considered (e.g. animal welfare or harvest quality). It is important to note that the algorithms used in the qualitative approach differ to the once used for quantitative forecasting. Ensemble Learning or Decision trees are the most commonly used methods, which can be subsumed in the area of regression for supervised learning. In the study of Liakos, the machine learning methods are analyzed according to their frequency of use, a qualitative analysis of the results is not considered. In addition, no information is provided on established methods for data preparation. Having analyzed the demand forecasts in agriculture, energy and finance, the following section applies the findings to the requirements of demand planning in the manufacturing industry.

4 Solution approaches and strategies for demand planning in manufacturing industry

The analysis of the different industrial sectors has shown that certain problems and challenges are common (e.g. data basis or algorithm selection). Due to the relatively low degree of digitalization in the manufacturing industry, it is useful to learn from more advanced industries and to use the experience gained (Demary, et al., 2016). Based on the available knowledge in those sectors, recommendations for action can be formulated for the application of machine learning methods for demand planning in the field of supply chain management in the manufacturing industry.

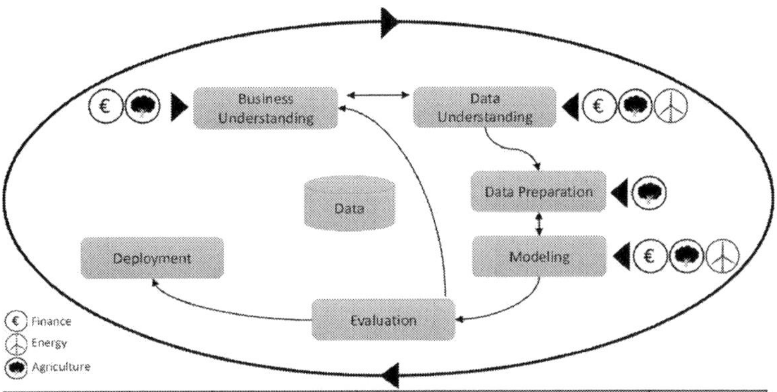

Challenges for using ML-Methods in demand planning	CRISP-Model	Possible Solution Approaches	Industry sector
Creation of a data base	Data Understanding/Data Preparation	▪ Historical data ▪ Preparation of the data base (use existing models and algorithm)	Energy, Finance
Definition of influencing parameters	Business Understanding/Data Understanding	▪ Use secondary sources ▪ Specific and general sources of information ▪ Take environmental and regional influences into account	Finance, Agriculture
Shortened product life cycles	Data Preparation	▪ Use data from similar products	Agriculture
Algorithm Selection	Modeling	▪ High effectiveness ANN and support vector machines	Energy, Finance
Definition of the Output-Parameter	Modeling	▪ Decision qualitative or quantitative statement influences the selection of algorithms	Agriculture

Figure 3: CRISP-Model solution approaches and strategies for using ML

Figure 3 summarizes the findings from the comparative analysis of other industries using the CRISP-Model (Cross-industry standard process) and provides recommenda-tions for the structured use of machine learning methods in the manufacturing industry. The challenges identified previously in chapter 2.2 are compared to the possible solutions from the three industrial sectors. The following section explains the approaches and findings of the analysis more in detail.

The data basis is essential for the success of machine learning methods and must be prepared for each application. Historical data sets are used in the energy industry to overcome this difficulty. In order to improve data quality, they can still be processed in order to eliminate missing data or data noise.

Afterwards a model must be designed for a specific target variable. This model takes into account both the data basis and the influencing factors that affect the target variable to be calculated. It is recommended to use secondary data (general or industry-specific) in addition to the primary data to enable a better integration of fluctuations (agriculture and finance sector). Publicly accessible data are of advantage, as they are available in a standardized form and with sufficient data quality, so that the process steps of data acquisition and data preparation can be reduced. In addition, during the selection of the algorithm, the use case must be taken into account, since the different algorithms depend on the type of the forecast value (qualitative or quantitative).

In the area of data exploitation in particular, successful approaches have already been developed in the financial sector, but the literature research also shows that challenges due to product behavior (shorter product life cycles) or in modelling can be solved by approaches from agriculture.

5 Conclusion

In this article, the authors have shown that machine learning methods are already used in several industries. This promises a high potential for the manufacturing industry. Requirements for the application in demand planning have been formulated and prepared in a structured way. It has been shown that many areas have not been sufficiently analyzed and described (e.g. Modelling, influencing factors (internal and external) and data preparation). This is especially valid for the modelling of the ML approach and the underlying data basis. There will be a need for future research in the area of demand planning in the context of the manufacturing industry. This includes the specification of necessary and influencing data sets. This would allow the creation of an initial model for successful demand planning. It is of utmost importance to focus further research in the field of the design of the data model to guarantee optimal input for the ML algorithms. Only if the required set of data is defined and available, the true value of machine learning can be evaluated.

References

Al-Alawi, S. M. and Islam, S. M., 1996. Principles of electricity demand forecasting. I. Methodologies. Power Engineering Journal, 10(3), pp. 139–143.

Alansari, Z., Anuar, N. B., Kamsin, A., Soomro, S., Belgaum, M. R., Miraz, M. H. and Alshaer, J., 2018. Challenges of internet of things and big data integration. International Conference for Emerging Technologies in Computing, pp. 47–55.

Arroyo, J., Espinola, R. and Maté, C., 2011. Different approaches to forecast interval time series: a comparison in finance. Computational Economics, 37(2), pp. 169–191.

Ashayeri, J. and Lemmes, L., 2006. Economic value added of supply chain demand planning: A system dynamics simulation. Robotics and Computer-Integrated Manufacturing, 22(5-6), pp. 550–556.

Backhaus, K., Erichson, B., Plinke, W. and Weiber, R., 2016. Multivariate Analysemethoden: Springer.

Bryanjolfsson, E. and Mcafee, A., 2014. The second machine age: Work, progress, and prosperity in a time of brilliant technologies.

Crone, S. F. and Preßmar, D. B., 2006. An extended evaluation framework for neural network publications in sales forecasting. Artificial Intelligence and Applications, pp. 179–186.

Demary, V., Engels, B., Röhl, K.-H. and Rusche, C., 2016. Digitalisierung und Mittelstand: Eine Metastudie: IW-Analysen.

Deutschländer, T., 2003. Über Möglichkeiten und Grenzen der statistischen Langfristprognose.

Döring, S., Döring, J., Borg, H. and Böttcher, F., 2011. Vergleich von Trockenheitsindizes zur Nutzung in der Landwirtschaft unter den klimatischen Bedingungen Mitteldeutschlands. Hercynia-Ökologie und Umwelt in Mitteleuropa, 44(2), pp. 145–168.

Egri, P., Csáji, B. C., Kemény, Z., Monostori, L. and Váncza, J., 2008. Komplexität der Bedarfsprognosen und ihre Wirkungen in kooperativen Logistiknetzwerken. pp. 455–469.

Engemann, C. and Sudmann, A., 2018. Machine Learning-Medien, Infrastrukturen und Technologien der Künstlichen Intelligenz: transcript Verlag.

Ester, M. and Sander, J., 2013. Knowledge discovery in databases: Techniken und Anwendungen: Springer-Verlag.

Feindt, M. and Kerzel, U., 2014. Prognosen bewerten: Statistische Grundlagen und praktische Tipps: Springer-Verlag.

Gajowniczek, K. and Ząbkowski, T., 2017. Two-stage electricity demand modeling using machine learning algorithms. Energies, 10(10), p. 1547–1547.

Géron, A., 2017. Hands-on machine learning with Scikit-Learn and TensorFlow: concepts, tools, and techniques to build intelligent systems: O'Reilly Media, Inc.

Gottschlich, J., Solar-Lezama, A., Tatbul, N., Carbin, M., Rinard, M., Barzilay, R., Amarasinghe, S., Tenenbaum, J. B. and Mattson, T., 2018. The three pillars of machine programming. Proceedings of the 2nd ACM SIGPLAN International Workshop on Machine Learning and Programming Languages, pp. 69–80.

Grömling, M., 2005. Konjunkturprognosen: Verfahren, Erfolgskontrolle und Prognosefehler.

Haasis, H.-D., Kopfer, H., Kotzab, H., Dethloff, J. and Schönberger, J., 2015. Logistics management. Products, actors, technology -- proceedings of the German Academic Association for Business Research, Bremen, 2013.

Habla, C., Driebel, R., Monch, L., Ponsignon, T. and Ehm, H., 2007. A short-term forecast method for demand quantities in semiconductor manufacturing. 2007 IEEE International Conference on Automation Science and Engineering, pp. 94–99.

Hammer, R. M., 1998. Unternehmensplanung: Lehrbuch der Planung und strategischen Unternehmensführung: Oldenbourg.

Hansmann, K.-W., 2013. Kurzlehrbuch Prognoseverfahren: Springer-Verlag.

Haykin, S. S., 2009. Neural networks and learning machines: Pearson education Upper Saddle River.

Hellingrath, B. and Kuhn, A., 2013. Supply chain management: Optimierte Zusammenarbeit in der Wertschöpfungskette: Springer-Verlag.

Helton, J. C., Johnson, J. D., Oberkampf, W. L. and Sallaberry, C. J., 2010. Representation of analysis results involving aleatory and epistemic uncertainty. International Journal of General Systems, 39(6), pp. 605–646.

Hornik, K., Stinchcombe, M. and White, H., 1989. Multilayer feedforward networks are universal approximators. Neural networks, 2(5), pp. 359–366.

Johnson, M. D., 2018. Kundenorientierung und Markthandlung: Walter de Gruyter GmbH & Co KG.

Koch, S., 2012. Logistik: Eine Einführung in Ökonomie und Nachhaltigkeit: Springer-Verlag.

Krollner, B., Vanstone, B. J. and Finnie, G. R., 2010. Financial time series forecasting with machine learning techniques: a survey. 2010. ESANN.

Küçükdeniz, T., 2010. Long term electricity demand forcesting: An alternative approach with support vector machines. Istanbul university of engineering sciences, pp. 45–53.

Kühnapfel, J. B., 2014. Vertriebsprognosen: Methoden für die Praxis: Springer-Verlag.

Lee, H. L., 2002. Aligning supply chain strategies with product uncertainties. California management review, 44(3), pp. 105–119.

Liakos, K., Busato, P., Moshou, D., Pearson, S. and Bochtis, D., 2018. Machine learning in agriculture: A review. Sensors, 18(8), p. 2674–2674.

Lüdeke-Freund, F. and Opel, O., 2014. Energie. Nachhaltigkeitswissenschaften: Springer, pp. 429–453.

Marsland, S., 2011. Machine learning: an algorithmic perspective: Chapman and Hall/CRC.

Mayr, S. and Moser, M., 2018. Produktvorlaufkosten im Blick haben. Controlling & Management Review, 62(5), pp. 42–47.

Mello, R. F. de and Ponti, M. A., 2018. Machine Learning: A Practical Approach on the Statistical Learning Theory: Springer.

Mertens, P. and Rässler, S., 2012. Prognoserechnung-Einführung und Überblick. 2012. Prognoserechnung: Springer, pp. 3–10.

Michie, D., 1968. "Memo" Functions and Machine Learning. Nature, 216(5136), pp. 19–22.

Murray, P. W., Agard, B. and Barajas, M. A., 2015. Forecasting supply chain demand by clustering customers. IFAC-PapersOnLine, 48(3), pp. 1834–1839.

Neef, A., 2017. Wie Künstliche Intelligenz die Wertschöpfung transformiert. Z-punkt The foresight company.

Onischka, M. and Orbach, T., 2008. Klima und Finanzmarkt. So investiert die Welt: Springer, pp. 77–96.

Papenfort, J., Frank, U. and Strughold, S., 2015. Integration von IT in die Automatisierungstechnik. Informatik-Spektrum, pp. 199–210.

Reder, B., 2018. Studie Machine Learning/Deep Learning.

Röniger, O., 2018. Self-Service Analytics Unlimited. Controlling & Management Review, 62(2), pp. 14–23.

Scheidler, A. A., 2017. Methode zur Erschließung von Wissen aus Datenmustern in Supply-Chain-Datenbanken: Cuvillier Verlag.

Schulthess, M., 2013. Ein Blick in fremde Branchen lohnt sich. Marketing Review St. Gallen, 30(1), pp. 98–107.

Setiawan, A., Koprinska, I. and Agelidis, V. G., 2009. Very short-term electricity load demand forecasting using support vector regression. 2009 International Joint Conference on Neural Networks, pp. 2888–2894.

Shalev-Shwartz, S. and Ben-David, S., 2014. Understanding machine learning: From theory to algorithms: Cambridge university press.

Stickel, E., Groffmann, H.-D. and Rau, K.-H., 2013. Gabler: Wirtschaftsinformatik Lexikon: Springer-Verlag.

Stoetzer, M.-W., 2017. Regressionsanalyse in der empirischen Wirtschafts-und Sozialforschung Band 1: Eine nichtmathematische Einführung mit SPSS und Stata: Springer-Verlag.

Tempelmeier, H., 2018. Begriff der Logistik, logistische Systeme und Prozesse: Springer.

Tsang, E., Yung, P. and Li, J., 2004. EDDIE-Automation, a decision support tool for financial forecasting. Decision Support Systems, 37(4), pp. 559–565.

Verein Deutscher Ingenieure e.V., 2018. Künstliche Intelligenz. VDI-Statusreport, 2018, p. 5–5.

Welc, J. and Esquerdo, P. J., 2018. Applied regression analysis for business: Springer.

Wesseler, B., 2018. Nichts verallgemeinern. iX Developer 2018-Machine Learning: Verstehen, verwenden, verifizieren, p. 38–38.

Wirth, R. and Hipp, J., 2000. CRISP-DM: Towards a standard process model for data mining. 2000. Proceedings of the 4th international conference on the practical applications of knowledge discovery and data mining, pp. 29–39.

Wittpahl, V., 2019. Künstliche Intelligenz: Springer.

Zell, A., 2000. Simulation Neuronaler Netze.

Zuboff, S., 2010. Creating value in the age of distributed capitalism. McKinsey Quarterly.

III.
Supply Chain Analytics and Blockchain

Blockchain Adoption at German Logistics Service Providers

Oliver Kühn[1], Axel Jacob[2] and Michael Schüller[2]

1 – Arvato Distribution GmbH

2 – Osnabrück University of Applied Sciences

Purpose: The study provides a recent overview of the diffusion of blockchain technology in the logistics industry. It reveals the adoption of blockchain technology at German logistics service providers (LSPs) and their expectations regarding the future relevance of the technology. Based on the TOE Framework adoption supporting and inhibiting factors are identified.

Methodology: In a first step, LSPs listed in the "Die Top 100 der Logistik 2016/2017" were contacted and questioned about their blockchain activities. Based on the responses, qualitative interviews were conducted with seven participants as part of a three-stage Delphi study.

Findings: In particular small and medium-sized German LSPs are currently hardly involved in blockchain technology. Larger LSPs are beginning to define their own use cases and are trying to develop them further in joint projects with partners. A systematic use is currently not taking place.

Originality: The study reveals the current discrepancy between rapidly evolving theoretical approaches for the use of blockchain technology in logistics, on the one hand, and the absence of the technology in everyday operations on the other hand. It also reveals a reluctant attitude of the management towards the technology.

Keywords: Blockchain Adoption, German Logistics Service Providers, TOE Framework, Delphi Method

First received: 18.May.2019 **Revised:** 28.May.2019 **Accepted:** 28.May.2019

1 Introduction

The blockchain technology is an emerging but also highly controversial innovation for the logistics industry that challenges LSPs (Hackius & Petersen, 2017). According to Tushman and Nadler (1986), innovations can create incremental, synthetic or discontinuous change. Incremental changes have a minor impact on an organization and involve only little risk. Synthetic changes result from existing ideas or technologies being reconstructed and combined in an innovative way. The highest impact on an organization with respect to change and risk arises from discontinuous change. Concerning the last-named category, in particular, the distinction between "competence-enhancing" and "competence-destroying" (Tushman & Anderson, 1986) innovations is crucial. Competence-enhancing innovations enlarge the expertise of companies and contribute to further development. Competence-destroying innovations may render existing technologies and expertise obsolete. As blockchain technology is in an early stage, the future will show which classification applies in the case of LSPs. However, even without a specific classification, it becomes clear that blockchain technology will influence logistics services anyway (Blossey, Eisenhardt, & Hahn, 2019; Dobrovnik, Herold, Fürst, & Kummer, 2018). LSPs that adopt blockchain technology will be able to exploit the benefits or to adjust their business models.

In consequence, this paper does not aim to evaluate the impact of blockchain on LSPs. Instead, the objective of this paper is to address a lack of research, pertinent to academics and practitioners by shedding light on the adoption of blockchain technology at LSPs. Thus, our research questions are:

RQ1: Which factors influence the adoption of blockchain technology at LSP?

RQ2: How advanced is the blockchain adoption at LSP?

By answering these questions, we respond to calls from Hughes et al. (2019) to research on the adoption of blockchain technology, especially on the benefit expectations. We also respond to Blossey et al. (2019) to identify success factors and challenges of implementing blockchain applications. Finally, we respond to Risius and Spohrer (2017) to research on how existing intermediary service providers position themselves towards blockchain technology.

The remainder of the paper is structured as follows. Section two lays the theoretical foundation for assessing the adoption of blockchain technology at German LSPs. Subsequently, in section three, the research methodology and the data sample are explained. The findings of our study are presented in section four. These are discussed and summarized in propositions in section five. In the final section, we summarize the contributions of our study and address implications for further research.

2 Theoretical Background

2.1 Blockchain in Logistics

The blockchain technology is a topic of steadily increasing interest for academics and practitioners (Li, Marier-Bienvenue, Perron-Brault, Wang, & Paré, 2018; Miau & Yang, 2018; Risius & Spohrer, 2017). A blockchain is a shared ledger that allows a decentralized verification and immutable storage of data and thus a trustworthy record of business activities (Nofer, Gomber, Hinz, & Schiereck, 2017). Because of its ability to create trust within a decentralized system with participants that do not necessarily trust each other, the application of blockchains seems to facilitate the elimination of a trustful third-party intermediary (Swan, 2015). Besides the execution of rather simple transactions such as currency or token transactions, blockchains also allow for more complex transactions via the application of smart contracts (Szabo, 1997). Thus, because of the blockchain characteristics and the possibility to execute smart contracts, the term "disruption" has frequently been used in recent years for the blockchain technology, in order to describe an estimated radical change of business processes and even business models in many industries (Mettler, 2016; Nofer et al., 2017; Peters & Panayi, 2016).

The logistics industry is expected to be one of the most impacted sectors. For example, Korpela et al. (2017) develop a concept for a blockchain-based digital supply chain that could enable an end-to-end data integration. Blossey et al. (2019) offer the most recent overview of blockchain applications in the field of logistics and supply chain management. According to them, blockchains can be applied in the fields of a) supply chain visibility, b) supply chain integrity, c) supply chain orchestration, d) supply chain virtualization, and e) supply chain finance. In most of the existing studies, blockchain technology is described as a competence-enhancing innovation for logistics services. However, it can also be a competence-destroying one. For example, Subramanian (2017) argues for blockchain-based decentralized marketplaces, inter alia for transportation services. Such a market place could seriously harm the existing business models of forwarders. It becomes clear that especially LSPs are well-advised to deal with blockchains, either in order to develop their business models or to be prepared for emerging new competitors.

The aforementioned studies are rather conceptual than empirical and only a few studies have been conducted which describe the adoption of the blockchain technology in logistics until now. Dobrovnik et al. (2018) offer a conceptual approach to discuss the adoption of blockchain technology in the logistics industry. Based on Rogers' attributes of innovation (relative advantage, compatibility, complexity, trialability, and observability) (Rogers, 2003), they hypothesize potential blockchain applications in the logistics industry (Dobrovnik et al., 2018). An empirical proof of the propositions does not take place. Hackius and Petersen (2017) conducted an expert survey of logistics professionals to explore potential applications and future prospects of blockchain technology in supply chain management. According to them, there are significant differences in regards to the estimated benefits and adoption barriers of blockchains between logisticians on the one hand, and consultants as well as scientists on the other. Especially logistics experts "have difficulties getting a clear idea of the benefits and use cases" (Hackius & Petersen, 2017, p. 16) and make reservations regarding an anticipated high level of collaboration and commitment. Thus, Hackius and Petersen ask logisticians to "engage in experiments to find out if and how Blockchain [sic!] could be of use for their own company" (2017, p. 16). We distinguish to the aforementioned literature in so far that we offer profound details on the antecedents for blockchain adoption of LSPs on an organizational level. To the best of our knowledge, there has been no other study on this topic so far. Thus, we intend to fill this gap with this paper and build on the findings of Hackius and Petersen (2017).

2.2 Innovation Adoption

Adoption of innovations takes place on different levels, e.g. individuals, groups, organizations or industries. The different entities on each level adopt the innovation, based on their attitude towards innovations. The more entities adopt an innovation, the higher is the possibility that adoption spills over to other entities and levels (Valente, 1995). Thus, adoption is a sub-process of diffusion according to Rogers (2003). There are several approaches that conceptualize the adoption process, e.g. the technology acceptance model (Davis, Bagozzi, & Warshaw, 1989), the theory of planned behavior (Ajzen, 1991), and the unified theory of acceptance and use of technology (Venkatesh, Morris, Davis, & Davis, 2003). However, in this paper, we apply the concept of Rogers (2003) as his approach seems to be applicable not only for individuals but also for organizations. Rogers divides the adoption process into five different stages: a) knowledge, b) persuasion, c) decision, d) implementation and e) confirmation. In the knowledge stage, a decision-making unit gets the awareness of an innovation, reaches a basal understanding of the innovation, and apprehends the principles of the innovation. During the persuasion stage, the decision-making unit forms an attitude toward the innovation, based on the perceived characteristics of the innovation, e.g. relative advantage, compatibility, complexity, trialability, and observability (cf. section 2.1). In consequence, the decision-making unit decides to adopt or to reject the innovation in the decision stage. The implementation stage describes the application of the innovation in routine processes. This stage often comes along with a degree of uncertainty and adaptions of the innovation (Rogers, 2003). After a distinct time, the decision-making unit evaluates the outcome of the decision stage within the confirmation stage. At this point in time, discontinuance, which

Rogers describes as "a decision to reject an innovation after having previously adopted it" (Rogers, 2003, p. 182) can occur.

In order to explain adoption on an organizational level, the TOE framework turned out to be very fruitful, because of its general applicability (Baker, 2012). According to the TOE framework (DePietro, Wiarda, & Fleischer, 1990), the organizational adoption of technological innovation is influenced by the context's technology, organization, and environment, which can be "constraints and opportunities for technological innovation" (1990, p. 154). As the contexts are not exclusively, Zhu and Kramer (2005, p. 63) describe TOE as a "generic" framework.

The technological context of the TOE framework refers to all relevant technology for an organization, whether it is equipment or processes. This includes technology that is already applied in an organization (Collins, Hage, & Hull, 1988) but also technology that is generally available but not yet applied in an organization (Kuan & Chau, 2001). The adoption of new technology is restricted by the significance and embeddedness of the already applied technology with respect to its ecosystem (Adner & Kapoor, 2016).

The organizational context represents all organizational characteristics and resources available for the adoption and operationalization of technological innovation. Elements of the organizational context include financial readiness, slack resources, top management support, the involvement of the employees, the degree of centralization, technology competence and, to a lesser degree, the size of the organization. Large organizations are considered to be more successful in adopting innovations (Damanpour, 1992) mainly because of higher financial resources (Iacovou, Benbasat, & Dexter, 1995) and more slack resources (Rogers, 2003).

The environmental context describes the setting in which an organization acts. Characteristics of this context are the size and structure of the industry, the technology support infrastructure, and the regulatory environment (DePietro et al., 1990). Because of the endogenous competitive pressure, the size and structure of an industry have been identified as highly important characteristics of the environmental context (Iacovou et al., 1995). However, not only competitors influence the adoption but also customers and partners in the value chain, such as suppliers, retail partners or service providers (Kuan & Chau, 2001). Especially vertical relations in value chains accelerate the adoption of new technologies (Alipranti, Milliou, & Petrakis, 2015).

3 Research Process

3.1 Methodological Approach

To be able to investigate the adoption behavior of LSPs, we conducted a three-stage Delphi study. Delphi studies use an iterative feedback technique with a group of experts (Dalkey & Helmer, 1963). In this study, we apply the procedure as proposed by Murry and Hammons (1995). According to them, a Delphi study consists of two main phases: a) Definition and selection of experts and b) multiple iterations of questionnaire rounds. In this study, we define experts as managers that are responsible for the technological or business development at German LSPs and have a job experience of at least three years. Based on these criteria, we contacted all companies listed in "Die Top 100 der Logistik 2016/2017" (Schwemmer, 2016) and asked for participation of relevant experts. A detailed description of the different experts follows in the next section.

In the first questionnaire round, we asked the participants for basic information about their companies and blockchain activities. Based on these results, we prepared a second round with guided qualitative interviews. The interviewees were informed about the intentions of the study before the start of the interviews and the guideline was made available to the interviewees one week before the interview. The interviews took place from June till August 2018. They were recorded on tape and later transcribed. Within the scope of qualitative content analysis, the statements of the interviewees were manually paraphrased and coded (Saldaña, 2009). The results of the qualitative interviews were condensed in a classification scheme with the help of the theory derived from the TOE framework (DePietro et al., 1990) and were used to create a survey to evaluate the blockchain efforts of the company. This helped to classify the companies even better within Rogers' adoption process (Rogers, 2003) and their assessment of blockchain. Finally, the findings were discussed with each participant. After the third stage, a consensus was reached among the participants and thus, the procedure came to an end (Murry & Hammons, 1995). In order to eliminate problems associated with face-to-face meetings, the names of the participating managers and companies were treated confidentially during the entire procedure (Murry & Hammons, 1995).

3.2 Sample Description

In total seven interviews with German LSPs were conducted. In the following, we name the seven cases A to G for the sake of anonymization. Company E is partly a logistics service provider. The investigated companies employ between 750 and 35,600 employees and their turnover was between 130 million and 15 billion in 2018. The interviewees job positions can be defined as follows: Site manager (A), CEO (B), CIO/Head of Global IT (C), Manager of Corporate Development/Head of Innovation (D), Blockchain-representative/employee in the field of digitalization (E), CIO (F) and Team-leader Innovation (G). The share of turnover in contract logistics referring to the full turnover of each company ranges from 18-97%.

4 Findings

4.1 Classification of Blockchain Engagement

The first questionnaire round revealed that all interviewees know the blockchain technology and attribute benefits to the technology. Cases A and B reached an awareness of the technology but in contrast to the cases C and D, they have not started to engage profoundly and deepen their knowledge on the technology yet. The cases E, F and G have done so and also developed blockchain use cases in order to gain specific knowledge and be able to evaluate the technology for their purposes. The procedure differs in so far as case F carries out an individual project, while cases E and G each cooperate with other companies. Figure 1 gives an overview of the companies' blockchain engagement

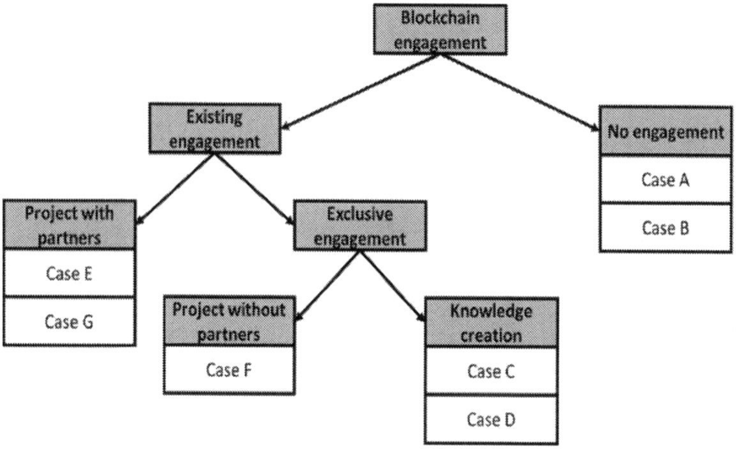

Figure 1: Classification of companies' blockchain engagements

4.2 Factors of Influence for Blockchain Adoption

During the qualitative interviews in the second stage of the Delphi procedure, factors, according to the TOE framework were identified that facilitate or inhibit blockchain adoption. The technological context comprises a high number of influencing factors in the categories perceived characteristics of the blockchain technology, existing IT infrastructure and data as well as IT knowledge. General perceived benefits of blockchain technology are more predictive supply chains, general optimization, and savings based on an end-to-end digitalization and a high level of privacy based on the applied encryption. The most mentioned perceived benefits of the blockchain technology are high data security based on the distributed ledger and transparency. In contrast, some interviewees expressed the fear of insufficient privacy and sharing data with all blockchain participants. Further perceived shortcomings are technological immaturity (i.e. performance, speed, scalability, and power consumption), the complexity of the technology and missing blockchain standards. Besides the perceived characteristics of the blockchain technology, the data quality and the maturity of the existing IT infrastructure and its compatibility with blockchain technology were named as adoption impediments. Likewise, the interviewees attributed to their companies a lack of relevant IT knowledge, especially in the fields of interface control and blockchain infrastructure.

The influencing factors of the organizational context were all considered to inhibit blockchain adoption. The main factors are the culture of the organization and financial considerations. The different cultures of the organizations were described as reluctant in most of the cases. The interviewees were apprehensive of the "curse of the first mover" (Rayna, and Striukova, 2009), called existing use cases "lobbying" and stated missing trust towards blockchain in general. The financial considerations represent anticipated high investments, sunk costs through a partial replacement of existing systems and in consequence an unclear cost-benefit ratio. Further organizational factors include strict IT governance rules and a high need for process harmonization.

Similar to the organizational context, the environmental context was predominantly described as inhibiting by the interviewees. The pressure from the customers as well as from competitors and other closely related industry partners seems to be low. The interviewees expected other companies are not willing to share data via blockchains and only very few customers asked for blockchain solutions. In addition, short contract periods impede investments into blockchain technology. Likewise, legal uncertainties in general, especially referring to smart contracts depict impediments. A facilitating factor is represented by the estimated ongoing technological development of the logistics industry.

Figure 2 gives an overview of the influencing factors for blockchain adoption based on the TOE framework.

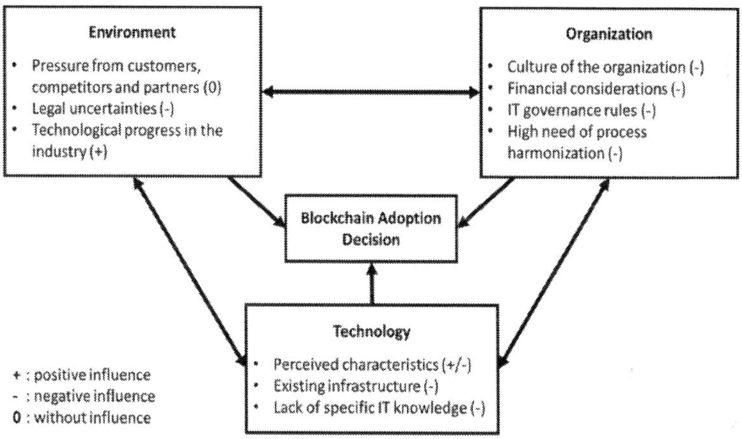

Figure 2: Factors that influence blockchain adoption at LSPs

4.3 Assessment of the Blockchain Adoption Process

During the third stage of our Delphi procedure, we conducted a second survey based on the blockchain and innovation adoption literature as well as the results of the qualitative interviews. The purpose of the survey was to determine the statuses of the different companies with regards to the adoption of blockchain technology. Thus, the interviewees first assessed their company's blockchain efforts. Based on the answers and the results of the analysis of the qualitative interviews, we determined the adoption statuses of the companies according to the adoption process of Rogers (Rogers, 2003). The results of our classification were discussed with the interviewees until consensus was reached. Table 1 depicts the final results of our classification.

Table 1: Adoption statuses according to Rogers (2003)

Case/Company	**Phase of the adoption process**
A	Knowledge phase (awareness knowledge)
B	Knowledge phase (how-to-knowledge)
C	Knowledge phase (how-to-knowledge)
D	Knowledge phase (principles-knowledge)
E	Begin of persuasion phase
F	End of persuasion phase
G	Decision phase

5 Discussion

5.1 On the Adoption of Blockchain Technology

Our research reveals the practitioners' perception of blockchain technology. Although more inhibiting than facilitating factors were mentioned, the interest in blockchains is rather high. Nevertheless, the reviewed LSPs act differently. Our results show that companies with solely theoretical blockchain knowledge have a positive attitude towards the technology. These companies acknowledged the future importance and advantages of the technology. For example, the interviewee in case D explained, if it is possible to conclude contracts soon without the parties knowing each other, the market will be very efficient through smart contract logistics operators. Cases C and D assume that in future automatic contract negotiations and conclusions will be possible through interlinked transport management systems and supply chain management systems. These estimated use cases correspond with the work of Blossey et al. (2019) who name the supply chain orchestration an emerging blockchain-driven innovation.

Companies with practical blockchain experience also highlighted the future relevance of blockchain technology, but in the same way, they learned about the difficulties of the technology. For example, the company of case F conducted various use cases but is still doubtful, which processes should be performed with the help of blockchains in the future. The company of case G initiated a blockchain consortium in which several companies are involved. The target of the consortium was to develop a standardized blockchain protocol for a specific tracking and tracing system that incorporates sensor data to initiate smart contracts for payments. However, the implementation is unclear, because of the uncertain stability of cryptocurrencies (Catania, Grassi, & Ravazzolo, 2019). The company of case E works on a use case in cooperation with a number other companies and investigates the application of blockchains for financial transactions, billing methods within the company, processes of order processing as well as track and trace of materials from production to consumption. Additionally, the company is a member of an international network of different companies that shares knowledge on blockchains and smart contracts e.g. for price determination and flexible adaption. These results add to the research of Hackius and Petersen (2017) who attested logisticians in 2017 difficulties in getting a clear idea of use cases. Two years later, several LSPs are able to identify and conduct first use cases but they have not made a final adoption decision so far. Furthermore, it confirms the practical utility of the so far hypothesized application of blockchain in logistics (Dobrovnik et al., 2018; Korpela et al., 2017).

The companies A and B, which had not engaged in blockchain technology behaved critically towards the technology. They observed the technology and market and found themselves mainly in a waiting position. Company A supposed benefits in the area of tracking and tracing, a transparent supply chain with live status messages and billing processes with the help of smart contracts. Company B assumed the same advantages. Additionally, it named predictive maintenance and network coordination with other companies as further prolific application areas.

5.2 Implications

Our research on the adoption of blockchain technology at LSPs has theoretical and practical implications, which will be summarized in this section. In general, we suggest that German LSPs should focus more on blockchain technology and define their own use cases. Companies should also cooperate more intensively to define certain blockchain rules and standards, e.g. for emerging networks.

Implication 1: The theory about blockchains in logistics is far more advanced than the practice. A lot of research has been conducted to conceptualize blockchain applications for logistics (Blossey et al., 2019; Dobrovnik et al., 2018; Korpela et al., 2017). However, our research reveals a gap between the existing literature and use cases in the reality of German LSPs. We identified facilitating and inhibiting factors according to the TOE framework (DePietro et al., 1990) that help to understand this development. In addition to these factors, the nature of the blockchain technology seems to impede its diffusion in practice. Blockchains can create trust in decentralized environments where the participants do not necessarily trust each other (Swan, 2015). However, this constellation makes it difficult for companies to set up comprehensive projects in order to gain experience, as they normally would prefer trustful partners for such kind of projects. This impediment should be addressed by researchers. For example, the game theory (Aumann & Hart, 1992) could be a prolific source for ideas. Moreover, future blockchain research in the field of logistics should apply more practice-oriented methodologies, e.g. action research (Lewin, 1946).

Implication 2: The size of the LSPs strongly influences the adoption of blockchain technology. All reviewed companies perceive the blockchain technology to be more positive than negative. Nevertheless, the willingness to change differs between larger and smaller LSPs. In our study, the larger LSPs are more engaged in the adoption of blockchain technology. They defined own use cases and developed them further in cooperative projects with partners. According to Rogers, this positive adoption behavior is the result of existing slack resources, which usually go along with a large size of an organization (Rogers, 2003). In contrast, we found also evidence that most of the companies do not want to change their (IT) structures, processes and work routines, Hannan and Freeman (1984) call this unwillingness to change structures and rejection of innovation "inertia", which is normally more likely to occur at large organizations. In our cases, inertia was more likely to occur at smaller companies. A possibility to overcome this inertia is the contact to external networks (Gilbert, 2005). However, in the case of German LSPs, external networks do not seem to positively influence the blockchain adoption behavior of smaller LSPs. The perceived pressure in the logistics industry as well as in partner companies was rated low and medium by the interviewees. Future research could focus on differences in the adoption behavior of smaller and larger LSPs.

Implication 3: The management, especially at small LSPs, is in torpor. Our study also reveals a reluctant attitude of the management towards the technology. All interviewees acknowledged the future importance of blockchain technology. Nevertheless, no one of the interviewees – even not in the cases A and B – felt threatened by the blockchain technology. Although numerous research revealed potential threats to LSPs (e.g. Subramanian, 2017), the management does not perceive the blockchain technology as a discontinuous change and competence-destroying technology in the sense of Tushman and Nadler (Tushman & Nadler, 1986). Especially small LSPs have not engaged in the technology so far. The managers of the small LSPs refer to technical issues of blockchain technology, problems with the own IT infrastructure, little slack resources, organizational inertia and no pressure from the stakeholders in order to justify the low engagement. We argue that most of the named inhibiting factors are under the control of the management and the management has not attempted to change these factors. In fact, the management seems to be as much affected by the organizational inertia as other members of the companies. However, this implication is limited as we only reviewed a few companies. Further research is necessary to prove this implication. Future research should also address the question, which kind of LSPs will be most affected by the blockchain technology.

Implication 4: LSPs fear transparency because it reveals weaknesses in their work processes. One interviewee mentioned that blockchain technology could make their company too transparent (De Cremer, 2016). Customers or competitors could become aware of failures or weaknesses and use this information against the company. However, this apprehension fades the more experiences were made with the blockchain technology. For example, all interviewees of companies with blockchain engagement evaluated the data disclosure in a blockchain positively. Future research could elaborate on this effect of the adoption of blockchain technology in logistics but also in general.

6 Conclusions, Limitations and Future Research

This study sheds light on the adoption of blockchain technology at German LSPs. Against the background of the rising importance of blockchain technology, we offer significant theoretical and practical implications for the further application of the blockchain technology in the logistics service industry. With the help of a Delphi study and the TOE framework (DePietro et al., 1990), we identified facilitating and inhibiting factors for the adoption of the blockchain industry from the perspective of managers at German LSPs. Furthermore, we evaluated the status of the blockchain adoption process according to Rogers (Rogers, 2003). Our results show that especially larger LSPs engage in the blockchain technology by deepening their knowledge or conducting projects. Nevertheless, none of the companies that we reviewed uses blockchains in their business processes. Smaller

LSPs are very reluctant adopters of the technology and remain in a waiting position.

However, our study as presented in this paper has certain limitations. The significance of our study is limited insofar as the sample size is rather small with only seven participants. Furthermore, it is conceivable that mainly companies participated in the interviews that expressed interest in the technology and tend to appraise it too positively. As a result, the real adoption engagement of LSPs may be lower. Although desktop research supports the findings, additional quantitative studies seem necessary.

Additionally to the directions for future research, which we already described in the implications, further research could also address the adoption behavior in different countries and cultural settings, because the entrepreneurial culture in other countries is different and thus could result in different findings. We also advocate for further research on the environmental factors of the TOE framework (DePietro et al., 1990) as we only focused on the evaluation from the viewpoint of the LSPs. It would be interesting to know if this perception also applies to the viewpoint of the different stakeholders.

References

Adner, R., & Kapoor, R. 2016. Innovation ecosystems and the pace of substitution: Re-examining technology S-curves. Strategic Management Journal, 37(4), 625–648. https://doi.org/10.1002/smj.2363

Ajzen, I. 1991. The theory of planned behavior. Organizational Behavior and Human Decision Processes, 50(2), 179–211. https://doi.org/10.1016/0749-5978(91)90020-T

Alipranti, M., Milliou, C., & Petrakis, E. 2015. On vertical relations and the timing of technology adoption. Journal of Economic Behavior & Organization, 120, 117–129. https://doi.org/10.1016/j.jebo.2015.10.011

Aumann, R. J., & Hart, S. (Eds.). 1992. Handbook of game theory with economic applications. /Handbooks in economics: Vol. 11. Amsterdam, New York, New York: North-Holland. Retrieved from http://www.sciencedirect.com/science/handbooks/15740005/3

Baker, J. 2012. The Technology–Organization–Environment Framework. In Y. K. Dwivedi, M. R. Wade, & S. L. Schneberger (Eds.), Integrated Series in Information Systems. Information Systems Theory (Vol. 28, pp. 231–245). New York, NY: Springer New York. https://doi.org/10.1007/978-1-4419-6108-2_12

Blossey, G., Eisenhardt, J., & Hahn, G. 2019. Blockchain Technology in Supply Chain Management: An Application Perspective. In 52nd Hawaii International Conference on System Sciences, HICSS 2019, Grand Wailea, Maui, Hawaii, USA, January 8-11, 2019 (pp. 1–9).

Catania, L., Grassi, S., & Ravazzolo, F. 2019. Forecasting cryptocurrencies under model and parameter instability. International Journal of Forecasting, 35(2), 485–501. https://doi.org/10.1016/j.ijforecast.2018.09.005

Collins, P. D., Hage, J., & Hull, F. M. 1988. Organizational and Technological Predictors of Change in Automaticity. Academy of Management Journal, 31(3), 512–543. https://doi.org/10.2307/256458

Dalkey, N., & Helmer, O. 1963. An Experimental Application of the Delphi Method to the Use of Experts. Management Science, 9(3), 458–467. https://doi.org/10.1287/mnsc.9.3.458

Damanpour, F. 1992. Organizational Size and Innovation. Organization Studies, 13(3), 375–402. https://doi.org/10.1177/017084069201300304

Davis, F. D., Bagozzi, R. P., & Warshaw, P. R. (1989). User Acceptance of Computer Technology: A Comparison of Two Theoretical Models. Management Science, 35(8), 982–1003. https://doi.org/10.1287/mnsc.35.8.982

De Cremer, D. 2016. When Transparency Backfires, and How to Prevent It. Retrieved from https://hbr.org/2016/07/when-transparency-backfires-and-how-to-prevent-it

DePietro, R., Wiarda, E., & Fleischer, M. 1990. The context for change: Organization, technology and environment. In L. G. Tornatzky & M. Fleischer (Eds.), Issues in organization and management series. The processes of technological innovation (pp. 151–175). Lexington, Mass.: Lexington Books.

Dobrovnik, M., Herold, D., Fürst, E., & Kummer, S. 2018. Blockchain for and in Logistics: What to Adopt and Where to Start. Logistics, 2(3), 18. https://doi.org/10.3390/logistics2030018

Gilbert, C. G. 2005. Unbundling the Structure of Inertia: Resource Versus Routine Rigidity. Academy of Management Journal, 48(5), 741–763. https://doi.org/10.5465/amj.2005.18803920

Hackius, N., & Petersen, M. 2017. Blockchain in logistics and supply chain: Trick or treat? In W. Kersten, T. Blecker, & C. M. Ringle (Eds.), Proceedings of the Hamburg International Conference of Logistics (HICL). Digitalization in supply chain management and logistics: Smart and digital solutions for an industry 4.0 environment (1st ed., pp. 3–18). https://doi.org/10.15480/882.1444

Hannan, M. T., & Freeman, J. 1984. Structural Inertia and Organizational Change. American Sociological Review, 49(2), 149. https://doi.org/10.2307/2095567

Hughes, L., Dwivedi, Y. K., Misra, S. K., Rana, N. P., Raghavan, V., & Akella, V. (2019). Blockchain research, practice and policy: Applications, benefits, limitations, emerging research themes and research agenda. International Journal of Information Management, 49, 114–129. https://doi.org/10.1016/j.ijinfomgt.2019.02.005

Iacovou, C. L., Benbasat, I., & Dexter, A. S. 1995. Electronic Data Interchange and Small Organizations: Adoption and Impact of Technology. MIS Quarterly, 19(4), 465–485. https://doi.org/10.2307/249629

Korpela, K., Hallikas, J., & Dahlberg, T. 2017. Digital Supply Chain Transformation toward Blockchain Integration. In : Proceedings of the Annual Hawaii International Conference on System Sciences. Hawaii International Conference on System Sciences. https://doi.org/10.24251/HICSS.2017.506

Kuan, K. K.Y., & Chau, P. Y.K. 2001. A perception-based model for EDI adoption in small businesses using a technology–organization–environment framework. Information & Management, 38(8), 507–521. https://doi.org/10.1016/S0378-7206(01)00073-8

Lewin, K. 1946. Action Research and Minority Problems. Journal of Social Issues, 2(4), 34–46. https://doi.org/10.1111/j.1540-4560.1946.tb02295.x

Li, Y., Marier-Bienvenue, T., Perron-Brault, A., Wang, X., & Paré, G. 2018. Blockchain Technology in Business Organizations: A Scoping Review. In T. Bui (Ed.): Proceedings of the Annual Hawaii International Conference on System Sciences. Hawaii International Conference on System Sciences. https://doi.org/10.24251/HICSS.2018.565

Mettler, M. 2016. Blockchain technology in healthcare: The revolution starts here. In 2016 IEEE 18th International Conference on e-Health Networking, Applications and Services (Healthcom) (pp. 1–3). IEEE. https://doi.org/10.1109/HealthCom.2016.7749510

Miau, S., & Yang, J.-M. 2018. Bibliometrics-based evaluation of the Blockchain research trend: 2008 – March 2017. Technology Analysis & Strategic Management, 30(9), 1029–1045. https://doi.org/10.1080/09537325.2018.1434138

Murry, J. W., & Hammons, J. O. 1995. Delphi: A Versatile Methodology for Conducting Qualitative Research. The Review of Higher Education, 18(4), 423–436. https://doi.org/10.1353/rhe.1995.0008

Nofer, M., Gomber, P., Hinz, O., & Schiereck, D. (2017). Blockchain. Business & Information Systems Engineering, 59(3), 183–187. https://doi.org/10.1007/s12599-017-0467-3

Peters, G. W., & Panayi, E. 2016. Understanding Modern Banking Ledgers Through Blockchain Technologies: Future of Transaction Processing and Smart Contracts on the Internet of Money. In P. Tasca, T. Aste, L. Pelizzon, & N. Perony (Eds.), Banking Beyond Banks and Money: A Guide to Banking Services in the Twenty-First Century (pp. 239–278). Cham: Springer International Publishing. https://doi.org/10.1007/978-3-319-42448-4_13

Risius, M., & Spohrer, K. 2017. A Blockchain Research Framework. Business & Information Systems Engineering, 59(6), 385–409. https://doi.org/10.1007/s12599-017-0506-0

Rogers, E. M. 2003. Diffusion of innovations (Fifth edition, Free Press trade paperback edition). Social science. New York, London, Toronto, Sydney: Free Press. Retrieved from http://www.loc.gov/catdir/bios/simon052/2003049022.html

Saldaña, J. 2009. The coding manual for qualitative researchers. London: Sage.

Schwemmer, M. (Ed.). (2016). Die Top 100 der Logistik: Marktgrössen, Marktsegmente und Marktführer in der Logistikdienstleistungswirtschaft (Ausg. 2016/2017). Hamburg: DVV Media Group Dt. Verkehrs-Verl.

Subramanian, H. 2017. Decentralized blockchain-based electronic marketplaces. Communications of the ACM, 61(1), 78–84. https://doi.org/10.1145/3158333

Swan, M. (2015). Blockchain: Blueprint for a new economy (1. ed.). Safari Tech Books Online. Beijing: O'Reilly. Retrieved from http://proquest.safaribooksonline.com/9781491920480

Szabo, N. (1997). Formalizing and securing relationships on public networks. First Monday, 2(9).

Tushman, M., & Nadler, D. 1986. Organizing for Innovation. California Management Review, 28(3), 74–92.

Tushman, M. L., & Anderson, P. 1986. Technological Discontinuities and Organizational Environments. Administrative Science Quarterly, 31(3), 439–465.

Valente, T. W. 1995. Network models of the diffusion of innovations. Quantitative methods in communication. Cresskill, NJ: Hampton Press.

Venkatesh, V., Morris, M. G., Davis, G. B., & Davis, F. D. 2003. User Acceptance of Information Technology: Toward a Unified View. MIS Quarterly, 27(3), 425–478.

Zhu, K., & Kraemer, K. L. 2005. Post-Adoption Variations in Usage and Value of E-Business by Organizations: Cross-Country Evidence from the Retail Industry. Information Systems Research, 16(1), 61–84.
https://doi.org/10.1287/isre.1050.0045

A Literature Review on Machine Learning in Supply Chain Management

Hannah Wenzel[1]*, Daniel Smit*[1] *and Saskia Sardesai*[1]

1 – Fraunhofer Institute for Material Flow and Logistics IML

Purpose: In recent years, a number of practical logistic applications of machine learning (ML) have emerged, especially in Supply Chain Management (SCM). By linking applied ML methods to the SCM task model, the paper indicates the current applications in SCM and visualises potential research gaps.

Methodology: Relevant papers with applications of ML in SCM are extracted based on a literature review of a period of 10 years (2009-2019). The used ML methods are linked to the SCM model, creating a reciprocal mapping.

Findings: This paper results in an overview of ML applications and methods currently used in the area of SCM. Successfully applied ML methods in SCM in industry and examples from theoretical approaches are displayed for each task within the SCM task model.

Originality: Linking the SC task model with current application areas of ML yields an overview of ML in SCM. This facilitates the identification of potential areas of application to companies, as well as potential future research areas to science.

Keywords: Supply Chain Management, Machine Learning, Literature Review, Predicitve Analytics

First received: 19.May.2019 **Revised:** 29.May.2019 **Accepted:** 11.June.2019

1 Motivation

In the age of information technology and increasingly complex technical and industrial processes, agile and efficient logistics processes play a central role. High logistic requirements such as reliability, transparency and flexibility combined with optimal economic conditions form the foundation for a successful supply chain (SC). Dynamically changing processes require a technology that is able to cope with the increasing complexity of supply chains.

The growth in data volume and diversity has led to data sets larger than ever before. Processing with conventional, practical management tools is often inefficient or impossible. To manage and evaluate these new and potentially valuable data sets, new methods and applications have been developed in the form of predictive analytics (Waller and Fawcett, 2013, p. 77). One of these predictive analytics methods is machine learning (ML). The success of this methodology is achieved through the invention of sophisticated ML models, the availability of large data sets ("big data") and the utilization of hardware architectures such as GPUs (Abadi et al., 2016, p. 256; Copeland, 2016; Steinbach, 2018, p. 32).

Current applications of ML focus on specific areas within SCM, while some areas remain unexploited. Thus the paper intends to create a link between current ML applications, actual research work and the SCM task model. This allows for a visualization of tasks wherein ML techniques are already applied and a deduction of potential areas of future research within the SCM task model.

The paper is based on a literature analysis by querying the "IEEE Xplore Digital Library", "Scopus" and "ScienceDirect" databases using the search term "Supply Chain Management [AND] Machine Learning". Due to the recent developments in ML, the results were restricted to publications from 2009-2019. Based on the abstracts, a total of 38 relevant papers have been identified.

Along with the full papers, the specific SCM task and the applied ML methods have been extracted. The papers have been categorized by the source of the datasets they used. 20 of the papers used real-world datasets acquired mostly from enterprises. 9 papers used synthetic datasets (e.g. generated by a simulation). For the remaining 9 papers, the types of datasets used were not mentioned. Furthermore, with the support of the Google search function several practical applications of ML within SCM were identified. This was done to provide a complete picture of current practical applications.

The paper is structured as follows. Firstly, the different tasks of SCM are explained, ensued by a general introduction to ML including the different types and methods, and explaining the process of applying ML techniques for solving real-world issues. This is succeeded by application examples of ML in SCM and a mapping to the SCM task model. The paper concludes with an overview on possible areas of future research within the SCM task model. The result of this paper addresses researchers in their future course of research work as well as Supply Chain Managers in applying ML in their field of work.

2 Supply Chain Management and its Task Model

According to Arnold et al. (2008) the primary goal of SCM is to fulfill customer needs while optimizing costs in terms of inventories, resources and processes in the network at the same time. To ensure that this goal can be achieved, it requires a number of subgoals e.g. improving customer orientation and satisfaction, increase of delivery capability and reduction of lead times (Arnold et al., 2008, 460 f.). The achievement of the preceding goals must be carried out by internal and cross-company tasks of SCM. A representation of these tasks is shown in Figure 1.

The three main tasks of the model are Supply Chain Design, -Planning and -Execution (Kuhn and Hellingrath, 2002, 142 f.). The inner circle of tasks is discussed briefly hereafter as it is of relevance for mapping ML applications.

Supply Chain Design

Supply Chain Design (SCD) deals with long-term planning in terms of location decisions, make-or-buy decisions, supply relationships, capacity dimensioning, logistics strategy and general tasks (Hompel and Wolf, 2013, p. 148). Another important task is cost-optimized structuring of the logistics and production processes. Consequently, this area involves a financial evaluation of changes in the logistics network (Kuhn and Hellingrath, 2002, 143 f.).

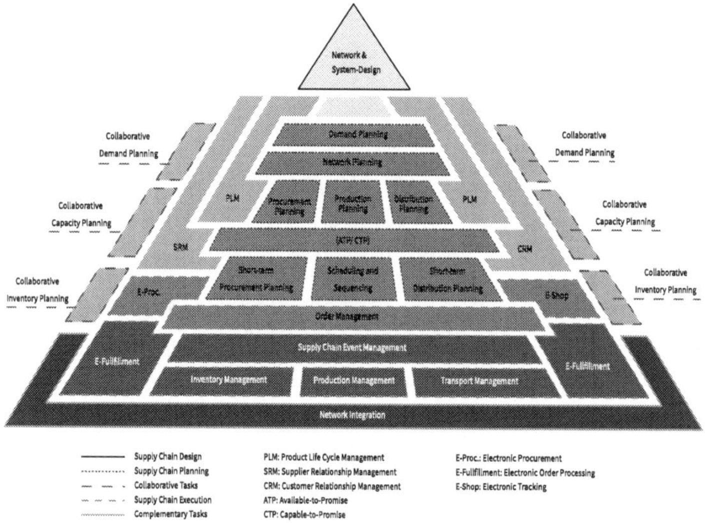

Figure 1: Supply Chain Management task model (Hompel and Wolf, 2013, pp. 146–147)

Supply Chain Planning

The goal of Supply Chain Planning (SCP) is medium- to long-term program planning across the entire SC. The area of SCP reflects the planning of production and logistics resources to fulfill customer orders. Hence, the basis of resource planning are forecasted customer requirements. The tasks for SCP include the areas of demand and network planning, which set the basis for procurement, production and distribution planning and enabling available-to-promise/ capable-to-promise checks. In the short-term, procurement planning, scheduling and sequencing and short-term distribution planning support SCP (Kuhn and Hellingrath, 2002, p. 144). Thus, demand information creates an important basis in order to coordinate decisions and planning along the logistics chain.

Tasks of demand planning are forecasting and visualization of short-term and long-term requirements. These are particularly important when it comes to fulfilling customer requirements, planning the utilization of the value chain, or carrying out inventory optimization (Kuhn and Hellingrath, 2002, 144 f.).

The next task, network planning, deals with the coordination between partners within the value chain. Through network planning, requirements as well as material and capacity resources are coordinated. It is possible to trace the fulfillment of the production program through network planning, due to its orientation towards production programs and sales plans (Kuhn and Hellingrath, 2002, 145 f.).

Procurement planning takes into account the preceding areas of demand and network planning and determines the planning of the parts supply. This is followed by production planning, which has the task of creating a production plan for each production unit in the SC. The aim is to achieve a high level of readiness for delivery while at the same time keeping inventory costs as low as possible (Kuhn and Hellingrath, 2002, 146 f.).

Distribution planning deals with the fulfilment of requirements, which are achieved through optimized planning and coordination of deliveries. It is based on the results of production planning and the specifications of network planning (Kuhn and Hellingrath, 2002, p. 147).

In order to satisfy customer requirements or customer orders, their feasibility is examined via available-to-promise/ capable-to-promise. Due to the gap in the task model between the anonymous planning tasks and the planning tasks with customer reference, this area represents an important task of SCM (Kuhn and Hellingrath, 2002, 147 f.).

Short-term procurement planning, scheduling and sequencing, as well as short-term distribution planning deal with more detailed tasks of procurement, production and distribution planning. For more detailed information, please refer to Kuhn and Hellingrath (2002).

Supply Chain Execution

The area of Supply Chain Execution (SCE) considers cross-company processes such as SC management and control. The aim of this area is to support decision-making at the operational level (Kuhn and Hellingrath, 2002, 152 f.). Its executive area includes the processing and monitoring of orders. Inventory management is also part of SC Execution, taking care of movements of stocks and materials including their documentation. Two further components are production management, which manages all information on production processes, and transport management, which handles transport orders on the procurement and distribution side.

An additional major task is Supply Chain Event Management. It controls the activities within the SC, announces plan changes and initiates corrective measures. The task aims to create transparency about parameters such as disruptions, inventories and requirements. Therefore, it includes all reactive risk managing activities (Kuhn and Hellingrath, 2002, p. 154).

3 Machine Learning

The following section interprets the term "Machine Learning" and covers types of ML methods. This supports a common understanding of the methods used in the different application areas of ML in SCM. Challenges are also addressed.

3.1 Interpretation of Machine Learning

ML is a sub-area of artificial intelligence and represents another way of programming. Example data replaces rigid calculation rules of a program. From the given example data, learning methods or algorithms extract statistical regularities, and represent those in the form of models. The models can react to new, unknown data and classify them into categories or make predictions (Hecker et al., 2017, p. 8).
ML deals with the computer-aided modelling and realization of learning phenomena (Görz, Schneeberger and Schmid, 2013, pp. 200–201). It is defined as a process that uses experience to improve performance or make concrete predictions. The experience refers to past information, which is provided to the procedure from an electronic data collection. ML involves the design of effective and precise algorithms (Mohri, Rostamizadeh and Talwalkar, 2012).

3.2 Types of machine learning

Several types of ML can be distinguished. Well-known ones are supervised, unsupervised and reinforcement learning. Further types can be found in e.g. Géron (2017).

Supervised Learning

The most widely used type of ML is Supervised Learning (Marsland, 2015, p. 6). Supervised Learning is a process in which a computer program is trained by using known example data. As the output is also known, this learning process aims at finding a connection in the form of rules, which relate input data to output data and finally apply the learned rules to new data. At this point the computer program is getting trained. With this newly gained knowledge it is now able to predict future input and output data. Two important tasks are classification and regression (Kirste and Schürholz, 2019, 25 f.).

Unsupervised Learning

Unsupervised learning describes a system that is able to discover knowledge. Correct answers are not provided in this type of learning, thus, there are no pre-labelled target values. This approach is also called "learning without a teacher". A well-known task of unsupervised learning is clustering. The method identifies similarities between the inputs to categorize inputs by common patterns. Similar tasks are e.g. Association Rules, Self-Organizing Maps, Multidimensional Scaling and Nonlinear Dimension Reduction (Marsland, 2015, p. 281; Russell and Norvig, 2016, p. 694; Hastie, Friedman and Tibshirani, 2017, 485 ff.).

Reinforcement Learning

With reinforcement learning, the optimal solution is unknown to the system at the beginning of the learning phase and therefore must be determined iteratively. In this process, sensible approaches are rewarded, and wrong steps tend to be punished. With this approach, it is possible that the system takes into account complex environmental influences and reacts accordingly. Hence, the system finds its own solutions autonomously through directional rewards and punishments (Gentsch, 2018, 38 f.).

3.3 Machine Learning Methods

Machine Learning methods can be categorized by the task they are solving. As described in 3.2 classification and regression are two tasks of supervised learning. Examples of tasks in unsupervised learning are clustering and association rule mining. Another common type of learning is reinforcement learning. Figure 2 shows selected methods of ML, grouped by the task they commonly solve. It is important to mention that this selection of methods is far from complete and some methods can be used to solve more than one task. Also, there exist several variants of algorithms for each method.

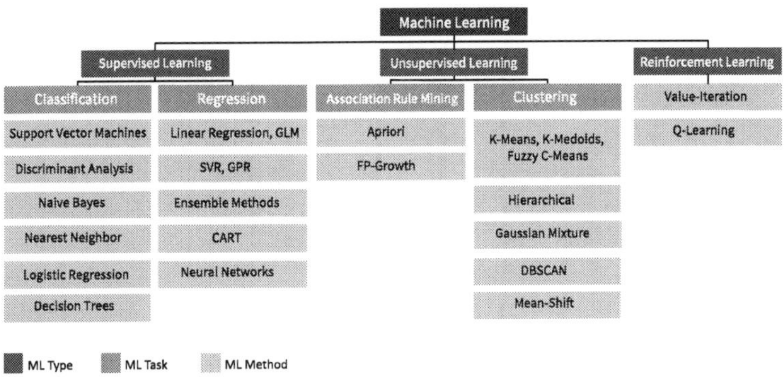

Figure 2: Overview of Machine Learning Methods based on (Witten, Eibe and Hall, 2011, pp. 116–216; The MathWorks, 2016, p. 7; Seif, 2018)

Part of the ongoing research in ML is the development of further methods, models and algorithms. Selecting a model and an algorithm depends on many factors and there is no one-method-fits-all-solution - also known as "There is no free lunch in statistics" (James et al., 2013, p. 29). The selection of an appropriate method for a given problem and dataset is one of the most challenging tasks in data analysis (James et al., 2013, p. 29).

When using ML methods and algorithms, some challenges may occur. Two different reasons can be mentioned when it comes to the development of error sources. On the one hand, the problem can be found in the data. The availability of a large amount of data which is necessary for the training of a model is still a frequent challenge, as well as insufficiently representative training data. Another challenge is poor data quality. In order to make patterns visible in data sets, they first must be detected. If the data contains errors or outliers, it is difficult to identify such patterns. Another issue is redundant features, which do not provide added value to the model. An important step for the success of a ML method is therefore to select suitable features for training (Géron, 2017, pp. 23–26). On the other hand, a poorly chosen algorithm can create difficulties in the form of overfitting and underfitting (Géron, 2017, p. 22). Overfitting means that ML algorithms can sometimes generalize incorrectly although the model works on a training data set. A model can be too complex, which means that the model follows noise or errors in the data too closely (James et al., 2013, p. 22). Underfitting is the opposite of overfitting and describes a model that is too simple for the structure of the data to be recognized and learned (Géron, 2017, p. 29).

3.4 Machine Learning Process

Real world machine learning applications are usually developed as part of a larger project. Several process model frameworks have been developed for such projects. One popular one is the Cross Industry Standard Process for Data Mining (Shearer, 2000; Piatetsky, 2014).

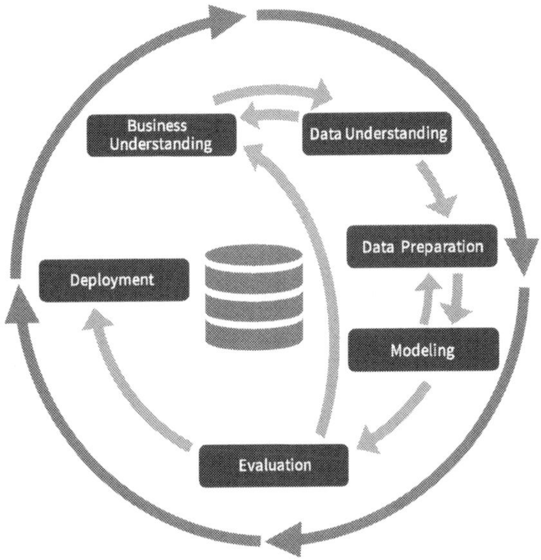

Figure 3: Cross Industry Standard Process for Data Mining (Shearer, 2000, p. 14)

The illustrated process in Figure 3 is also used for ML projects and acts as a common standard for ML (Cohen, 2017). The process defines six phases with specific tasks. Starting from "Business Understanding" to "Data Understanding", "Data Preparation", "Modeling", its "Evaluation" and ending with "Deployment". The goal of the framework is to improve results while reducing costs and time for the project. Shearer (2000) describes the specific tasks in more detail.

The paper is focused on presenting ML methods that are usually used during the modeling phase, as this was also the focus of the examined papers.

4 Machine Learning Application Areas in Supply Chain Management

This chapter portrays practical applications of ML by using three examples. Each of these examples represents one of the three main tasks of the SCM Model and is intended to show the reader how to use ML in practice. The three applications are selected as each of them represents a concise structure, a comprehensible explanation and gives a first impression of ML application.

4.1 Selecting Supply Chain Partners

In the field of SCD, one of the major challenges remains to find suitable business partners such as customers or suppliers to exploit new opportunities. This is especially valid regarding the globalization and the fast development of technology. To simplify finding new plausible business partners, Mori et al. (2010) use ML techniques based on company profiles and transaction relationships.

For data collection, the authors obtained company data of suppliers and customers via the "Teikoku Data Bank". A total of 30,660 companies from the manufacturing industry could be retrieved from the databank, all of them located in the Kanto and Koshinetsu areas of Japan. The training data consisted of 34,441 supplier and customer pairs from the existing inter-firm network. Those pairs are used as positive examples. Negative examples are defined as randomly chosen supplier and customer pairs, which are not related.

A Support Vector Machine (SVM) was applied to model the existing relationships by using features inherent to the company, such as number of employees and capital, as well as features defined by the interlinkage of firms, like customers of a supplier and common Industrial categories. The accuracy of the SVM-model reached up to 85%.
The most important determinants of customer-supplier relationships were identified as number of employees or industry category. The authors explain the relevance of these categories with country specific characteristics such as hierarchical structures. This shows that different determinants might be relevant when applying the method elsewhere. Along with the SVM, the ML model provides a tool to predict plausible candidates for future customer-supplier relationships (Mori et al., 2010).

4.2 Demand forecast for DM

One of Germany's largest drugstore companies, dm, uses machine learning algorithms to predict future demands. With 3,350 stores worldwide, the dm drugstore chain runs six distribution centers to ensure the availability of goods. The distribution centers have to ensure that the incoming goods of the industry partners are arranged for the individual stores in such a way that there is no product deficit and customer satisfaction remains high. Frequently, stores require a delivery of products within a short time frame, but the manufacturers have relatively long delivery times. This remains the biggest challenge for the distribution centers. Planning well in advance is accompanied by long-term orders, resulting in high storage costs and tied up capital.
To overcome this issue, artificial intelligence algorithms are used to create weekly demand forecasts based on SKU level over a time horizon of half a

year. For training of the algorithm, the data from the last 2.5 years of the particular distribution center has been used and the seasonality factor has been taken into account.

As a result of the artificial intelligence algorithms, the demand forecasts became so precise over a period of six months that a significant improvement in forecast quality could be achieved and industrial partners are now able to plan much earlier. Delivery reliability and product availability can thus be significantly optimized, which is supported by an automated information flow of future requirements to industrial partners. The advantages of using the algorithms are obvious for the industrial partners as well as for the dm drugstore market. The industrial partners can count on increased planning and ordering security through valid forecasts and the drugstore chain has secure product availability and less excess stocks. The results are satisfied customers and lower costs (JDA Software, 2019, pp. 8–9).

4.3 Detecting False-Positive RFID Tag Reads in Transport Management

One of SC's key processes is the shipment of goods from distribution centers. An error-free shipment depends on several factors, be it the picking process or the assignment of pallets to the allotted truck. To integrate a control mechanism, the METRO Group Cash & Carry implemented RFID portals to check on outgoing pallets. In its distribution center in Unna, Germany, loading ramps are equipped with RFID portals to automatically detect goods leaving the warehouse. For the detection of a pallet, it carries a transponder. As soon as the pallets pass the portals, the reading device records goods leaving the warehouse and automatically adjusts the available inventory via direct communication with the warehouse management system.

The RFID identification supports the shipment of goods and the automatic inventory adjustment but carries one major issue. The antennas of the portals have ranges of several meters and therefore identify all pallets in range, pallets to be shipped and pallets placed for intermediate storage. This means that during a collection cycle it is not possible to distinguish between a moving pallet loaded into a truck and a static pallet that is not relevant for the current shipment and hence, has been accidentally read. The non-relevant pallets represent a false-positive reading of the RFID portals.

In order to differentiate between the moving and static pallets, METRO Group Cash & Carry applied ML techniques based on the recorded attributes by each pallet. The attributes include the reception of the signal strength indicator (RSSI), the time stamp of each tag event and the number of reads per antenna based on the low-level reader data. It became apparent, that the tags of static and moving pallets submit different RSSI values, based on frequencies submitted to the portals. This is where ML comes in, as classifying two characteristics according to their threshold value is a well-known ML task. If the RSSI average value is below -58.1dBm, it is a static tag (false positive); otherwise it is a moving tag.

To automatically evaluate the attributes by ML, a very large data set is required for the training phase. The distribution center installed RFID portals on all 70 loading ramps and monitored them over a period of 7 months. A total of 53,998 pallets were monitored, of which 40,743 were static and 13,245 were moved through the outgoing goods portal.

After completion of the data acquisition, the used algorithm could correctly classify more than 95.5% of the data. Thus the pallets can be quickly identified and false-positive pallets can be captured directly. This minimized faulty inventory adjustments and deliveries (Keller et al., 2010).

4.4 Mapping of Use Cases to the Supply Chain Task Model

The analysis of the papers resulted in an assignment of the ML use cases to all three main task areas of the SCM model, which are outlined below. Out of illustration purposes in Figure 4, a number is assigned to each paper.

In the area of SCD, all retrieved papers deal with supplier selection. The authors apply different ML methods. While Kong and Xue (2013) utilize Radial Basis Function (RBF) neural networks [1], Zhang et al. (2017) use ensemble learning, decision trees and logistic regression [2]. Similar to the case of Mori et al. (2010) [3], the authors apply SVM, too. Another paper by Cheng et al. (2017) integrates Adaboost on expert knowledge basis to deal with supplier selection [4].

In the area of SCP, ML topics address the task of demand planning and procurement planning. The number of papers in relation to other tasks is very comprehensive. Based on the literature analysis carried through, fourteen use cases could be assigned to this area.

The authors Xue et al. (2018) use machine learning to solve the dynamic forecasting problem of the supply of goods in the event of a catastrophe [5]. Gamasaee, Zarandi and Turksen (2015) [6] and Sarhani and El Afia (2014) deal with approaches to demand forecasts of SC. It should be noted that all three papers use Support Vector Regression (SVR) as the ML method [7]. Ap-

plication examples for artificial neural networks are provided by the authors Gaur, Goel and Jain (2015), who apply a data set from Walmart to forecast demand for the SC [8], and Yang and Sutrisno (2018), who distribute perishable goods among franchise companies [9]. Adhikari et al. (2017) contribute with an ensemble technique to combine different ML methods in order to obtain demand forecasts [10]. The described practical application of the drugstore chain dm can be assigned to SCP, too, as the methodology creates a planning basis for a more effective SC to the stores. JDA Software (2019) mentions that the goal is to achieve high store availability [11]. Souichirou (2015) deals with commodity demand forecasting and systematization of SCM by using heterogenous mixture learning, and uses this to simulate sales measures [14]. As one of the rare studies so far, Cui et al. (2018) [12] and Lau, Zhang and Xu (2018) [13] integrate social media sources to forecast sales. The last three authors base their analysis on real data.
In the area of procurement planning, Hogenboom et al. (2009) apply RBF Networks [15] to deal with product pricing as well as Kiekintveld et al. (2009) who use a Bayesian Network and artificial neuronal network [16]. In addition to the two authors mentioned above, Lee and Sikora (2016) also deal with product pricing by using Q-learning softmax function [17]. In the same context, Ketter et al. (2009) use gaussian mixture models [18]. It is interesting to mention that all three authors deal with the same topic based on the same synthetic data, but use different methods.
Within distribution planning Mokhtarinejad et al. (2015) outline vehicle routing and scheduling problems in cross-docking systems by using Bi-Clustering, K-means and artificial Neural Networks [19].

In the third, operationally structured SCE, the literature addresses the areas of Order Management, Supply Chain Event Management, Inventory, Production and Transport Management.

Order management contains a total of four use cases. Based on synthetic data, the authors Sun and Zhao (2012) present a multi-agent coordination mechanism using reinforcement learning to derive an optimal ordering strategy for the entire SC with several levels [20]. Wang, Ng and Ng (2018) group SKUs according to their demand and performance attributes by unsupervised learning [21]. Real data is used in Wang and Liu (2009) to train a model using neural networks, which represents an index system for evaluating order priority [22]. Zhu, Ma and Zhang (2014) determine the manufacturing priority of an order with the help of RBF neural networks, Kriging and SVR [23].

There are four use cases in the area of Supply Chain Event Management. Among them, Arumugam, Umashankar and Narendra (2018) create an intelligent logistics solution that negotiates contracts, and includes logistics planning and condition monitoring of the facilities [24]. The early detection of supplier risks, Frerichs (2018) [25], as well as the identification of fraud and deceptive practices Zage, Glass and Colbaugh (2013) [26] are topics addressed in the context of risk management. Both applications rely on real data. Hiromoto, Haney and Vakanski (2017) use neural networks to identify vulnerabilities within SC's and mitigate the consequences of unforeseen risks [27].

A total of four use cases could be assigned to inventory management. The authors Barbosa de Santis, Pestana de Aguiar and Goliatt (2017) use ML classifiers to identify material residues within an inventory management system that have negative effects on inventory management [28]. Furthermore, Inprasit and Tanachutiwat (2018) use neural networks to optimize safety stocks and reorder points for products, taking into account various influencing factors such as performance and lead time [29]. The applications outlined above rely on real data. Another application is the identification of technically obsolete spare parts in the warehouse. Supervised learning methods were used to solve this problem (Cherukuri and Ghosh; Inprasit and Tanachutiwat, 2018) [30]. The authors Priore et al. (2018) use decision trees on the basis of synthetic data to determine the optimal replenishment rule [31].

Production management contains four use cases, while all of the papers had access to real data. Ali Ahmadi et al. (2016) use machine learning approaches to differentiate between the manufacturing locations of products. ML distinguishes products that were manufactured in a ratified factory from the ones which originate from unknown sources [32]. In the paper of Tirkel (2011) predictive models of ML and data mining are used to determine the cycle time [33]. By using linear regression and tree based methods, Dávid Gyulai et al. (2018) and Lingitz et al. (2018) estimate production lead times [34], [35].

The described use case of Keller et al. (2010), whose authors use ML to distinguish between static and moving pallets, is assigned to transport management [36]. In addition, the authors Gulisano et al. (2018) formulate a challenge to predict the arrival time and destinations of vessels [37].

An interesting approach is taken by Kao, Niraula and Whyatt (2018) who analyze part identification to enable data curation. Hence, the paper can cover several areas within the supply chain task model and is a rather general topic.

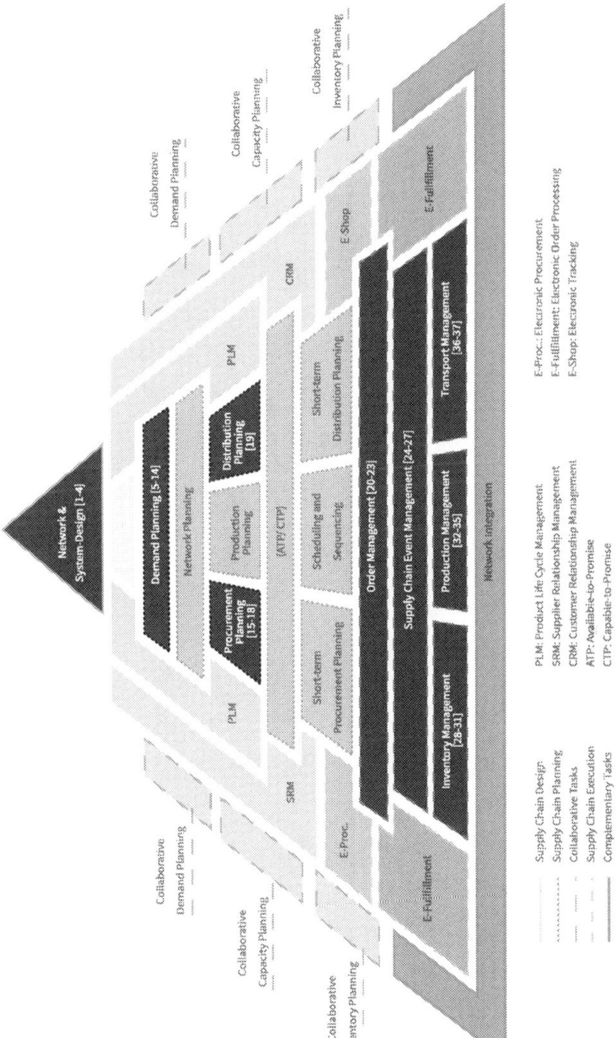

Figure 4: Illustration of the ML use cases in the SC model with numbers of the use cases based on (Hompel and Wolf, 2013, pp. 146–147)

Figure 4 depicts the ML application cases described above sorted into the respective supply chain tasks. The papers have been integrated into the task model along with their assigned numbering. All tasks with ML use cases are highlighted.

It shows that with four papers, SCD contains the lowest number of ML applications, all of them in the area of supplier selection. SCP has a total number of 15 use cases in the area of demand, procurement and distribution planning, most of them allocated to demand planning. Within SCE, each component of the SCM task model has at least one use case assigned. The highest number of applications, 18, could be found in SCE. Many of the applications are based on real data, in some cases synthetic data serves to verify the applicability of ML methods.

5 Conclusion

This paper assigns use cases of machine learning to the task model of Supply Chain Management, resulting in an overview of ML applications within the different supply chain tasks. It was demonstrated that in the SCM task model a single area could have different ML methods applied for a common goal. A large portion of research focused on demand planning, with 10 out of 38 papers handling this task. An investigation of ML methods for inter-company areas such as SRM and CRM could be promising for SCM. Future research should review relevant literature to suggest which methods apply best to certain SCM tasks.

With regard to the standard process model, the research focus lies in the modeling phase. Further research emphasizing on and explaining the concepts used in Business, Data Understanding and Evaluation is required. The examined papers cover these phases, but hardly present useful insights and concepts for SCM. The Deployment phase is not treated at all.

References

Abadi, M., Barham, P., Chen, J. and Chen, Z., 2016. Proceedings of OSDI '16: 12th USENIX Symposium on Operating Systems Design and Implementation. [e-book]. Berkeley, CA. <https://www.usenix.org/system/files/conference/osdi16/osdi16-abadi.pdf>.

Adhikari, N. C., Garg, R., Datt, S., Lalit Das, Deshpande, Srinivas and Misra, A., 2017. Proceedings of the International Conference on Intelligent Sustainable Systems (ICISS 2017). Piscataway, NJ.

Ali Ahmadi, Mohammad-Mahdi Bidmeshki, Amit Nahar, Bob Orr, Michael Pas and Yiorgos Makris, 2016. A machine learning approach to fab-of-origin attestation.

Arnold, D., Isermann, H., Kuhn, A., Tempelmeier, H. and Furmans, K., 2008. Handbuch Logistik. [e-book]. Berlin, Heidelberg. <https://link.springer.com/content/pdf/10.1007%2F978-3-540-72929-7.pdf>.

Arumugam, S. S., Umashankar, V. and Narendra, C. N., 2018. IOT Enabled Smart Logistics Using Smart Contracts. [online].

Barbosa de Santis, R., Pestana de Aguiar, E. and Goliatt, L., 2017. Reordering Point Determination using Machine Learning Technique for Inventory Management. [online]. Piscataway, NJ [Accessed 3 May 2019].

Cheng, Y., Zhou, Z., Gu, X. and Liu, W., 2017. A Hybrid DEA-Adaboost Model in Supplier Selection for Fuzzy Variable and Multiple Objectives. [online].

Cherukuri, M. B. and Ghosh, T., 2016. Control Spare Parts Inventory Obsolescence by Predictive Modelling. In: X. Liu, ed. 2016. iThings-GreenCom-CPSCom-Smart Data 2016. Piscataway, NJ, pp. 865–869.

Cohen, S., 2017. What is the most important step in a machine learning project? [online]. Available at: <https://www.kdnuggets.com/2017/08/most-important-step-machine-learning-project.html>.

Copeland, M., 2016. What's the Difference Between Artificial Intelligence, Machine Learning, and Deep Learning? [online]. Available at: <https://blogs.nvidia.com/blog/2016/07/29/whats-difference-artificial-intelligence-machine-learning-deep-learning-ai/>.

Cui, R., Gallino, S., Moreno, A. and Zhang, D. J., 2018. The Operational Value of Social Media Information. Production and Operations Management, 27(10), pp. 1749–1769.

Dávid Gyulai, András Pfeiffer, Gábor Nick, Viola Gallina, Wilfried Sihn and László Monostori, 2018. Lead time prediction in a flow-shop environment with analytical and machine learning approaches.

Frerichs, P., 2018. Welche Chancen künstliche Intelligenz dem Supply Chain Manage-ment bietet. <https://www.hermes-supply-chain-blog.com/welche-chancen-kuenstliche-intelligenz-dem-supply-chain-management-bietet/>.

Gamasaee, R., Zarandi, M. F. and Turksen, I. B., 2015. A type-2 fuzzy intelligent agent based on sparse kernel machines for reducing bullwhip effect in supply chain. [online] [Accessed 3 May 2019].

Gaur, M., Goel, S. and Jain, E., 2015. Comparison between Nearest Neighbours and Bayesian Network for Demand Forecasting in Supply Chain Management. New Delhi, India.

Gentsch, P., 2018. Künstliche Intelligenz für Sales, Marketing und Service. Mit AI und Bots zu einem Algorithmic Business - Konzepte, Technologien und Best Practices. [e-book]. Wiesbaden: Springer Gabler. <https://link.springer.com/content/pdf/10.1007%2F978-3-658-19147-4.pdf>.

Géron, A., 2017. Hands-on machine learning with Scikit-Learn and TensorFlow. Concepts, tools, and techniques to build intelligent systems. [e-book]: O'Reilly. <http://shop.oreilly.com/product/0636920052289.do>.

Görz, G., Schneeberger, J. and Schmid, U., 2013. Handbuch der Künstlichen Intelligenz. [e-book]. München: Oldenbourg. <http://dx.doi.org/10.1524/9783486719796>.

Gulisano, V., Jerzak, Z., Smirnov, P., Strohbach, M., Ziekow, H. and Zissis, D., 2018. The DEBS 2018 Grand Challenge. In: Unknown, ed. 2018. DEBS'18. Proceedings of the 12th ACM International Conference on Distributed and Event-Based Systems : June 25-29, 2018, Hamilton, New Zealand. New York, pp. 191–194.

Hastie, T., Friedman, J. and Tibshirani, R., 2017. The Elements of Statistical Learning. New York, NY: Springer New York.

Hecker, D., Döbel, I., Petersen, U., Rauschert, A., Schmitz, V. and Voss, A., 2017. Zukunftsmarkt künstliche Intelligenz. Potenziale und Anwendungen. [online]. Available at: <http://publica.fraunhofer.de/eprints/urn_nbn_de_0011-n-4976615.pdf> [Accessed 16 April 2019].

Hiromoto, R. E., Haney, M. and Vakanski, A., 2017. A Secure Architecture for IoT with Supply Chain Risk Management. Romania.

Hogenboom, A., Ketter, W., van Dalen, J., Kaymak, U., Collins, J. and Gupta, A., 2009. Product Pricing using Adaptive Real-Time Probability of Acceptance Estimations based on Economic Regimes. New York, NY: ACM.

Hompel, M. ten and Wolf, O., eds., 2013. IT in der Logistik 2013/2014. Marktübersicht & Funktionsumfang: Enterprise-Resource.Planning, Warehouse-Management, Transport-Management & Supply-Chain-Management-Systeme. Stuttgart.

Inprasit, T. and Tanachutiwat, S., 2018. Reordering Point Determination Using Machine Learning Technique for Inventory Management. In: . 2018. "Exploring innovative solutions for smart society". Piscataway, NJ: IEEE, pp. 1–4.

James, G., Witten, D., Hastie, T. and Tibshirani, R., 2013. An Introduction to Statistical Learning. New York, NY: Springer New York.

JDA Software, 2019. Erfolgreicher Einsatz von Bedarfsprognosen bei dm. Wie künstliche Intelligenz bei der Drogeriemarktkette Mehrwert schafft. [online] [Accessed 26 April 2019].

Kao, A., Niraula, N. B. and Whyatt, D., 2018. PANDA - Discovering Part Name in Noisy Text Data. In: . 2018. 2018 IEEE International Conference on Prognostics and Health Management (ICPHM). Piscataway, NJ, pp. 1–6.

Keller, T., Thiesse, F., Kungl, J. and Fleisch, E., 2010. Using Low-Level Reader Data to Detect False-Positive RFID Tag Reads. In: . 2010. Internet of things (IOT), 2010. Papers presented at the Second International Internet of Things Conference ; Tokyo, Japan, 29 November - 1 December 2010. Piscataway, NJ: IEEE, pp. 1–8.

Ketter, W., Collins, J., Gini, M., Gupta, A. and Schrater, P., 2009. Detecting and forecasting economic regimes in multi-agent automated exchanges. Decision Support Systems, 47(4), pp. 307–318.

Kiekintveld, C., Miller, J., Jordan, P. R., Callender, L. F. and Wellman, M. P., 2009. Forecasting market prices in a supply chain game. Electronic Commerce Research and Applications, 8(2), pp. 63–77.

Kirste, M. and Schürholz, M., 2019. Entwicklungswege zur KI. In: V. Wittpahl, ed. 2019. Künstliche Intelligenz. Entwicklungswege zur KI. Berlin, Heidelberg: Springer Berlin Heidelberg, pp. 25–26.

Kong, Z.-j. and Xue, J.-B., 2013. A comparative study of supplier selection based on support vector machine and RBF neural networks.

Kuhn, A. and Hellingrath, B., 2002. Supply Chain Management. Optimierte Zusammenarbeit in der Wertschöpfungskette. [e-book]. Berlin, Heidelberg, s.l.: Springer Berlin Heidelberg. <http://dx.doi.org/10.1007/978-3-662-10141-4>.

Lau, R. Y. K., Zhang, W. and Xu, W., 2018. Parallel Aspect-Oriented Sentiment Analysis for Sales Forecasting with Big Data. Production and Operations Management, [e-journal] 27(10), pp. 1775–1794. http://dx.doi.org/10.1111/poms.12737.

Lee, Y. and Sikora, R., 2016. Design of Intelligent Agents for Supply Chain Management. In: V. Sugumaran, V. Yoon, and M. J. Shaw, eds. 2016. E-Life: web-enabled convergence of commerce, work, and social life. 15th Workshop on e-Business, WEB 2015 ; Fort Worth, Texas, USA, December 12, 2015 ; Revised Selected Papers: Springer International Publishing, pp. 27–39.

Lingitz, L., Gallina, V., Ansari, F., Gyulai, D., Pfeiffer, A., Sihn, W. and Monostori, L., 2018. Lead time prediction using machine learning algorithms: A case study by a semiconductor manufacturer. Procedia CIRP, [e-journal] 72, pp. 1051–1056. http://dx.doi.org/10.1016/j.procir.2018.03.148.

Marsland, S., 2015. Machine learning. An algorithmic perspective. Boca Raton, FL.

Mohri, M., Rostamizadeh, A. and Talwalkar, A., 2012. Foundations of machine learning. [e-book]. Cambridge MA. <https://pdfs.semanticscholar.org/e923/9469aba4bccf3e36d1c27894721e8dbefc44.pdf>.

Mokhtarinejad, M., Ahmadi, A., Karimi, B. and Rahmati, S. H. A., 2015. A novel learning based approach for a new integrated location-routing and scheduling problem within cross-docking considering direct shipment. Applied Soft Computing, pp. 274–285.

Mori, J., Kajikawa, Y., Sakata, I. and Kashima, H., 2010. Predicting customer-supplier relationships using network-based features. In: . 2010. IEEE International Conference on Industrial Engineering and Engineering Management (IEEM), 2010. Piscataway, NJ, pp. 1916–1920.

Piatetsky, G., 2014. CRISP-DM, still the top methodology for analytics, data mining, or data science projects. <https://www.kdnuggets.com/2014/10/crisp-dm-top-methodology-analytics-data-mining-data-science-projects.html>.

Priore, P., Ponte, B., Rosillo, R. and La Fuente, D. de, 2018. Applying machine learning to the dynamic selection of replenishment policies in fast-changing supply chain environments. International Journal of Production Research, 27(11).

Russell, S. J. and Norvig, P., 2016. Artificial intelligence. A modern approach. Boston, Columbus, Indianapolis.

Sarhani, M. and El Afia, A., 2014. Intelligent system based support vector regression for supply chain demand forecasting. In: . 2014. 2014 Second World Conference on Complex Systems (WCCS): IEEE, pp. 79–83.

Seif, G., 2018. The 5 Clustering Algorithms Data Scientists Need to Know. [online]. Available at: <https://www.kdnuggets.com/2018/06/5-clustering-algorithms-data-scientists-need-know.html>.

Shearer, C., 2000. The CRISP-DM model: the new blueprint for data mining. J Data Warehouse, 5.

Souichirou, N., 2015. Application of Big Data Technology in Support of Food Manufacturers' Commodity Demand Forecasting. NEC Technical Journal, (10).

Steinbach, P., 2018. Wo noch nie eine CPU gewesen ist. In: . 2018. iX Developer 2018 - Machine Learning: Verstehen, verwenden, verifizieren, pp. 32–37.

Sun, R. and Zhao, G., 2012. Analyses about efficiency of reinforcement learning to supply chain ordering management. [online].

The MathWorks, I., 2016. Introduction to Machine Learning. [online] [Accessed 30 April 2019].

Tirkel, I., 2011. Cycle time prediction in wafer fabrication line by applying data mining methods. In: S. Lantz, ed. 2011. 2011 22nd Annual IEEE/SEMI Advanced Semiconductor Manufacturing Conference (ASMC). Piscataway, NJ, pp. 1–5.

Waller, M. A. and Fawcett, S. E., 2013. Data Science, Predictive Analytics, and Big Data: A Revolution That Will Transform Supply Chain Design and Management. Journal of Business Logistics, (34(2)), pp. 77–84. <https://pdfs.semanticscholar.org/9c1b/9598f82f9ed7d75ef1a9e627496759aa2387.pdf>.

Wang, F., Ng, H. Y. and Ng, E.T., 2018. Novel SKU Classification Approach for Autonomous Inventory Planning. [online] [Accessed 6 May 2019].

Wang, X.-F. and Liu, B., 2009. Study on the method of order optimization and arrangement based on neural network technology. In: . 2009. International Conference on Machine Learning and Cybernetics, 2009. Piscataway, NJ, pp. 2697–2701.

Witten, H. I., Eibe, F. and Hall, A. M., 2011. Data Mining: Practical Machine Learning Tools and Techniques: Elsevier.

Xue, H., Jiang, C., Cai, B. and Yuan, Y., 2018. Research on demand forecasting of retail supply chain emergency logistics based on NRS-GA-SVM. In: . 2018. Proceedings of the 30th Chinese Control and Decision Conference (2018 CCDC). Piscataway, NJ, pp. 3647–3652.

Yang, C.-L. and Sutrisno, H., 2018. Short-Term Sales Forecast of Perishable Goods for Franchise Business. In: . 2018. "Cybernetics in the next decades". Piscataway, NJ, pp. 101–105.

Zage, D., Glass, K. and Colbaugh, R., 2013. Improving supply chain security using big data. In: K. Glass, ed. 2013. IEEE International Conference on Intelligence and Security Informatics (ISI), 2013. Piscataway, NJ, pp. 254–259.

Zhang, H., Cheng, C. B., Wang, H., Yang, D., Liu, J., Nakayama, A., Cen, F. J., Liu, P., Jia, X., Zhang, H., Li, W., Pang, Y., HUANG, D., Xu, H. and uU, H., eds., 2017. Supplier credit evaluation model based on ensemble learning. Riverwood, NSW: Aussino Academic Publishing House.

Zhu, L.-y., Ma, Y.-z. and Zhang, L.-y., 2014. Ensemble model for order priority in make-to-order systems under supply chain environment. In: H. Lan, ed. 2014. International Conference on Management Science & Engineering (ICMSE), 2014. Piscataway, NJ, pp. 321–328

Impact and Beneficiaries of Blockchain in Logistics

Thomas Twenhöven[1] *and Moritz Petersen*[1]

1 – Kühne Logistics University

Purpose: Blockchain in logistics is slowly moving beyond the hype. Against this background, we investigate the current expectations concerning the impact and possible beneficiaries of Blockchain applications in the logistics services industry.

Methodology: We conduct an online survey among logistics professionals to understand their expectations regarding specific use cases, potential issues, and general developments. Specifically, we ask the respondents to evaluate impacts and beneficiaries of three actual Blockchain projects from the logistics domain.

Findings: We find that industry professionals are still optimistic about Blockchain and expect it to make an impact on both communication effectiveness and costs in the industry. However, we also find that the expected impact and beneficiaries strongly depend on the underlying use case.

Originality: While there is a fast-growing body of research on Blockchain in the logistics industry, the specific impacts and beneficiaries of Blockchain usage as they relate to different use cases have received little attention to date.

Keywords: Blockchain, Logistics, Survey, Use Cases

First received: 08.June.2019 **Revised:** 19.June.2019 **Accepted:** 19.June.2019

1 Introduction

As an emerging technology, Blockchain has received an increasing amount of attention in recent years. After being brought into the general consciousness by the cryptocurrency Bitcoin, the technology was initially perceived as a phenomenon in the financial industry. However, a Blockchain is a general-purpose decentralized database with cryptographic assurances about integrity and write access. Monetary systems such as Bitcoin are only one possible application of this structure (Tapscott and Tapscott, 2018). Several sectors could benefit from this technology (e.g., Carson et al., 2018; Gentemann, 2019). One industry in the spotlight of Blockchain considerations is logistics, characterized by a large number of interacting organizations without efficient information exchange processes (Dobrovnik et al., 2018). The Blockchain as a distributed data structure is uniquely suited for this purpose. Consequently, it is widely believed that Blockchain will have a profound transformational effect on the industry as a whole (e.g., Hackius and Petersen, 2017; Treiblmaier, 2018).

Digitally enabled technologies, in general, are increasingly used to improve logistics processes. Additive manufacturing, artificial intelligence, and autonomous driving are just three examples of technologies that promise to change how logistics is done today (Hoberg et al., 2019). Blockchain is also one of these technologies, raising hope and fear at the same time. After a period of inflated expectations with little substance, the industry is slowly figuring out how Blockchain could be utilized for its logistics processes (e.g., Dobrovnik et al., 2018; Groenfeldt, 2017; Petersen et al., 2018). Also, companies increasingly scrutinize the actual value Blockchain can bring beyond a technology savviness signaling effect. To further illuminate the value proposition of Blockchain for logistics, we seek to provide insights into the impact that Blockchain might have on the logistics process as a whole and the role of different actors. Two research questions emerge from this perspective: (1) “Which positive impacts might Blockchain in logistics entail?" and (2) "Who might benefit from Blockchain in logistics?" The subsequent sections of this paper have the following structure: We begin with a theoretical analysis of Blockchain with a focus on logistics. Following this, we introduce three projects as examples of the emerging Blockchain logistics industry. Then, we present the setup and results of an expert survey that we conducted among industry professionals. We discuss the results in the subsequent section, followed by limitations and further research, and a conclusion.

2 Theoretical Background

In this section, we introduce the theoretical underpinnings of Blockchain and provide a short overview of recent research in the emerging subfield of Blockchain application for the benefit of logistics.

A Blockchain is a decentralized, verified, and immutable ledger for the storage of information (Figure 1). To implement these key characteristics, the data is fully replicated across all nodes in a peer-to-peer network. By storing this data itself, each node can verify that the rest of the network is storing the data correctly and not tampering with the database. Moreover, all transactions must be signed with a private key cryptographically linked to the sending account. This allows the access management to be verified across the network as well. Through a consensus algorithm, the nodes can then establish a common understanding of the current state of the database, making it immutable (Tapscott and Tapscott, 2018).

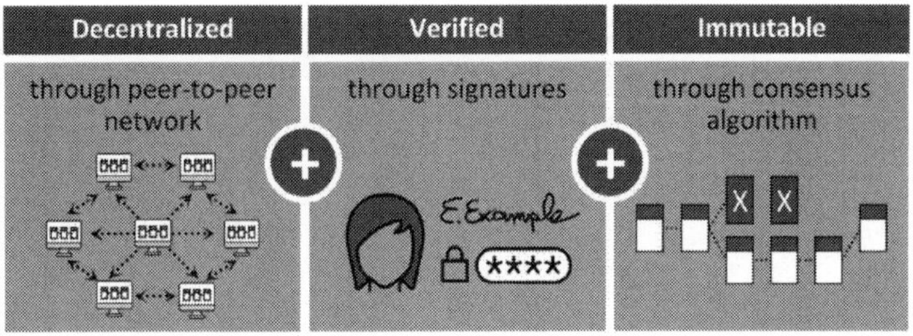

Figure 1: Characteristics of Blockchain (Hackius and Petersen, 2017)

What this mechanism allows for is the creation of a database without the central authority that is usually required to run that database. The most obvious consequence would be the development of databases that store information important to multiple stakeholders in a business process. However, the previous requirement for a central database operator has traditionally enabled the business models of intermediaries and other third parties seeking to insert themselves between business partners and extract revenue from their role in the process. Removing the need for these intermediaries potentially has dire consequences for their business models and potentially positive implications for the industry as a whole (Avital et al., 2016). This close link to the structure of the industry also means that any analysis of the implications of Blockchain must consider the affected industry and not just general characteristics of the technology (Nowiński and Kozma, 2017).

A key characteristic of the distributed database is that all nodes in the network have access to all transaction data. This is a fundamental aspect of Blockchain because it is necessary for them to verify the correctness of the transactions. It also enables the nodes to conduct data analysis to draw conclusions on the real-world business associated with the stored data (Kalodner et al., 2017). Therefore, the Blockchain poses challenges regarding the privacy of people associated with the data and regarding confidential internal information by the involved organizations.

Regarding the state of research on Blockchain in logistics, we survey some of the existing literature in the following. Note that most of these contributions discuss Blockchain in both logistics and supply chain management, suggesting that the fields are closely related. Since our focus on Blockchain in logistics is indeed on its impact on the whole supply chain, such papers are well-suited to shed light on our research area. In an early practitioner survey, Hackius & Petersen (2017) found that while high hopes are on Blockchain in logistics and supply chain, scientific publications on the subject are still scarce. This no longer holds. In their more recent literature review, Queiroz et al. (2019) and Wang et al. (2019) found an increasing number of relevant publications. However, despite this increase, the number of peer-reviewed publications is found to be still below 30 papers in total, possibly illustrating a misfit of long review cycles and fast technological developments.

Wang et al. (2019) distinguish between enhancements of transparency and associated characteristics and the associated gains in efficiency. The first aspects relate to tracking information on the Blockchain. This way, the information is authentic and can no longer be altered, and since it is visible to all participants in the network, it makes the process transparent. This is useful for information about the contracting, the state of the transport process itself (which would essentially replace Bills of Lading and other forms of paper-based documentation), and other metrics about the freight such as location information or temperature (Poszler et al., 2019). The immutability also means that unauthorized movement of freight (such as theft) cannot be represented on the Blockchain, potentially easing the recovery of stolen merchandise (Queiroz et al., 2019). Utilizing this information can help increase the efficiency of the overall network. Since the information is available in digital form, it can be processed automatically and the paper form is no longer required. This also means it is instantly available to all interested parties without handling physical documents (Wang, Singgih, et al., 2019). Dobrovnik et al. (2018) propose that a critical mass of users is necessary to make applications viable through the network effect. Thus, coordination between many parties is required. However, they focus on diffusion within the industry without considering external parties, such as the customers of the industry. Wang et al. (2019) follow a similar approach to analyze the diffusion of Blockchain in Supply Chain Management. Gallay et al. (2017) additionally highlight the importance of decentralization – if a central party has influence over or even control of the network, other organizations are unlikely to adopt it.

Overall, the focus of the Blockchain logistics literature is set to improving efficiency. Thus, it is in stark contrast to publications like Flint et al. (2005), who find that customer value drives successful innovation in the field of logistics. Perboli et al. (2018) also propose that Blockchain projects in the industry should focus on understanding business processes to be able to deliver customer value.

3 Industry Analysis

In this section, we present industry projects to highlight the potential of Blockchain for logistics. In their analysis, Poszler et al. (2019) determine the value-add of different Blockchain startups in logistics. They list use cases, partners, revenue, and cost structure of the startups, illustrating the breadth of the industry. Our overview also includes projects undertaken by established companies. The Blockchain in logistics industry consists of a large number of diverse organizations with different backgrounds and approaches (e.g., Bita, 2019). Thus, this section is not intended to be an exhaustive list, but rather a sample of well-known projects selected by the authors. The intent is to capture projects that have already moved beyond the conceptualization stage and have developed real applications and to capture different use cases and implementation mechanisms.

3.1 CargoX

Our first example is CargoX, a Slovenian startup that has developed a Blockchain-based Bill of Lading for containerized ocean freight. The startup is focused on the Bill of Lading but intends to include all other freight documentation in their product as well through an iterative development process. In a trial with just the Bill of Lading, they claim to have reduced processing costs by 85% (Wee, 2018). The startup partners with external logistics companies for the deployment of the technology, but it aims to be a neutral platform for all actors in the market instead of a venture by any particular market participant (such as the TradeLens platform headed by Maersk and IBM). CargoX believes that an open platform is necessary for succeeding in the marketplace (Rajamanickam, 2018).

3.2 dexFreight

dexFreight is a startup building a freight exchange on the Blockchain that allows shippers and carriers to negotiate directly. dexFreight is also looking to provide an open platform and partners with established logistics companies. Trucking companies can store various company-related documents on the platform to speed up the onboarding process of subcontractors. Thus, the focus for the shippers can be on negotiating terms related to the specific freight rather than administrative tasks. The negotiation of specifics such as rates, delivery and pickup times, and other stipulations also occurs on the platform. Ultimately, the entire contracting part of freight transportation is both handled and documented on the Blockchain (Prevost, 2018). dexFreight has recently partnered with CargoX to provide documentation for both contracting and the actual transportation process within one platform and remove paper from the entire process (Rajamanickam, 2019).

3.3 GS1

A specific part of the logistics process is the target of a pilot project for the digitization of pallet notes. Most companies that require pallets for the movement of freight already participate in a pallet-sharing program in which pallets trade flexibly among all involved parties. This requires an accounting structure for the ownership, state, and flow of the pallets. The systems currently rely on paper-based pallet notes, but the project aims to bring Blockchain to this field. Through a smartphone application, the operators can access a Blockchain backend to store information about the transfer of pallets between the different parties. Since there is no single pallet owner who could operate a traditional digital ownership-tracking infrastructure, Blockchain has enabled the digitization of this process (Uhde, 2018). After the trial period, the project partners consider the project to be a success and intend to continue developing this system (Nallinger, 2018).

4 Survey about the Impacts and Beneficiaries of Block-chain in Logistics

The three project exemplars introduced above illustrate that the market for Blockchain in logistics extends across multiple use cases and organizational backgrounds. The projects' impacts and beneficiaries are likely depending on the use case - warranting a more diversified approach to evaluating Blockchain's potential benefits for logistics. We decided to run an expert survey to shed light on these issues. Parts of the survey update the findings from a similar study we ran in 2017 (i.e., Hackius and Petersen, 2017). This section describes the survey's setup and its most relevant findings.

4.1 Setup and Data Collection

We conducted a four-section survey using Typeform. First, we asked about the general industry role of the participant and their general opinion and knowledge of IT in logistics and Blockchain in general. Secondly, we introduced the three projects as described in the previous section, and inquired about the participants' opinion of them. Following, we asked more specific questions about Blockchain as a technology. We then concluded the survey with some more questions about the background of the participants.

For the evaluation of the projects, we adopted the approach proposed by Bienstock et al. (1997) and added dimensions recommended by Qureshi et al. (2008) and Yeung (2006). Hence, we used the criteria reliability, timeliness, cost, sustainability, and information sharing to evaluate how the logistics process might benefit from each of the three projects.

Data collection started April 10th and ran until May 1st, 2019. The study was distributed among respondents of the previous survey, known industry partners, and various social media channels (Twitter, Xing, and Reddit) related to the topic of supply chain and logistics. The survey link was also included in email newsletters by two industry associations in Hamburg and three online bulletin boards. We incentivized participation through small donations. After data collection was concluded, we donated 50 Euro to "Zeit für Zukunft" (Hamburg-based mentoring program) and to "Ingenieure ohne Grenzen" (Berlin-based aid organization) on behalf of our participants. In total, we received 81 qualified responses.

Especially for anonymous internet surveys, it is recommended to identify and exclude potentially careless responses (Meade and Craig, 2012). However, none of the data sets exhibited any unusual answer pattern. All 81 collected data sets were prepared for statistical analysis following the guidelines of Hair et al. (2009). We used the software jamovi for data analysis.

4.2 Findings

In the following, we present the results of our survey. In addition to looking at the descriptive statistics, we ran statistical analyses (ANOVA) to explore differences between respondent groups. Such differences are presented only if they are found to be statistically significant at the 5%-level.

Figure 2 gives an overview of the 81 respondents: A clear plurality works in logistics services, with consulting and manufacturing being the other large groups of respondents. German participants represent the majority, followed by individuals from the USA, India, and a large number of countries with a small number of participants. In total, 70% of the participants are from Europe. As expected, company headcount and company turnover yielded similar overall distributions. A slight majority of respondents work in small organizations and about a quarter in medium-sized companies.

Figure 3 summarizes the results pertaining to the company's stance towards Block-chain. 20% of the respondents indicate to not be involved with Blockchain in any fashion. Together with respondents observing the development from a distance, they represent the majority of companies.

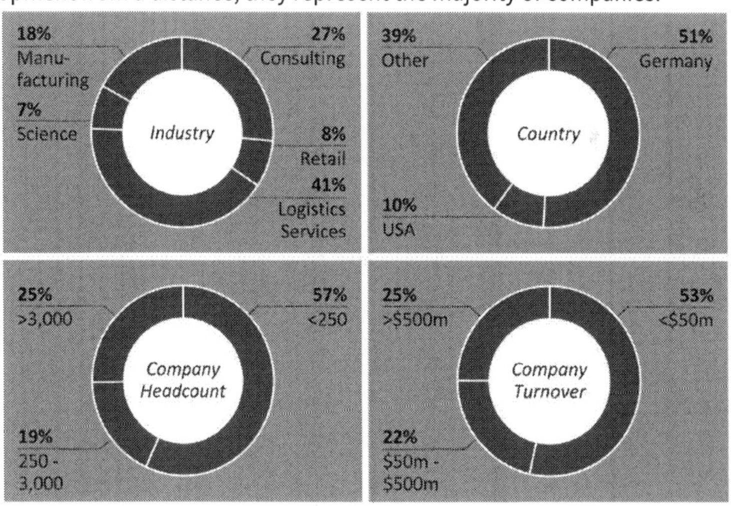

Figure 2: Industry and Organizational Background

Only 14% are actively working with Blockchain, around 30% state to be investigating use cases. Overall, these results indicate a hesitant approach to the technology. We also found that big companies are significantly more likely to be implementing Blockchain solutions than small companies (as measured by company turnover).

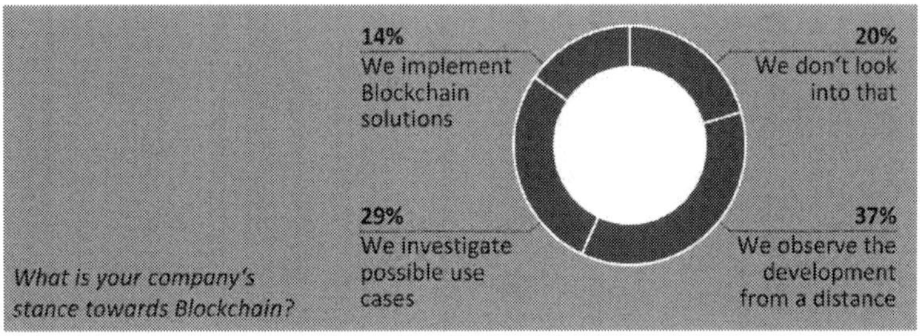

Figure 3: Stance towards Blockchain

As shown in Figure 4, we asked our respondents for their opinion on two statements on Blockchain. They show a tendency to believe that Blockchain has real use cases and offers tangible benefits for business (average rating of 6.68). This is particularly true for the executives among them, who have given an average response of 8.4, a significantly higher value than the one provided by both middle management (5.95) and operational personnel (6.43). Also, we asked about the maturity of the technology and got mixed results (average rating of 5.48). Again, the executives in the sample show a significantly more positive attitude towards the technology (with a score of 7.2) than their fellow respondents (4.80 and 5.22).

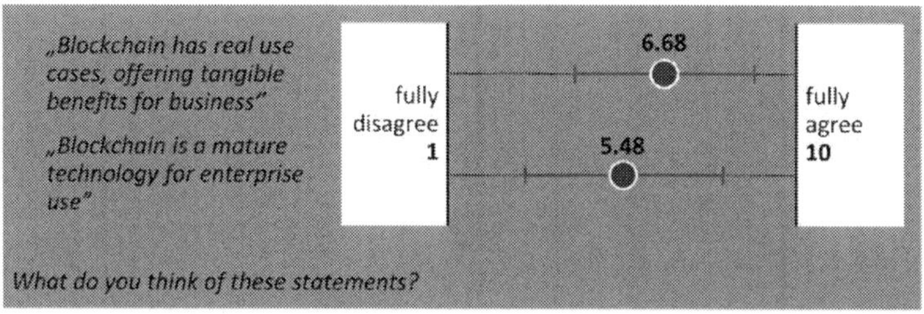

Figure 4: Statements on Blockchain

Respondents working in logistics services have rated the maturity of the technology more than 1.5 points lower than their colleagues in other sectors. For both statements, it has to be kept in mind that the standard deviations are high. Thus, our respondents are far from agreeing on these issues.

Next, we inquired about the potential beneficiaries of Blockchain usage. Figure 5 shows that multiple groups have received a large percentage of votes. 84% of respondents believe that Blockchain would benefit logistics service providers, followed by 77% for technology providers and similar responses of slightly above 60% for both senders and receivers. In the 2017 survey, we asked the same question and got a more positive result for senders and receivers. Thus, the customers received ratings similar to those of logistics service providers and technology providers. This may be indicative of a shift in the perception of Blockchain in the industry. However, the 2017 sample size and structure differ from this survey and, thus, this inter-temporal relationship cannot be established reliably. Still, we consider it noteworthy.

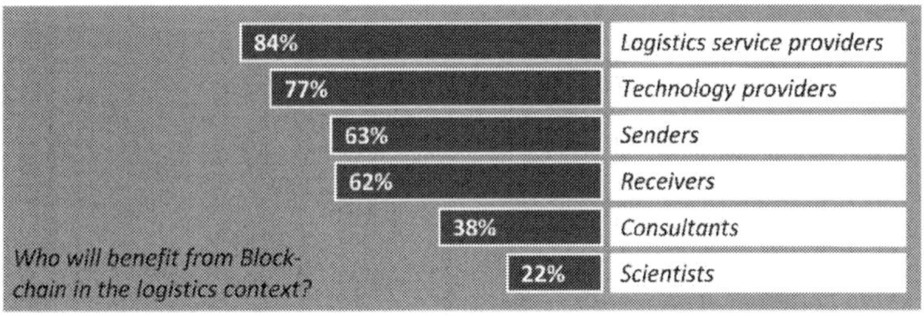

Figure 5: Beneficiaries of Blockchain in Logistics

In the question depicted in Figure 6, we asked how our respondents expect Blockchain to change the logistics industry. All options presented in this multi-option question have been reasonably popular with the respondents, with even the least popular option being selected 41% of the time. The most frequently selected option at 85% was the stronger IT integration with customers, suggesting that customers can also benefit from Blockchain in the logistics context.

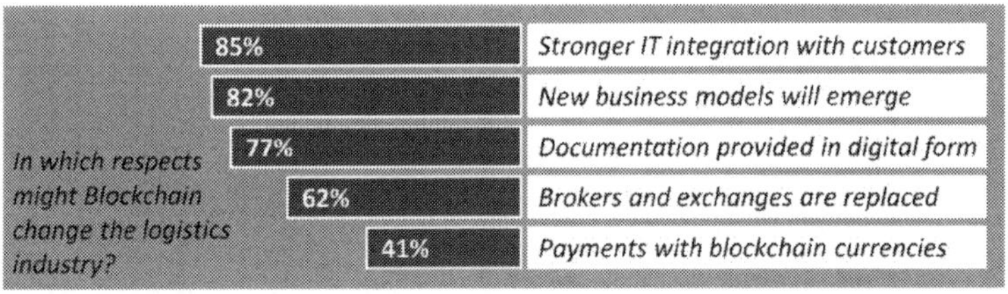

Figure 6: Blockchain-induced Industry Changes

The emergence of new business models came in a close second, followed by digital documentation. The replacement of brokers and exchanges and

Blockchain payments received fewer votes than the leading group of options.

We found that respondents from companies with more Blockchain implementation experience anticipate more different kinds of change than their less experienced counterparts. We measured experience through the responses shown in Figure 3; to achieve this, we created a binary "less experience/more experience" measure by splitting the four response options into two groups of two options each.

To enable a more detailed analysis of the expected impacts and beneficiaries of the technology, we asked participants to evaluate the three projects presented previously (Figure 7). Our participants received a short introduction to each project, based on the information in Chapter 3, to allow them to respond meaningfully.

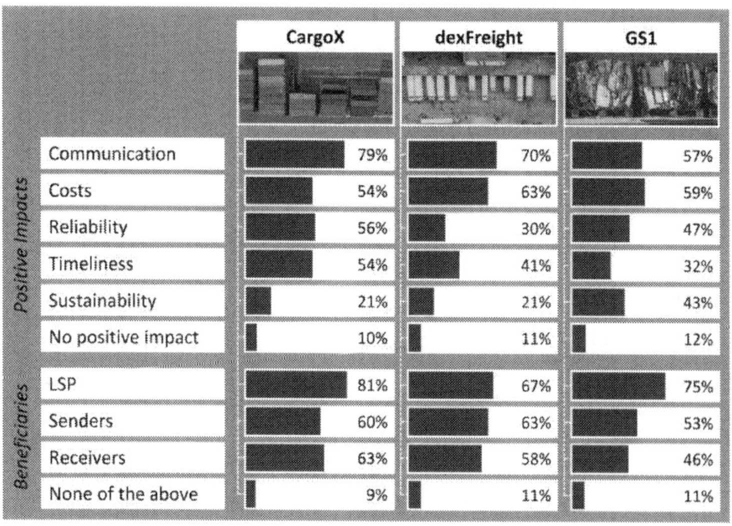

Figure 7: Evaluation of Specific Blockchain Projects

Overall, the results show that Blockchain is expected to make the strongest impact on communication between stakeholders, followed by an impact on cost. For the GS1 project, cost is slightly ahead of communication. The technology is also expected to have an impact on reliability and timeliness. For the CargoX project, the two have received almost the same percentage of votes. However, timeliness leads for dexFreight, and reliability leads for GS1. Sustainability is the least popular potential impact area with just above 20% of the respondents expecting an improvement for CargoX and dexFreight. However, this figure jumps to 43% for GS1. In general, the evaluation of GS1 has yielded different results than the evaluation of CargoX and dexFreight. The impact areas of the latter two are similar in their order and overall values (except for a large difference in reliability impact), and the large spread between the impact areas. For GS1, costs rank ahead of communication due to a big drop in communication impact, and sustainability beats timeliness due to a significant increase in sustainability impact. Additionally, the spread between the biggest and the smallest impact areas is much smaller.

Regarding the beneficiaries of these projects, the logistics service providers themselves rank ahead of their customers (senders and receivers). However, there is a difference between the projects – CargoX benefits the logistics service providers significantly more than dexFreight. Overall, senders and receivers have received similar responses, with senders ahead of receivers for dexFreight and GS1, but a slight lead for the receivers with CargoX. Both have received values between 60% and 70% for CargoX and dexFreight, but only around 50% for GS1. For both CargoX and GS1, the logistics service providers rank about 20 percentage points ahead of their customers as potential beneficiaries. Only the dexFreight project is expected to benefit the customers on a similar (but still lower) level. Overall, the technology is expected to have a strong impact on all participants, with only one value dropping slightly below 50%, and most far above.

The beneficiaries of the CargoX project have received very similar percentages to the beneficiaries of the overall technology (Figure 5), while both dexFreight and GS1 deviate from the result. This might suggest that the value proposition of CargoX is the use case most commonly associated with the technology, while the other projects implement solutions that are not as well-known to the respondents.

The specific use case associated with each project is paramount when comparing the results between the projects. It is noticeable that while senders and receivers are expected to benefit similarly from CargoX and dexFreight, the freight exchange has scored worse for the logistics service providers themselves. One may expect that easier negotiations between customers and logistics service providers would make it easier for customers to receive multiple offers for each item of freight. Thus, competition between logistics service providers might increase, ultimately hurting their ability to maintain profit margins. For the GS1 project, logistics service providers receive a higher result again, but senders and receivers fare lower. This would be expected for innovation in the logistics process itself, but the sender usually provides the pallets along with the freight. One may speculate that respondents wrongly assumed the pallet notes to be a logistics process innovation not relevant to the customers.

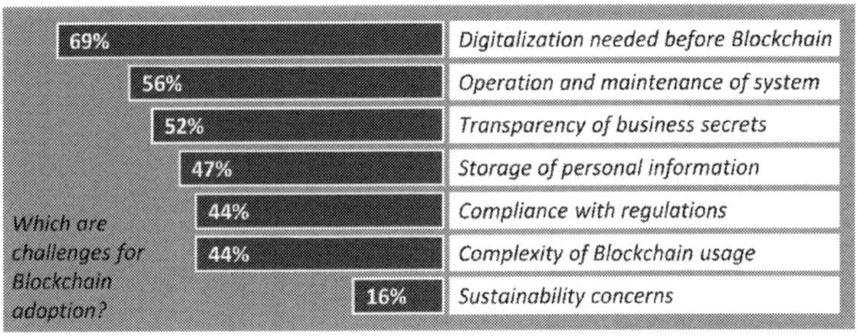

Figure 8: Challenges for Blockchain Adoption

Finally, regarding barriers to the adoption of Blockchain, 69% of respondents believe that digitalization is necessary before Blockchain can be introduced, as shown in Figure 8. This option has received the most votes by a large margin. This means that most respondents agree that many companies still have to do their homework concerning digitalization - a fact that can act as a decisive roadblock for any digital technology. The three options of storage of personal information, usage complexity, and regulatory compliance have received almost the same amount of votes at close to 50%. Sustainability, however, has merely received 16%, suggesting that such concerns are not perceived as a significant barrier to adoption. Similar to what has been found for the impacts of the technology, more experienced respondents anticipate significantly fewer different challenges than their less experienced counterparts.

5 Discussion and Implications

Our study reveals that most respondents and their companies are still cautious about implementing Blockchain even though they expect many benefits in different areas of the logistics industry. Overall, general questions about the technology have received more positive responses than those asking about the current state. This may indicate that the technology has not yet reached its expected potential.

Remarkably, respondents working in companies with more Blockchain experience tend to see more positive impact areas and fewer challenges. One may expect that more experienced companies possess better information about the technology than their less experienced peers. Interestingly, this may point to a positive relationship between past decisions to conduct Blockchain projects (which we use to measure technology experience) and future Blockchain projects (which we expect to be based on the general attitude towards the technology by people within the company). This may lead to the emergence of a divide between companies which keep on adding to their experience with the technology, and other companies stagnating at a low level of experience.

While this is an assumption, it highlights that companies should seek to develop an understanding of Blockchain and use it according to their findings, not based on general skepticism towards an unknown technology. This is an important finding from a managerial perspective. However, the positive attitude by executives suggests that management is well-equipped to utilize this effect.

Due to its background in cryptocurrencies such as Bitcoin, these currencies are still a major theme in the industry. Our study confirms that this holds for the logistics industry as well, with 41% of respondents expecting cryptocurrencies to be used for payments.

Regarding problems for Blockchain, the biggest one appears to be the lack of digitalization in the industry. This is a positive result for Blockchain, but it also presents a valuable lesson to both technology providers and logistics services providers – Blockchain projects cannot just move digitized assets to a different kind of database, but they must digitize the physical world first. Therefore, they have to understand the real-world logistics processes, interactions, and paperwork in order to be successful, as previous research has suggested. However, technology-related problems also rank high, highlighting the lack of maturity.

This may be viewed as a statement regarding the difference between startups and Blockchain projects started within the logistics industry. Indeed, of our three example projects, the two startups were rated differently compared to the industry project. Overall, the number of selected impact areas is similar, but for the industry project, the spread between the different impact areas is much smaller. However, at a sample size of three, we cannot make any reliable statement comparing startups and projects by established companies.

One key stakeholder of the logistics process is the customer that benefits from the freight being moved. Our results show that the logistics companies themselves, rather than their customers, are seen as the primary beneficiaries in Blockchain projects. The technology providers as the secondary beneficiary also rank ahead of senders and receivers. However, the logistics process as a whole is ultimately undertaken for the benefit of senders and receivers of freight. A technology that primarily benefits the logistics service providers and technology providers might not be a good fit for such a market. As a technology is expected to benefit mostly the logistics service providers, Blockchain may generate a positive impact on their customers through cost savings, but this impact area only ranks second. The primary impact area is customer communication. It is surprising to see that this impact area ranks higher than the customer as a beneficiary, highlighting that customer communication does benefit the logistics service providers as well as the customer.

6 Limitations and Further Research

The survey underlying this publication is subjected to some limitations. The sample is relatively small, not demographically representative, and concentrated in the geographic area of Germany. Additionally, we expect our sample to be biased towards those who actively network in social media channels and those who are part of the industry groups contacted by the authors. These limitations represent a starting point for future research, as others may conduct similar surveys within other subgroups.

Regarding the differences between the three Blockchain projects, a larger sample of projects would be necessary to conclude the differences between the associated company structures or use cases. Further research may also focus on the disruption of the established industry by new players, which is a common expectation in our study. Our conjecture of a divide between users and non-users may also be a starting point for future research.

7 Conclusion

In this contribution, we present the results of our study regarding the state of Block-chain in logistics and its impact on the supply chain. We find that the technology may have a large impact on the industry in the future, but for now, companies are rather cautious about implementations. We find that those who have implementation experience it are significantly more optimistic about it – a cautious approach may be associated with unwarranted caution in the future. We also find that a lack of digitalization is the biggest hurdle for Blockchain adoption in the industry, ahead of any technical problems related to Blockchain technology itself.

While our research shows that Blockchain in logistics is expected to have the strongest impact on logistics services itself, its customers (i.e., senders and receivers of freight) have also received strong results regarding a potential impact. In terms of impact areas, communication and costs dominate. Further research is required to understand the precise nature of this impact on the logistics process and the value chain as a whole, and the differences between startups and projects by established companies. While our results in this area are ultimately inconclusive, they suggest that differences exist.

Financial Disclosure

The research reported in this paper was carried out as part of the HANSE-BLOC project, funded by the German Federal Ministry of Education and Research under the funding code 03VNE2044J. Responsibility for the content of this publication is with the authors.

References

Avital, M., Beck, R., King, J., Rossi, M. and Teigland, R., 2016. Jumping on the Blockchain Bandwagon: Lessons of the Past and Outlook to the Future. The 37th International Conference on Information Systems ICIS 2016, AIS Electronic Library, Dublin.

Bienstock, C.C., Mentzer, J.T. and Bird, M.M., 1997. Measuring Physical Distribution Service Quality. Journal of the Academy of Marketing Science, Vol. 25 No. 1, pp. 25–31.

Bita., 2019. Members of BiTA. BiTA: Blockchain in Transport Alliance, available at: https://www.bita.studio/members (accessed 6 June 2019).

Carson, B., Romanelli, G., Walsh, P. and Zhumaev, A., 2018. Blockchain Beyond the Hype: What is the Strategic Business Value? June, available at: https://www.mckinsey.com/business-functions/digital-mckinsey/our-insights/blockchain-beyond-the-hype-what-is-the-strategic-business-value (accessed 24 May 2019).

Dobrovnik, M., Herold, D.M., Fürst, E. and Kummer, S., 2018. Blockchain for and in Logistics: What to Adopt and Where to Start. Logistics, Vol. 2 No. 3.

Flint, D.J., Larsson, E., Gammelgaard, B. and Mentzer, J.T., 2005. Logistics Innovation: A Customer Value-oriented Social Process. Journal of Business Logistics, Vol. 26 No. 1, pp. 113–147.

Gallay, O., Korpela, K., Tapio, N. and Nurminen, J.K., 2017. A Peer-to-Peer Platform for Decentralized Logistics. in Kersten, W., Blecker, T. and Ringle, C.M. (Eds.), Digitalization in Supply Chain Management and Logistics, epubli, Hamburg, pp. 19–34.

Gentemann, L., 2019. Blockchain in Deutschland – Einsatz, Potenziale, Herausforderungen. Bitkom e.V., available at: https://www.bitkom.org/sites/default/files/2019-04/190423_bitkom_studie_blockchain_2019.pdf (accessed 13 May 2019).

Groenfeldt, T., 2017. IBM And Maersk Apply Blockchain To Container Shipping. Forbes.Com, March, available at: https://www.forbes.com/sites/tomgroenfeldt/2017/03/05/ibm-and-maersk-apply-blockchain-to-container-shipping/#7d43507a3f05 (accessed 8 February 2019).

Hackius, N. and Petersen, M., 2017. Blockchain in Logistics and Supply Chain: Trick or Treat?. in Kersten, W., Blecker, T. and Ringle, C.M. (Eds.), Digitalization in Supply Chain Management and Logistics, epubli, Hamburg, pp. 3–18.

Hair, J.F., Black, W.C., Babin, B.J. and Anderson, R.E., 2009), Multivariate Data Analysis, Prentice Hall, Upper Saddle River.

Hoberg, K., Petersen, M. and Heinen, J., 2019. Die Implikationen digitaler Technologien für die Supply Chain 4.0. In Obermaier, R. (Ed.), Handbuch Industrie 4.0 Und Digitale Transformation: Betriebswirtschaftliche, Technische Und Rechtliche Herausforderungen, Springer Gabler, Wiesbaden.

Kalodner, H., Goldfeder, S., Chator, A., Möser, M. and Narayanan, A., 2017. BlockSci: Design and Applications of a Blockchain Analysis Platform. ArXiv.

Meade, A.W. and Craig, S.B., 2012. Identifying Careless Responses in Survey Data. Psychological Methods, Vol. 17 No. 3, pp. 437–455.

Nallinger, C., 2018. Hat der Palettenschein ausgedient? 17 December, available at: https://www.eurotransport.de/artikel/palettentausch-mithilfe-der-blockchain-hat-der-palettenschein-ausgedient-10634469.html (accessed 3 April 2019).

Nowiński, W. and Kozma, M., 2017. How can Blockchain Technology Disrupt the Existing Business Models?. Entrepreneurial Business and Economics Review, Vol. 5 No. 3, pp. 173–188.

Perboli, G., Musso, S. and Rosano, M., 2018. Blockchain in Logistics and Supply Chain: A Lean Approach for Designing Real-world Use Cases. IEEE Access, Vol. 6, pp. 62018–62028.

Petersen, M., Hackius, N. and von See, B., 2018. Mapping the Sea of Opportunities: Blockchain in Supply Chain and Logistics. It - Information Technology, Vol. 60 No. 5–6, pp. 263–271.

Poszler, F., Ritter, A.-C. and Welpe, I., 2019. Blockchain Startups in the Logistics Industry: The Technology's Potential to Disrupt Business Models and Supply Chains. In Schröder, M. and Wegner, K. (Eds.), Logistik Im Wandel Der Zeit, Springer, Wiesbaden, pp. 567–584.

Prevost, C., 2018. dexFreight Completes First Truckload Shipment Using Blockchain. 24 October, available at: https://www.freightwaves.com/news/blockchain/technology/dexfreight-completes-first-truckload-shipment-using-blockchain (accessed 25 March 2019).

Queiroz, M.M., Telles, R. and Bonilla, S.H., 2019. Blockchain and Supply Chain Management Integration: A Systematic Review of the Literature. Supply Chain Management: An International Journal.

Qureshi, M., Kumar, D. and Kumar, P., 2008. An Integrated Model to Identify and Classify the Key Criteria and their Role in the Assessment of 3PL Services Providers. Asia Pacific Journal of Marketing and Logistics, Vol. 20 No. 2, pp. 227–249.

Rajamanickam, V., 2018. CargoX Introduces an Open and Neutral Blockchain Platform for the Maritime Industry. November, available at: https://www.freightwaves.com/news/technology/cargox-blockchain-smart-open-neutral-platform (accessed 22 March 2019).

Rajamanickam, V., 2019. CargoX's Blockchain-based Smart Bill of Lading Solution is now on dexFreight's Platform. March, available at: https://www.freightwaves.com/news/blockchain/cargoxs-blockchain-based-smart-bill-of-lading-solution-is-now-on-dexfreights-platform (accessed 13 May 2019).

Tapscott, D. and Tapscott, A., 2018), Blockchain Revolution: How the Technology behind Bitcoin and Other Cryptocurrencies Is Changing the World, Penguin, New York et al.

Treiblmaier, H., 2018. The Impact of the Blockchain on the Supply Chain: A Theory-Based Research Framework and a Call for Action. Supply Chain Management: An International Journal, Vol. 23 No. 6, pp. 545–559.

Uhde, T., 2018. Blockchain-Technologie und Architektur im Pilot-Projekt. December, available at: https://www.gs1-germany.de/innovation/blockchain-blog/das-passende-system-blockchain-technologie-und-architek/ (accessed 25 March 2019).

Wang, Y., Han, J.H. and Beynon-Davies, P., 2019. Understanding Blockchain Technology for Future Supply Chains: A Systematic Literature Review and Research Agenda. Supply Chain Management: An International Journal, Vol. 24 No. 1, pp. 62–84.

Wang, Y., Singgih, M., Wang, J. and Rit, M., 2019. Making Sense of Blockchain Technology: How Will it Transform Supply Chains? International Journal of Production Economics, Vol. 211, pp. 221–236.

Wee, V., 2018. CargoX Runs First Successful Electronic Bill of Lading Shipment. August, available at: http://www.seatrade-maritime.com/news/europe/cargox-runs-first-successful-electronic-bill-of-lading-shipment.html (accessed 22 March 2019).

Yeung, A.C., 2006. The Impact of Third-party Logistics Performance on the Logistics and Export Performance of Users: An Empirical Study. Maritime Economics & Logistics, Vol. 8 No. 2, pp. 121–139.

Prototype for a Permissioned Blockchain in Aircraft MRO

Jakob Schyga[1], Johannes Hinckeldeyn[1] and Jochen Kreutzfeldt[1]

1 – Technische Universität Hamburg

Purpose: The paper aims to show strengths, weaknesses and challenges in the application of Blockchain technology for aircraft maintenance, repair and overhaul (MRO). The analysis is based on a prototype of a permissioned Blockchain for the maintenance documentation of aircraft and their components.

Methodology: A prototyping approach is used to gain a deeper understanding of the underlying problems and proposed solutions of applying Blockchain technology to aircraft MRO. The open source platform 'Hyperledger Fabric' and the toolset 'Hyperledger Composer' are used to develop the permissioned Blockchain application. Based on devised use cases the prototype is validated and evaluated.

Findings: Examining the prototype presents opportunities to increase data credibility and efficiency within a decentralized business network through increased transparency, traceability and the use of Smart Contracts. The Blockchain technology could help address challenges in the aircraft MRO industry such as the use of counterfeit spare parts or costly documentation verifications.

Originality: In-depth descriptions of Blockchain use-cases can rarely be found. These are necessary in order to understand the requirements, strengths and weaknesses of using Blockchain in certain business contexts. The development of a prototype furthermore shows a deep insight in the conceptual design, development and usage of Blockchain applications.

Keywords: Blockchain, Hyperledger, Aircraft, Maintenance

First received: 29.May.2019 **Revised:** 12.June.2019 **Accepted:** 28.June.2019

1 Introduction

Blockchain as a technology and as a trend has drawn a lot of attention since its publication by a Satoshi Nakamoto in 2008 (Satoshi Nakamoto, 2008). Although Blockchain has given rise to a myriad of virtual currencies, summarized under the term cryptocurrencies, it has become more than just another means of payment. Basically, the Blockchain is a shared electronic ledger for digital data records that are managed by a number of participants of a distributed peer-to-peer network without the initial requirement of confidence among the members. This means, the innovation of the Blockchain is that transactions between not fully trusted parties can be carried out securely without the necessity of a central institution (Frigden, et. al., 2017). Smart Contracts use this feature to enable the secure and automatic execution of programcode between transaction parties (Buterin, 2013). Currently it is difficult to estimate how big the impact of this technology will ultimately be. Since missing out on the entry into a disruptive technology can lead to a rapid displacement in the market, it is suggested to include Blockchain technology in strategic considerations (Frigden, et al., 2017). Although a number of studies concerning the application of Blockchain technology exists, such as Boyle, et al. (2018), Herwljer, et al. (2018) or Hackius and Petersen (2017), in-depth descriptions of use cases can rarely be found. These, however, are necessary in order to understand the requirements, strengths, weaknesses and challenges of applying Blockchain Technology in certain business contexts. Therefore, this study aims at contributing to this gap through a detailed description of a particular Blockchain use case.

Aircraft maintenance has certain characteristics, due to which the application of Blockchain is suitable, such as a high complexity of processes, networking of several organizations and the demand for mechanisms to increase the efficiency of communication as well as transparency and credibility of data. Resulting from a lack of those aspects, industry-specific issues range from high costs due to missing maintenance records (Canaday, 2017) to an impairment of aviation safety due to the use of counterfeit spare parts (Locatory, 2012). A range of different companies collaborate for maintenance, repair and overhaul (MRO) services of aircraft, starting from maintenance service job shops (backshops) to original equipment manufacturers (OEM) and public organizations for certification and regulation. The International Air Transport Association (IATA) expects substantial benefits from the application of Blockchain to aircraft MRO processes, but also admits the difficulty of realizing this improvement potential (Goudarzi, et. al., 2018). The communication of a large number of companies between the respective often paper-based systems leads in practice to asynchronous data and a relatively small volume of data made available to each company. Data can be used in MRO, for instance for demand forecasting, risk analysis or process improvement. Large companies, in particular OEMs, which also carry out MRO activities, aggregate most of the data and thus gain a strong competitive advantage over small businesses, with the consequence of monopolizing data power (Elliot, 2018). Blockchain technology could help to overcome these issues by offering solutions for a decentralized, comprehensible and immutable database.

This article presents the application of Blockchain and Smart Contracts in aircraft MRO through the development and evaluation of a prototype for the registration of aircraft and their components in a Blockchain. Maintenance activities and ownership transfers are registered in order to provide a complete and transparent documentation by and for the organizations participating in the network. The next section summarizes the current state of Blockchain, while section 3 provides a short overview on the aircraft MRO industry and describes the conceptual model of this research study. The open source platform ‘Hyperledger Fabric‘ and the toolset ‘Hyperledger Composer‘ are used to develop the permissioned Blockchain application. The implementation of the application in Hyperledger Fabric and its way of functioning are described in section 4. Based on devised use cases the prototype is validated and evaluated in section 5. The evaluation contains a discussion of the strengths and weaknesses of the application, its potential impact in the industry and challenges that need to be overcome. The paper closes with some general remarks and further demand of development and research in the area of Blockchain technology.

2 Current State of Blockchain Technology

The Blockchain technology was developed for the original purpose of enabling value transactions between users of a decentralized network securely, directly and trustless and therefore without the need of an intermediary. For the first time the technology was described 2008 in the Bitcoin white paper (Satoshi Nakamoto, 2008) authored by a person or a group under the pseudonym 'Satoshi Nakamoto'. The identity behind the pseudonym has been unknown to the public, in spite of multiple speculations. This original type of Blockchain technology for money transactions is sometimes called first generation of Blockchains (Cummings, 2019). The Blockchain technology is not a monolithic invention as such, but rather a meta-technology combining a number of different technologies, which were already invented and known at the time the Bitcoin whitepaper was published. It integrates previously known technological components such as peer-to-peer technology, asymmetric cryptography, digital signature, Merkle-trees and a consensus algorithm such as proof-of-work into a common concept (Narayanan and Clark, 2017). Figure 1 shows an abstract of a generic Blockchain data structure. The Blockchain data structure links transactions between two parties efficiently and in a verifiable way (Iansiti and Lakhani, 2017). Transactions are linked by so-called hash references inside of a block. The blocks additionally contain a timestamp with the time of block creation, its own cryptographic hash value (Block ID) and the Block ID of the previous block, therefore building a chain of blocks. The transactions and blocks are created and validated by the nodes communicating through a peer-to-peer network (Drescher, 2017). In Bitcoin technology the proof-of-work algorithm is used to build a common consensus. If the proof-of-work algorithm is applied, a block contains additionally the difficulty for block creation and a random number called nonce. Before Bitcoin, it has not been possible to conduct transactions between distributed individuals without a verifying central control instance, such as a central bank. Today, also other consensus algorithms are used such as proof-of-stake or practical byzantine fault tolerance (Hinckeldeyn, 2019).

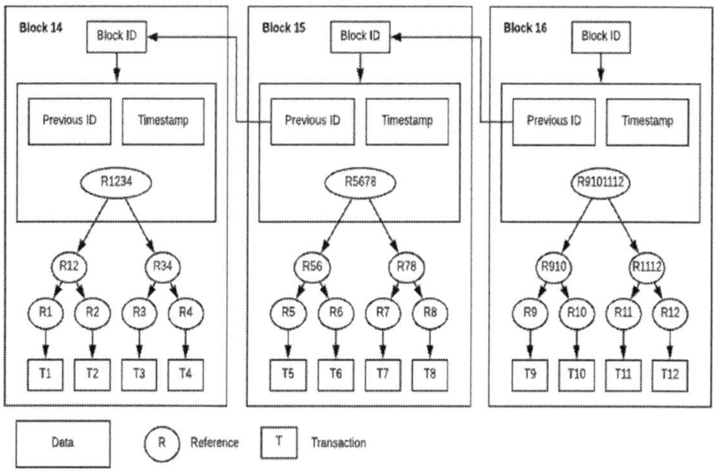

Figure 1: Abstract from a Blochain data structure

Since the creation of Bitcoin, Blockchain technology has evolved. A second generation of Blockchain emerged with partially extended functionality. In 2013 Vitalik Buterin proposed Ethereum, which expanded the Bitcoin technology by the functionality of Smart Contracts (Buterin, 2013). These contracts were proposed already in 1996 by Nick Szabo (1996) but were first made possible by Ethereum in 2015 as software scripts, running on a Blockchain (Tual, 2015). This means that the transactions run a computer program with the same certainty and trust like a Bitcoin transaction. Hence, it allows the automatic execution of programmed agreements between distributed transaction parties within the Blockchain. The innovation of Smart Contracts created a large number of use cases that go far beyond cryptocurrencies. Various companies in industry-specific consortia develop use

cases and solutions, suitable to be implemented using Blockchain technology (Frauenhofer-Gesellschaft, 2017, Chamber of Digital Commerce, 2016). Bitcoin and Ethereum are public and permissionless Blockchain networks, which means that anyone can join the network with more than one account, view its information and use its functions. However, for the business contexts there are usually requirements to comply with confidentiality of data, identification of the participants and network performance.
This led to a third generation of Blockchain technology, which integrates better the requirements and systems of existing companies. In contrast to public Blockchains, these platforms are not necessarily open to everyone and provide other architectures and consensus algorithms for better scalability and data protection. For example, Hyperleger Fabric is a private, permissioned Blockchain platform, developed especially for the use in business networks, offering a modular architecture and a high degree of flexibility, confidentiality and reliability (Hyperledger Fabric, 2019). Several fields have been identified for a suitable application, such as internet of things, energy supply, the insurance sector or the supply chain (Herwljer, et al., 2018. Brandt, et al., 2018, Schlatt et al. 2016).

3 Conceptual Model

Aircrafts are the most complex machines produced in series and therefore represent substantial investment and operational costs for the owner. An aircraft can only make profit while in operation. It is therefore necessary to limit ground time to the bare essentials and to allocate aircraft a tight flight schedule. Since maintenance activities are performed only on ground they

are planned in detail and if possible follow the operational breaks according to the flight schedule. In the event of unplanned downtime, it may be a problem to provide a replacement aircraft at the respective airport. This results in irregularities or failures in flight planning, which entail high follow-up costs. The objective of aircraft MRO is to maintain the airworthiness of aircraft or aircraft components through planned activities of regular maintenance, inspection and overhaul, or to restore them through repair.
As the survival of passengers is directly ensured by the safety of the systems, the processes of aircraft maintenance are strictly regulated and monitored by regulatory authorities such as the European Aviation Safety Agency (EASA). Within the European Union, the requirements and procedures for design, manufacture and maintenance are set by EASA as part of the so-called Implementing Rules. Maintenance operations are defined by the Implementing Rules Continuing Airworthiness (IRCA) in Part 145 (EASA, 2014). The American equivalent to ICAR Part 145 is the Federal Aviation Regulation for Repair Stations (FAR 145). Since the American structures were adopted or used by EASA, the regulations of the USA are very similar to those of the European Union (Hinsch, 2017). Due to the strict regulation of the aviation industry, the business processes differ only slightly. Based on the IRCA, the generic process diagram in Figure 2 is considered in order to illustrate business processes and how they are interconnected between organ-izations involved in aircraft maintenance.

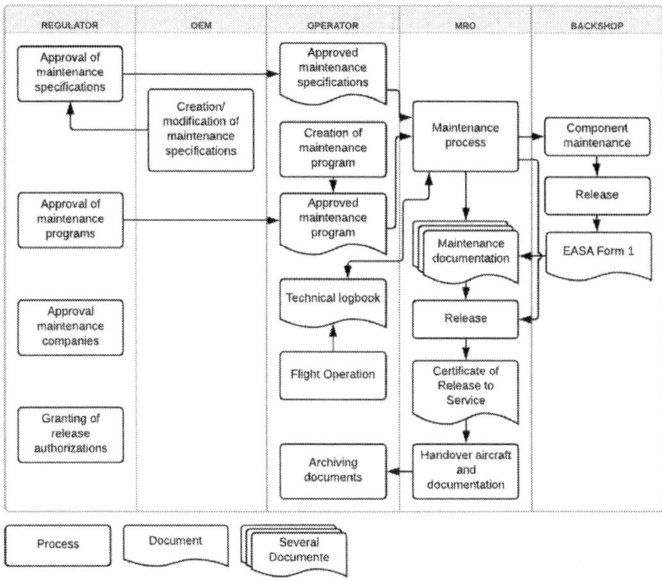

Figure 2: Processes in aircraft MRO

OEMs, MROs and backshops must be certified by the regulatory authority and maintenance activities may only be carried out in accordance with approved maintenance specifications. Planned maintenance activities are scheduled by the operator in a maintenance program, which is also to be approved by the regulatory authority. Furthermore, the regulatory authority grants authorizations for the release of components and aircraft. The OEM prepares maintenance specifications for each aircraft model by describing the maintenance activities and defining the maintenance intervals for the individual components. The aircraft operator is obliged to create a maintenance program based on information provided by the OEM. In the maintenance program, all maintenance activities of aircraft components with their planned intervals are listed. Functional failures called findings as well as their rectification are documented in the technical log book of an aircraft. The aircraft operator is also responsible for archiving the maintenance documents. The maintenance process is triggered either by specifications of the maintenance program or by findings that have occurred. Due to the high complexity of the products various parties are involved in the maintenance of an aircraft. Maintenance companies (MROs) are responsible for the maintenance of the operator's aircraft. In particular, the maintenance of complex components such as engines maintenance is carried out by specialized third-party workshops called backshops. For repaired aircraft and certain components, approval must be granted again by a certifying officer approved by the regulatory authority. As a result of a successful release, the so-called EASA Form 1 will be issued for components and the Certificate of Release to Service (CRS) for aircraft. The aircraft is then handed over to the operator, including all incurred maintenance documents. (Hinsch, 2017, EASA, 2014)

The Blockchain network of the conceptual model is mapping the maintenance network consisting of an arbitrary number of directly or indirectly in the maintenance involved organizations. In the network participating organizations have to be categorized according to their roles in order to enable different permissions. In addition to the regulatory authorities, OEMs, aircraft operators, MROs and backshops, component suppliers are included into the model. Each organization provides one or more participants to the network. The participants are grouped according to their role into Supplier,

OEM, Operator, MRO, Backshops and Regulator. The possibilities of interaction between participants within a maintenance network differ depending on their role. Figure 3 shows the section of the UML class diagram referring directly to the participants.

Aircraft and aircraft components (Parts) are registered in the Blockchain data structure and status changes such as the transfer of ownership, execution of flight operation or conduction of MRO activities are recorded. Due to the properties of the Blockchain technology a ledger of transactions is created containing the full history of the objects conditions without the need for a central instance. The states of the objects, such as the certification can be verified by the immutable history data and filtered through database queries. Methods are completely transparent within the network, but can only be processed by authorized network participants.

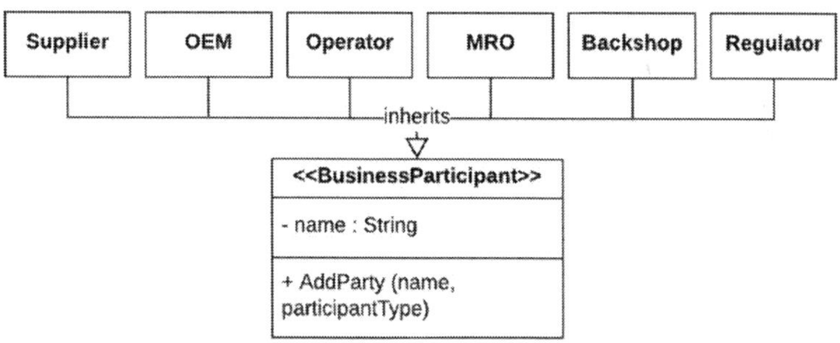

Figure 3: Class diagram participants

In particular, for registering and changing the state of assets, the secure allocation of the physical objects to the digital entries in the Blockchain data structure must be ensured. In order to prevent deceptive action, the responsible person has to digitally sign each transaction. Manipulation at the entry of data cannot be ruled out on the program side even with centralized systems.

The tangible assets Aircraft and Part define two basic levels of the program, while Aircraft are composed of Parts. Figure 4 shows the corresponding section of the class diagram, with the methods and attributes being hidden for reasons of clarity. The classes on the left are assigned to the Aircraft level and the ones on the right to the Part level of the program. Participants are enabled to create and delete, send and receive objects, to change property rights and issue or withdraw airworthiness. Furthermore, the installation and removal of parts in an aircraft can be registered. In order to fulfill the main function of documenting the maintenance history, the classes Finding and Report are defined.

A Finding contains information on damage or anomalies that have occurred to the physical object and must therefore always be assigned to an object in the program, whereby several Findings can be assigned to one object. The attributes store information about the documenting organization and person, the date of occurrence and a description.

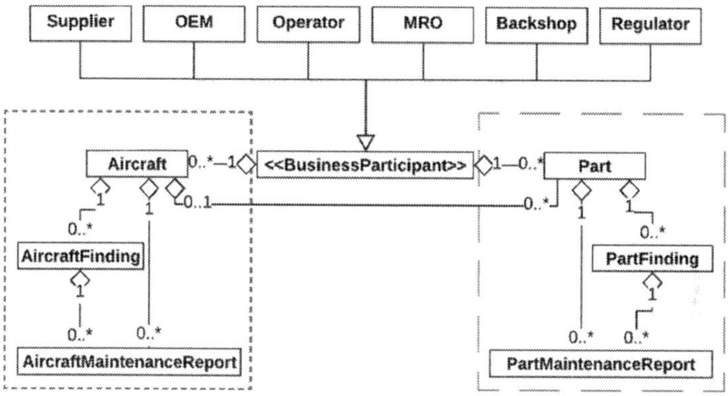

Figure 4: Abstracted class diagram

A Report describes the execution of maintenance activities. Similar to Findings, a Report is always assigned to one tangible object, whereby one object can be assigned to several reports. A Report can be declared as either a routine task or a non-routine task, while in the case of a non-routine task, a Report usually also refers to a Finding. A Report contains the performing organization and person, execution date and a description of tasks. In addition, the mileage of the object at the time of execution is stored.

4 Implementation

The use case of MRO possesses characteristics, which constrain the selection of an appropriate Blockchain framework. The industry-specific requirements for data confidentiality and safety make it particularly important to draw attention to the identification and permission of network participants. In conventional Blockchain technology like Bitcoin or

Ethereum different branches of the Blockchain data structure occur and the proof-of-work algorithm serves to agree on a common version. However, this type of consensus algorithm is not capable of reaching finality on transactions, which means that it is increasingly unlikely over time but possible to change transactions after execution. As Hyperledger Fabric uses a different transaction flow and consensus mechanism it allows for a higher performance in lower scaled networks and it is capable of reaching finality of transactions, making them impossible to be changed or reversed (Hinckeldeyn, 2019). Since maintaining consensus between the ledger copies in Fabric does not require proof-of-work, no cryptocurrency is needed as an economic incentive for block creation (Sajana, 2018). Furthermore, Fabric is providing a modular architecture that allows for high resiliency and flexibility to adapt the network according to the specific use case requirements (Hyperledger Fabric, 2019).

The modular design of Hyperledger Fabric for business context is leading to great popularity of the platform. Forbes published a list of 50 companies with a minimum revenue of one billion dollars that are leading the way in adapting decentralized ledgers to their operating needs. 22 of the 50 Companies, from Amazon to Walmart are dealing with Hyperledger Fabric (Del Castillo, 2019).

Due to its modularity, confidentiality and performance characteristics, Hyperledger Fabric is currently the most suitable Blockchain technology capable of running Smart Contracts to implement the previously described model. Furthermore, all Hyperledger projects are open source software under open governance of the Linux Foundation, so everyone can see, use, copy and contribute to the program code (Dhillon et al., 2017).

A ledger in Fabric is composed of a database called World State and a Blockchain data structure. The database stores the states of participants and assets. Assets are tangible goods, such as aircraft, or intangible goods, such as maintenance records. A state of the asset Aircraft for instance could be the name of the owner. The Blockchain data structure is containing the preceding transactions, which usually lead to changes of the World State such as send Aircraft to initate the transfer of an aircraft's ownership. The World State can be calculated at any time from the Blockchain data structure and thus be verified. Figure 5 illustrates the relationship between the ledger, World State, and the Blockchain data structure in Fabric. The ledger is composed of the World State and the Blockchain data structure, whereby the Blockchain data structure determines the World State. The ledger is managed by Smart Contracts called Chaincode in Fabric. These contain asset definitions as well as the business logic for changing those assets. Nodes in Fabric are assigned to an organization and are distinguished according to their function in peers, ordering-service-nodes called orderer and applications. Typically, every business would be a separate organization in Fabric and would contribute to the network with their own peers, ordering-service-nodes and applications. In order to save resources, it is also conceivable that several, in particular smaller companies join together to form a common organization in Fabric.

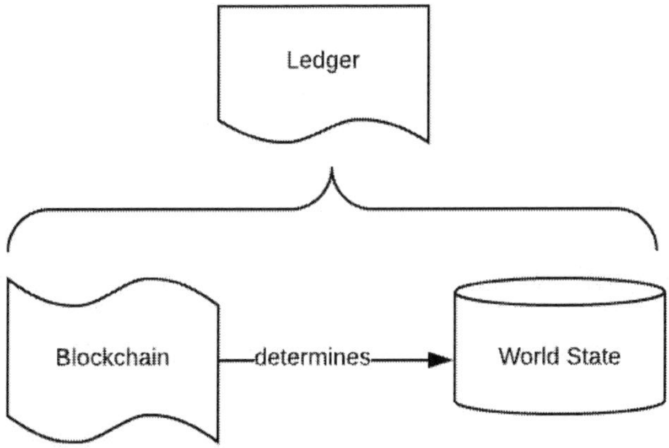

Figure 5: Components of a Ledger in Fabric (Hyperledger Fabric, 2019)

Peers are the most fundamental building block of the network because they manage the copies of the ledger and Chaincode. Applications are users of the network which communicate with it through transactions while the orderer is responsible for arranging transactions into blocks. The only mechanism of interaction with Chaincode is through transactions invoked by applications.

The transaction flow is illustrated in Figure 6. After connecting to the peer (1) the application initiates the transaction flow by creating a transaction proposal. The application sends the transaction proposal to the peers of the network (2), which execute it independently by invoking the Chaincode to generate a transaction response (2.1 and 2.2). The proposal response is not applied directly to the ledger, but only digitally signed by each peer and returned to the application (3). As soon as the application has received sufficient identical transaction responses, they are sent to the orderer (4), who

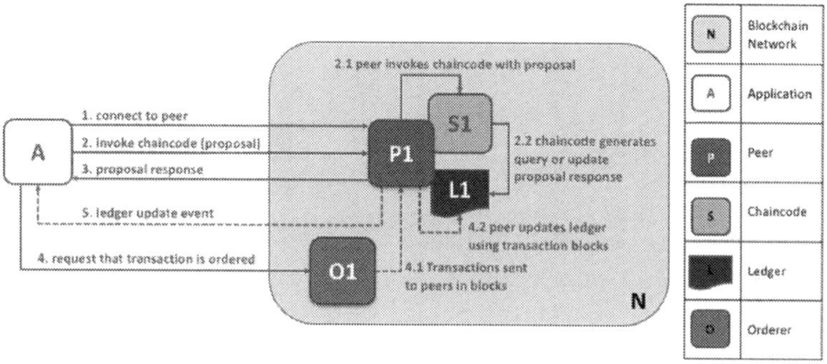

Figure 6: Transaction flow in Fabric (Hyperledger Fabric, 2019)

is then sorting the transactions mechanically into a block and forwarding it again to the peers (4.1). The peers process the obtained block independently. After each transaction has been verified, the block is added to the Blockchain data structure and the World State updated (4.2). Failed transactions are marked as such and are also added to the ledger as part of the block, but do not affect the World State. Finally, the applications are notified of the success or failure of the transactions (5).

To identify them, network participants in Fabric are issued a digital certificate, and a cryptographic key pair by an organization-owned component called Certificate Authority. Another component called Membership Service Provider assigns roles and organizations to the identities and authenticates them as members for the use of the network. MSPs provide dynamic membership by adding and removing access permissions in a decentralized way to ensure long-lasting network integrity.

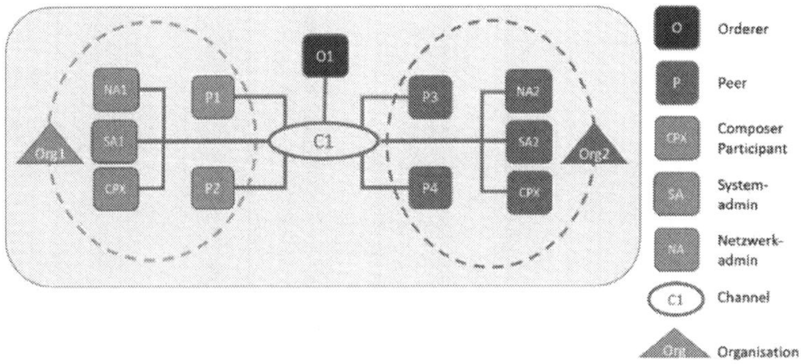

Figure 7: Structure of the designed fabric network

To support the development of Blockchain applications with Fabric, Hyperledger Composer was developed as part of the Hyperledger project. Composer is a collection of tools allowing the creation of Chaincode in a structured and clear logic as well as the subsequent provision in a Fabric network (Dhillon et al., 2017). One of the tools is the browser-based user interface Hyperledger Composer Playground, which is used to test the functionalities of the program. The programming in Composer is carried out on the basis of assets, participants and transactions. The program code is written in several files with different purposes. In the model files the fundamental components are defined as participants, assets and transactions. Methods are written as transactions in script files using JavaScript. Restrictions are defined in an access control file to ensure that only authorized network participants can initiate appropriate transactions. Furthermore, separate query files are used for database queries. From the separate files of a program, a Business-Network Archive (BNA)-file is generated, which is then provided as Chaincode to an existing Fabric network.

The designed Fabric network is illustrated in Figure 7. For a real use case, the Fabric network would be formed by a consortium of several companies, and optimally each participating company would form an organization in Fabric. The designed Fabric network is formed by the two organizations Org1 with the peers P1, P2 and Org2 with P3, P4 and a central orderer O1. The affiliations of the network components to the organizations are illustrated by the dashed circles. System administrators SA1 and SA2 were added to deploy the Chaincode on each organization's peers and to create the network administrators NA1 and NA2. The organizations can then use the rights specified in the deployed Chaincode to create further network participants CPX of any role according to Figure 3 and link them with identities. The nodes NA, SA and CPX represent the users of the network who in practice communicate with it via client applications. Regarding the given use-case CP1 could be the MRO-company Lufthansa Technik and CP2 the Airline Lufthansa, forming a common organization Org1 and contributing to the network by providing peers and orderer. A second organization Org2 could be formed by the OEM-company Airbus and the Supplier General Electric.

In the present Fabric network, there is a single orderer and thus a partial centralization of the network. A more decentralized approach could be realized by each organization providing an ordering-node, which form a common consensus on the Kafka protocol. Kafka uses Apache Kafka, an open-source stream processing platform that enables processing of continuous data streams of structured data (Apache, 2017). All orderers would send incoming transactions to Kafka and receive transactions in the same order as the others. (Hyperledger Fabric, 2019)

5 Evaluation

In order to validate the investigation model, case studies were designed and carried out. Subsequently the model's strength and weaknesses were evaluated based on the conducted case studies. Furthermore, current issues in aircraft MRO industry to which the application of a Blockchain network could contribute to the solution and challenges that need to be overcome to reach a use in practice are shown.

5.1 Validation

To consider the network size and complexity of processes in the evaluation, four case studies with two different numbers of participants and two different maintenance scenarios were designed.
Network A is a small network, consisting of one operator, one MRO, one regulator and three backshops without direct competition. The main purpose of the network is to record the history of maintenance activities rather than tracking property rights. Figure 8 illustrates the business network on the application level. The whole Fabric network infrastructure is built by the nodes previously described, while the participants such as MRO A or Operator A are the Composer participants CPX according to Figure 7 and are assigned to one of the organizations Org1 or Org2.
Network B is a medium-sized network consisting of a total of 18 participants for the tracking of maintenance activities and property rights. For each role, with the exception of the regulator, there are competing participants.

Based on the same base configuration of assets (one Aircraft containing four Parts), two hypothetical maintenance scenarios were designed. The scenario routine maintenance describes the execution of maintenance activities as part of an A-check. First, the aircraft to be inspected is handed over by an aircraft operator to the responsible maintenance company, which detects damage by inspecting the structural elements winglet and flaps. The findings of the winglet will be fixed during the check and the repair of the flaps will be postponed to the following check. Subsequently, the aircraft is returned to the original aircraft operator.

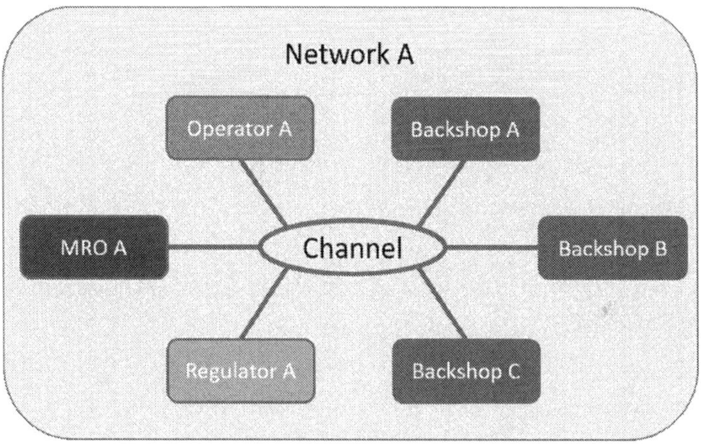

Figure 8: Illustration of business network A

The second investigated scenario is a non-routine maintenance in which a bird strike during flight operation results in findings and the need to carry out maintenance. The aircraft operator ascertains that the aircraft has been taken out of operation, declares it non-airworthy and hands it over to a maintenance company. The maintenance company detects damage to a turbine and to a hydraulic pump. The hydraulic pump has to be replaced while the turbine is declared non-airworthy and transferred to a backshop for repair. Since the maintenance operation does not keep the hydraulic pump in stock, it is ordered by a supplier and then reinstalled by the maintenance company. The turbine is repaired by the backshop, airworthiness is granted by the regulatory authority and the turbine is sent back to the maintenance organization for installation. After a test flight is carried out, the regulatory authority issues a new airworthiness certificate. Ultimately, the aircraft is returned to the original aircraft operator.

The pairwise combination of the scenarios with the networks A and B result in four case studies. The designed case studies were successfully carried out with the developed prototype and therefore mapped in the Blockchain data structure. For the scenario of routine maintenance, eight and for non-routine maintenance 24 transactions were executed in the correct sequence by the different participants. The executed transactions are stored in the Blockchain data structure resulting in a new configuration of objects and their states. Figure 9 shows the configuration of the objects before and after the non-routine scenario was carried out. Each block represents an object in the business network and the connections between the blocks correspond to the assignment between the assets. Within the blocks, attributes of the objects relevant for the representation are entered. Figure 10 shows the transactions performed by the participants in order to carry out the scenario.

The fulfillment of the network requirements for identification and authorization of network participants and dynamic admission restriction could be confirmed. The decentralization of the network infrastructure was limited by the central ordering service and the fact that the network infrastructure was formed by only two organizations, but it was shown in which way a decentralized network could be realized. The transaction confirmation time and transaction throughput cannot be appropriately quantified using the developed model. However, with a similar Fabric setup a benchmark analysis from Baliga et al. (2018) shows a linear transaction throughput with a transaction confirmation delay time of up to 2 seconds at an increasing input load of up to 1000 transactions per second. These performance parameters are more than sufficient for the considered use case. Furthermore, the fulfillment of the application level requirements for the digital registration of aircraft and components to store their states and status history, the transparency of the data and querying of data in a Blockchain network was successfully demonstrated. On the Smart Contract-layer the same business logic could be implemented using other programmable Blockchain platforms such as Ethereum or Hyperledger Sawtooth. However, the consequences for the properties of the network would change based on the chosen Blockchain technology.

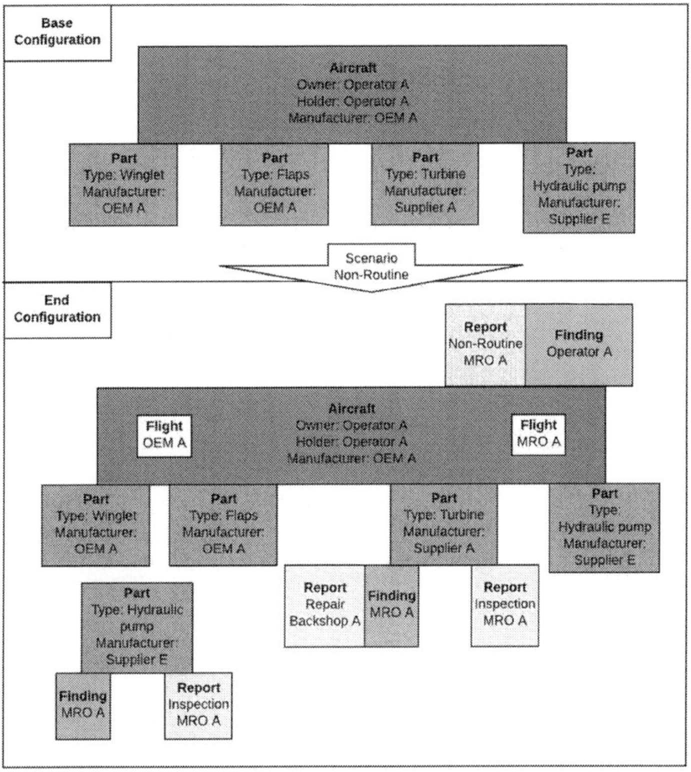

Figure 9: Base and end configuration of the objects after the non-routine scenario

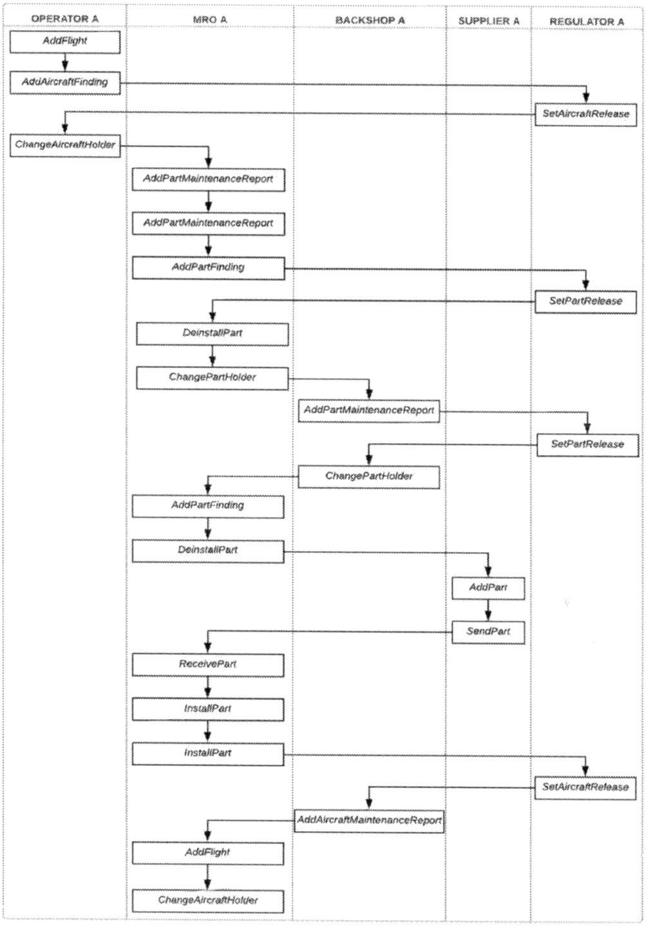

Figure 10: Transactions performed for the non-routine maintenance

5.2 Strengths and weaknesses

In the following, the strengths and weaknesses of using a decentralized network according to the developed prototype will be discussed.

The power over the data in a decentralized network is distributed among the participating organizations. The developed Fabric infrastructure is built on two organizations and a single orderer and is thus partially centralized. A more decentralized approach could be realized by assigning each participant of the network to a separate organization, providing its own peers and orderers. Nevertheless, the distributed power over the data leads to a higher willingness of the organizations to share their data in order to benefit from the network and from the data of other organizations. The additional data an organization obtains from participating in the network may contribute for instance to demand forecasting of spare parts, risk analysis, or product and process improvement. Each network participant is required to contribute data to the network, as one participant would otherwise not benefit others, but increase their risk. The larger the network and the higher the amount of competing participants the bigger is the risk of sharing data. This is particularly evident when comparing network A and network B from the carried out case studies.

Fabric offers the functionality of different channels to keep transactions in a shared network secret. A channel is formed by peers, applications and orderers and serves for the complete decapsulation of ledgers and Chaincode. The nodes can be assigned to multiple channels. Only the nodes of a channel are involved in the consensus building of the respective ledger and can thus see Chaincode, ledger and transactions. Channels could therefore protect data between competing organizations and only provide it to the participants desired. For smaller networks channels are easy and logical to implement. With reference to the use cases, sensitive data from competing backshops Backshop A and Backshop B could keep their data secret from each other and still make it available to the rest of the network. Figure 11 illustrates the use of channels for a given network. All network participants form a common channel (Channel A). The competing backshops also each form a separate channel (Channel B and Channel C) with the other network participants. To maximize the overall value of the network, without penalizing individual network users by the risk of external data usage, compromises must be made between the participating organizations in order to agree on channel configuration and sharing of data. In practice, cooperation between often competing companies with different interests must be formed in order to define common goals and requirements.

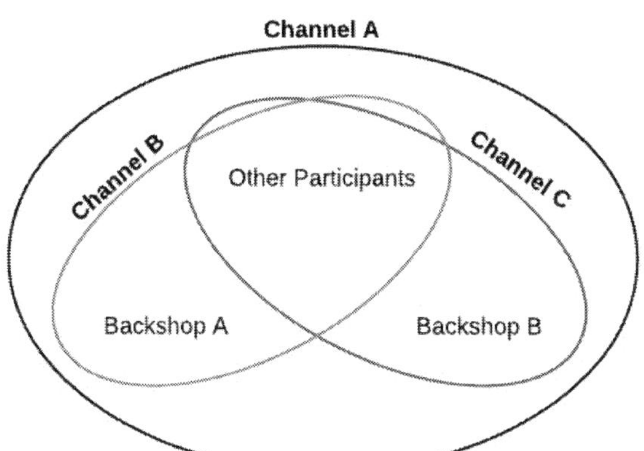

Figure 11: Confidentiality of sensitive data through the use of channels

Applying a Blockchain network according to the prototype, certain intermediaries could be avoided and as a result process time and costs be saved. By reducing process times, the entire maintenance process can be accelerated, thereby reducing the cost-intensive downtime of aircraft. Intermediaries, in particular, are auditors of documentation for the maintenance network, e.g. custodial employees of the regulatory authorities or document inspectors of an aircraft operator at an aircraft overtake.

Due to the distributed data and consensus building, there is a very low risk of data loss due to malfunction or hacker attacks and a high synchronicity of data between the organizations. On the other hand, the process of distributed consensus building requires a high time for transaction confirmation and low transaction throughput, compared to centralized systems, especially for large networks.

From the properties of the Blockchain data structure and the distributed consensus building results a high change inertia of the network and the irreversibility of transactions. This effect increases with the size and decentralization of the network. It follows a high security against data manipulation, since once entered transactions cannot be removed or changed without a common consensus. Together with the data transparency within the network, there is a good traceability of all activities of the participant in the networks. Traceability in the case of aircraft MRO can for instance contribute to facilitate rapid problem identification and elimination in case of premature component failure. On the other hand, the irreversibility of transactions limits the flexibility of organizations, as erroneous or erroneously submitted transactions generally persist. The program code of the developed prototype therefore provides transactions to reverse the effects of registered transactions in the World State.

The implementation of Smart Contracts in a distributed network enables the automation of process execution and thus also a shortening of process times by avoiding human labor with a high system availability and reliability of function execution. Due to the inertia of change and the fact that the programmed code of Smart Contracts is executed exactly as it was instructed, great care must be taken to understand how the code can affect network subscribers. Therefore, experts and adequate tools for the programming of Smart Contracts are needed.

5.3 Issues in Aircraft MRO

In the following, two selected issues in the aircraft MRO industry are shown and explained in which way a decentralized network according to the developed prototype could contribute to their solution.
Due to the complexity of aircraft and the high costs of spare parts and certifications arise risks of flight safety through the use of counterfeit spare parts. Counterfeit spare parts are components that have not been approved by the aviation authorities often made of inferior materials or with exceeded lifetime. Monitoring the authenticity of spare parts is problematic because of the component complexity, especially when the maintenance processes have been outsourced to MRO operations abroad (Locatory, 2012). In 1989, Partnair Flight 394 crashed due to counterfeit components on the route from Oslo to Hamburg, leading to the deaths of all 50 passengers and 5 crew members (Luedemann, 1996). To guarantee the authenticity of a part it is essential to ensure that physical parts are correctly linked to their digital counterparts. An object therefore has to be linked to a unique identifier, which is printed, embossed or attached as a tag to the

object. Balagurusamy et al. (2019) explain how an identifier, such as the object's surface structure, can be used to securely link an object to its properties asserted in a trusted database. The traceability of the transaction history with full data transparency could help to reduce the use of counterfeit spare parts. If a counterfeit component were to be installed in an aircraft, then the entry of the component is already to be falsified. All parties in the network would be able to review the transactions and assign responsibilities. If the component turns out to be a counterfeit, the guilty party could be easily and quickly identified. A prerequisite for this is that the storage and certification of the aircraft and components in the Blockchain data structure is recognized by the legislator.

The cost of transferring ownership of aircraft and components between aircraft operators in commercial aviation is approximately one billion dollars per year according to Canaday (2017). In addition to the storage, management and transport of records, there is a significant cost component in the verification of maintenance records. The verification of maintenance records is necessary to ensure the safety of the aircraft and to meet regulatory requirements. Missing, incorrect or incomprehensible records lead to costly rework and thus delays. (Canaday, 2017)

The use of a decentralized network according to the developed prototype could increase the credibility and traceability of maintenance records. During the transfer an aircraft's ownership, the inspection effort and therefore the costs of document verification could be reduced. Furthermore, the distributed nature of the Blockchain minimizes the risk of data loss, which could also reduce rework due to lacking records. If the old and new owners are located in a common network, seamless data transmission and utilization could be realized.

5.4 Challenges

In order to enable an efficient application of the Blockchain technology, interoperability is required to easily integrate the various systems (Banerjee, 2018). The largest producer of software solutions for enterprise software SAP is actively involved in the integration of Blockchain applications into their existing systems and has been a premium member of the Hyperledger Foundation since March 2017 (Hyperledger, 2017). In October 2018, Hyperledger's subproject Hyperledger Burrow enabled the provision of Ethereum Smart Contracts on Hyperledger Fabric (Ledger Insights, 2018). Another recent project of the Hyperledger Foundation is Hyperledger Quilt for the interoperability of Blockchain systems through the application of a payment protocol (Hyperledger Foundation, 2018).

Furthermore, challenges arise from uncertainties due to a lack of regulations. Currently, the legal implications of Blockchain data are not clearly defined. It is necessary to explore how existing contract law affects Smart Contracts and to adjust legal procedures to properly manage Smart Contracts. A prerequisite for the development of legal regulations and the creation of interoperability is the introduction of standards for designations and descriptions in order to facilitate a smooth communication and to prevent misunderstandings (Banerjee, 2018, Chamber of Digital Commerce, 2016).

For the creation of standards and legal regulations and the overcoming of technological challenges the inter-organizational collaboration of companies, research institutes and legislators is required. While these challenges are likely to be overcome in the foreseeable future, the need to co-operate with competing companies in a common network presents a challenge with an uncertain outcome.

Further uncertainties arise because there are hardly any practical applications in the industry or experts whose application experience could benefit a business network. The introduction of novel systems within the organizations is generally associated with a great deal of time and expense. The systems therefore have to last for a longer period of time to recoup their costs. However, there is a great deal of long-term planning uncertainty due to the changing nature of the industry, the participating organizations with their relations and needs, legal regulations and in particular the technology. A technology implemented today could already be obsolete in a few years' time.

6 Conclusion

Blockchain technology was first described in the end of 2008 and the Hyperledger project is existing since december 2015. Therefore, it is comprehensible that at the time there are various technological, regulatory and organizational challenges ahead. It is important to educate about the technology, its application and potentials without fanning false hopes. It will be necessary to further identify opportunities and risks through the development of investigation models and testing of applications. The present work contributes to this by providing a prototype solution for the MRO-industry, demonstrating its development and analyzing the applicability.

Based on an analysis of the processes in aircraft MRO a conceptual model was developed. The program code was then authored using the Hyperledger Composer toolset, deployed on a Hyperledger Fabric network, and validated by case studies. The analysis of the study model shows a considerable potential of the technology to increase the efficiency of an MRO network through credibility, traceability and transparency of the data, with high reliability and system availability, especially for larger networks with a high complexity of process flows. The usual high costs in the aviation industry due to warehousing, document checking and downtime of aircraft could potentially be reduced and the safety of the aircraft increased by avoiding the use of counterfeit spare parts.

Businesses could start using the technology for non-sensitive data to demonstrate the security and operational efficiency of the systems by first deploying applications within their enterprise and then extending them to cooperating companies. Finally, the development of the Blockchain's full potentials as an infrastructure technology requires a critical mass of key players in the market. While the MRO industry has lagged behind other industries in terms of IT deployment, it could be a trailblazer in the application of Blockchain technology because of the highly regulated and interconnected nature of maintenance for small-scale, high-complexity products. Adoption of an application for aircraft MRO is conceivable in similar industries that rely on complex fleet maintenance with several parties involved, such as shipping or rail.

References

Apache, 2017. Apache Kafka Documentation. [online] Available at: <https://kafka.apache.org/documentation/> [Accessed 15 May 2019].

Balagurusamy, V., Cabral, C., Coomaraswami, S., Delamarche, E., Dillenberger, D., Dittmann, G., Friedman, O., Hinds, N., Jelitto, J., Kind, A., Kumar, A.D., Libsch, F., Ligman, J., Munetoh, S., Narayanaswami, C., Narendra, A., Paidimarri, A., Delgado, M.P., Rayfield, J., Subramanian, C. and Vaculin, R., 2019. Crypto Anchors. IBM Journal of Research and Development.

Baliga, A., Chaterjee, S., Kamat, P., Pednekar, A., Solanki, N. and Verekar, S., 2018. Performance Characterization of Hyperledger Fabric. Crypto Valley Conference on Blockchain Technology.

Banerjee, A., 2018. Integrating Blockchain with ERP for a Transparent Supply Chain. [online] Available at: <https://www.infosys.com/Oracle/white-papers/Documents/integrating-blockchain-erp.pdf> [Accessed 4 May 2019].

Boyle, Brandon, B., Brenner, A., Steger, S., Janssen, S. and Rasamoela, M., 2018. The Blockchain bandwagon. Is it time for automotive companies to start investing seriously in Blockchain? Roland Berger GmbH.

Buterin, V., 2013. Ethereum Whitepaper: A next generation Smart Contract & Decentralized Application Platform. [online] Available at: <http://blockchain-lab.com/pdf/Ethereum_white_paper-a_next_generation_smart_contract_and_decentralized_application_platform-vitalik-buterin.pdf> [Accessed 18 Apr. 2019].

Canaday, H., 2017. Blockchain in MRO Could Happen Sooner Than You Think. [online] MRO Network. Available at: <https://www.mro-network.com/big-data/blockchain-mro-could-happen-sooner-you-think> [Accessed 12 May 2019].

Chamber of Digital Commerce, 2016. Smart Contracts: 12 Use Cases for Business & Beyond. [online] Available at: <https://digitalchamber.org/wp-content/uploads/2018/02/Smart-Contracts-12-Use-Cases-for-Business-and-Beyond_Chamber-of-Digital-Commerce.pdf> [Accessed 17 Apr. 2019].

Cummings, S., 2019. The Four Blockchain Generations. [online] Available at: <https://medium.com/altcoin-magazine/the-four-blockchain-generations-5627ef666f3b> [Accessed 12 Apr. 2019]. Dhillon, V., Metcalf, D. and Hooper, M., 2017. Blockchain Enabled Applications: Understand the Blockchain Ecosystem and How to Make it Work for You. 1st ed. New York, NY: Apress.

Del Castillo, M., 2019. Blockchain 50: Billion Dollar Babies. Forbes. Available at: <https://www.forbes.com/sites/michaeldelcastillo/2019/04/16/blockchain-50-billion-dollar-babies/#18cce6a157cc> [Accessed 12 Jun. 2019].

Drescher, D., 2017. Blockchain Basics: A Non-Technical Introduction in 25 Steps. 1st ed. Berkeley, California: Apress.

EASA, 2014. Acceptable Means of Compliance and Guidance Material - Annex II to Regulation No 1321/2014.

Elliott, J., 2018. MROs Battle OEMs for Commercial Aviation Aftermarket. [online] Available at: <https://www.aviationtoday.com/2018/06/21/mros-battle-oems-commercial-aviation-aftermarket/> [Accessed 15 Jun. 2019].

Frauenhofer-Gesellschaft, 2017. Blockchain und Smart Contracts: Technologien, Forschungsfragen und Anwendungen. [online] Available at: <https://www.fraunhofer.de/content/dam/zv/de/forschung/artikel/2017/Fraunhofer-Positionspapier_Blockchain-und-Smart-Contracts_v151.pdf> [Accessed 13 Apr. 2019].

Gilbert Frigden, Sven Radszuwill, André Schweizer and Nils Urbach, 2017. Entwicklung disruptiver Innovationen mit Blockchain: Der Weg zum richtigen Anwendungsfall. Wirtschaftsinformatik des Fraunhofer-Instituts für Angewandte Informationstechnik FIT. Wirtschaftsinformatik & Management, 9(5), pp.52–59.

Goudarzi, H., Martin, J. and Warren, S., 2018. IATA Whitepaper: Blockchain in Aviation. [online] Available at: <https://www.iata.org/publications/Documents/Blockchain-in-aviation-white-paper.pdf> [Accessed 12 May 2019].

Hackius, N. and Petersen, M., 2017. Blockchain in Logistics and Supply Chain: Trick or Treat? In: Digitalization in supply chain management and logistics. Hamburg International Conference of Logistics.

Herwljer, C., Waughray, D. and Warren, S., 2018. Building Block(chain)s for a Better Planet. World Economic Forum.

Hinckeldeyn, J., 2019. Blockchain-Technologie in der Supply Chain: Einführung und Anwendungsbeispiele. 1st ed. Springer Vieweg.

Hinsch, M., 2017. Industrielles Luftfahrtmanagement: Technik und Organisation luftfahrttechnischer Betriebe. 3rd ed. Berlin, Germany: Springer Vieweg.

Hyperledger, 2017. Hyperledger Welcomes SAP as Premier Member. Available at: <https://www.hyperledger.org/announcements/2017/03/22/hyperledger-welcomes-sap-as-premier-member> [Accessed 7 May 2019].

Hyperledger Composer, 2019. Hyperledger Composer Documentation. [online] Available at: <https://hyperledger.github.io/composer/latest/introduction/introduction.html> [Accessed 2 Apr. 2019].

Hyperledger Fabric, 2019. A Blockchain Platform for the Enterprise - Hyperledger ReadTheDocs. [online] Available at: <https://hyperledger-fabric.readthedocs.io/en/release-1.4/> [Accessed 2 Apr. 2019].

Hyperledger Foundation, 2018. Hyperledger Quilt. Available at: <https://www.hyperledger.org/projects/quilt> [Accessed 22 Apr. 2019].

Iansiti, M. and Lakhani, K., 2017. The Truth About Blockchain. Harvard Business Review. [online] Available at: <https://hbr.org/2017/01/the-truth-about-blockchain> [Accessed 12 May 2019].

Ledger Insights, 2018. Hyperledger Fabric integrates Ethereum Smart Contracts. Available at: <https://www.ledgerinsights.com/hyperledger-fabric-integrates-ethereum-smart-contracts-evm-blockchain/> [Accessed 17 Mar. 2019].

Locatory, 2012. Stolen and counterfeit spares is still one of the major issues in the aviation industry. [online] locatory.com. Available at: <https://www.locatory.com/en/news-media/press-releases/2012/stolen-and-counterfeit-spares-still-one-major-issues-aviation-industry/> [Accessed 18 Apr. 2019].

Luedemann, R., 1996. Flying Underground: The Trade in Bootleg Aircraft Parts. Journal of Air Law and Commerce. 1(62).

Narayanan, A. and Clark, J., 2017. Bitcoin's Academic Pedigree The concept of cryptocurrencies is built from forgotten ideas in research literature. ACM Queue [online] Available at: <https://queue.acm.org/detail.cfm?id=3136559> [Accessed 25 Apr. 2019].

Sajana, P., Sindhu, M. and Sethumadhavan, M., 2018. On Blockchain Applications - Hyperledger Fabric and Ethereum. [online] International Journal of Pure and Applied Mathematics (118). Available at: <https://pdfs.semanticscholar.org/7674/10f40ed2ef1b8b759fec3782d8a0f2f8ad40.pdf> [Accessed 17 Apr. 2019].

Satoshi Nakamoto, 2008. Bitcoin: A Peer-to-Peer Electronic Cash System. [online] Available at: <https://bitcoin.org/bitcoin.pdf> [Accessed 3 Apr. 2019].

Schlatt, V., Schweizer, A., Urbach, N. and Fridgen, G., 2016. Blockchain: Grundlagen, Anwendungen und Potenziale. Wirtschaftsinformatik des Fraunhofer-Instituts für Angewandte Informationstechnik FIT. [online] Available at: <https://www.fit.fraunhofer.de/content/dam/fit/de/documents/Blockchain_WhitePaper_Grundlagen-Anwendungen-Potentiale.pdf> [Accessed 17 Apr. 2019].

Szabo, N., 1996. Smart Contracts: Building Blocks for Digital Markets. [online] Available at: <http://www.fon.hum.uva.nl/rob/Courses/InformationIn-Speech/CDROM/Literature/LOTwinter-school2006/szabo.best.vwh.net/smart_contracts_2.html> [Accessed 15 May 2019].

Tual, S., 2015. Ethereum Launches. Ethereum Blog. Available at: <https://blog.ethereum.org/2015/07/30/ethereum-launches/> [Accessed 15 May 2019].

Design of Self-regulating Planning Model

Maria Paula Espitia Rincon[1]*, David Alejandro Sanabria Martínez*[1]*,*
Kevin Alberto Abril Juzga[1] *and Andrés Felipe Santos Hernández*[1]
1 – Escuela Colombiana de Ingeniería Julio Garavito

Purpose: This research aims to develop a dynamic and self-regulated application that considers demand forecasts, based on linear regression as a basic algorithm for machine learning.

Methodology: This research uses aggregate planning and machine learning along with inventory policies through the solver excel tool to make optimal decisions at the distribution center to reduce costs and guarantee the level of service.

Findings: The findings after this study pertain to planning supply tactics in real time, self-regulation of information in real time and optimization of the frequency of the supply.

Originality: An application capable of being updated in real time by updating data by the planning director, which will show the optimal aggregate planning and the indicators of the costs associated with the picking operation of a company with 12000 SKU's (Stock Keeping Unit), in which a retail trade of 65 stores is carried out

Keywords: Linear Programming, Linear Regression, Aggregate Planning, Cost Minimization

First received: 23.May.2019 **Revised:** 24.June.2019 **Accepted:** 01.July.2019

1 Introduction

The prognosis is made to be aware of the demand and the future conditions of the market, the companies use planning as a strategic tool, which transforms the prognosis into plans that allow satisfying the consumer's requirements. The aggregate plan is thus an important informative actor along the supply chain which affects the demand of suppliers and customers (Chopra, 2008).

That is why some of the engineering tools that most impact the production processes, are those for programming and planning. These tools are implemented with the aim of reducing process times and associated costs, giving clear guidelines on what to do, how to do it, where to do it and who does what, with the objective of maximizing the efficiency of both operators and the lines of processes (Borissova, 2008).

In Colombia, the mixed market of drugstores has a ratio of one drugstore for every 100,000 inhabitants which is 43.1% in Latin America, followed by Brazil, Mexico and Chile. With a growth of 12.6% between October 2017 and the same month in 2018 of the platform studied, which manages four product categories, these include medicines, personal care, beauty, and baby care and where business is comprised 50% of medicines, and the other 50% is distributed in the other categories mentioned above. The categories that have the best performance are analgesics, cleaning products, personal care, and natural products, the latter of which have a growth of 6.8%, the frequency of purchase for Colombian consumers is every 34 days, of which 52% of buyers plan their purchase (Dinero, 2019).

The current market forces all companies to use technology along the value chain, not only for its internal processes but also external ones, such as

sales through virtual channels, which represent 20% of the platform studied. This generates a logistical-level challenge because it requires satisfying the demand in shorter periods of time, generating high logistical costs.
The percentage of the logistics cost of sales in Colombia, according to the National Logistic Survey of 2018, represents 13.5%, which is composed of the following: cost of storage (46.5%), transport cost (35.2%), cost of administration and customer service (11.1 %) and other costs (7.2%); the cost of storage being an important factor within the total weighting, therefore the need to concentrate their efforts on reducing this type of cost is evident (Alonso, Martínez, Dorado,Páez, Lota, 2018).
This industry has created the need to optimize its planning in accordance with the logistical costs, as is intended with this project, which consists of the preparation of an aggregate planning, whose objective it is to calculate the real time, the optimal quantity to be ordered, and the reorder point - using the information found in the demand history and the support tools for linear programming solutions, such as the extension "Microsoft Excel solver".

2 Machine Learning and Logistics

At present, the advanced analytics, or Big Data, is very important as a support to the planning of an organization, since it is employed to manage large volumes of information and obtaining working data of different sources and formats. All of this is done in order to apply statistical calculations or different mathematical analysis tools that allow one to obtain current data, to identify flaws and to improve operations.

However, advanced analytics is not able to learn or make decisions independently. This is how the concept of artificial intelligence (AI) arises: it is the capacity of a machine to carry out processes without any intervention, having access to the data supplied by the Big data. The machine is able to learn patterns or characteristics provided by the information, and thanks to the acquired learning it creates knowledge, which generates intelligent actions that become more precise over time (Pereira, 2018).

Making use of this technology in the context of the industry 4.0 allows the understanding of data from different sources along the value chain through AI, and to make support decisions in real time, managing to optimize the quality of production, energy-saving, equipment performance (Rüssmann, 2015).

The supply chain in the industry 4.0 uses a branch of the AI called machine learning, which allows the discovery of patterns in the data through algorithms; these identify which elements that have greatest influence in the supply network by means of continuous learning. The algorithms find new patterns every day without human intervention or some previously established classification. This allows orienting the search, making use of models based on restrictions they find. A set of x elements with high predictive accuracy identify patterns that influence supplier quality, inventory levels, demand forecasts, purchasing processes, production planning, and transportation management. Through the entire value chain, companies have the freedom to evolve considerably, resulting in them being more competitive thanks to automatic learning skills (Columbus, 2018).

A machine algorithm learning is a linear regression, which is the most rudimentary algorithm of the branch of machine learning. This method models an objective value, which is based on independent predictors, it's main use concentrates on making forecasts and discover cause-and-effect relationships between dependent and independent variables.

In simple linear regression to a linear relationship between the independent variable (x) and dependent (y), were to quantify this relationship an objective function is used, which allows one to find the best possible values by minimizing the sum of used square errors in the set of variables. This is the best fit line of the data involved and the best data found that these problems can minimize the error between the predicted value and the current (Gandhi, 2018).

The linear regression was interpreted through linear programming to project the future supply in an optimal and timely way to the shelves of the picking area for each SKU (Stock Keeping Unit).

3 Linear Programming and Added Planning

Linear programming is the first instance of modeling reality problems at a computational level, however, to solve the problem is only the first step, since the objective of this investigation is to create the basis of a program that solves problems by using new data. In order to accomplish this, it needs to create a relationship between linear programming (as a subset of the convex programming, which has a linear objective function f (x) linked to a set of linear equations and inequalities) and the general principle of optimization applied to artificial intelligence through machine learning (Julian, 2016).

Planning is a tool that allows satisfying the demand by optimizing the resources of the system. An aggregate plan is characterized by considering the time horizon, both in the medium and the short term. The criteria of decision are based on the maximization of profit, which is understood as the difference between the income and the costs, this is why it can be seen as the criteria decision for the minimization of the total costs.

Given the free modeling features of linear programming which involve a lot of restrictions, the solution of the added plans or problems that seek to minimize not only costs, but also resources, can be achieved thanks to the help of an extension for Microsoft Excel called "Solver", which is suitable for solving these types of problems. This application is very popular in the academic and professional field, because of its versatility and ability to give optimal answers according to the reality of the system (Anon., 2017).

In researching of (Granja, 2014), the object is focused on reducing the waiting time of patients, which has generated a problem in regard to the public sector versus the private sector's inability to withstand the growing demand. The results obtained by the investigation, give to understand the importance of the use of linear programming, showing reductions of 38% in the total waiting time of patients.

In another study, the planning area of the mining industry has made use of classical linear programming based on goal programming systems, using multiple objective functions that reduce variations, reserves, and mixtures, ensuring compliance with the tonnage specifications in short and long term planning, reaching the best operational scenario that respects the cost restrictions present in the situation valued (Souza, 2018).

Another industry shows another example, a mixed linear programming model was designed, which presents a multi-objective function that aims to minimize production costs, employee turnover and, in turn, maximize sales while maintaining or improving quality service (Gholamian, 2015).

4 Impact on the social responsibility of employees

Customer satisfaction is reflected in the response times given by the market. For this, the distribution centers deliver products to the sales channels correctly and quickly. The automation of the processes through the use of new technologies of the industry 4.0 has allowed the logistical activities to achieve greater speed in the answer of the orders. However, not everything is positive, given the improvements in speed of the system, the employees of these distribution centers must intensify their work in order to keep up with the rhythms imposed by the automated areas, generating job insecurity and dissatisfaction, as evidenced by the investigation of (Fernández, 2011) in the distribution center of Inditex (Zara).

Any imbalance between human capacity and technology, creates dissatisfaction, either by internal customers or external customers. Unlike in the previous case, where the dissatisfaction of the employees was due to the excess of technology, in this investigation, the lack of technology has generated dissatisfaction of the employees for their long working hours.

The lack of adequate planning generates not only costs associated with the inventory, but also overtime per employee (€ 1.61 / hour) and a high rate of staff turnover. These two consequences have been recognized, given the increase in work outside of regular shifts, resulting in widespread discontent in the workers, which correlates to the monotony of the work and the high workload of more than 10 hours a day.

Keeping in mind the new planning in the process of shelf assortment, the workforce will not intervene in the supply of the products, which contributes to the elimination of delays caused by reprocessing, which is represented in an increase of overtime, of up to 4 hours daily. This will allow the daily work to be accomplished within the established schedule, avoiding additional costs for overtime.

5 Case Study

5.1 Company and Process Background

The regulations on health, the arrival of foreign competition with new sales formats has forced the traditional drugstore chains to reinvent themselves. They sell €1,537 million approximately each year, according to Euromonitor data.

Its new business plan is to strengthen medicine sales, representing 50% of the total sales and the inclusion of OTC (over the counter) or over-the-air products.

In the case of large areas, as is the case with the drugstore platform involved in this research, their annual sales represent €135 million approximately in 2018, according to Euromonitor data.

However, according to ASOCOLDRO (Asociación Colombiana de droguistas y detallistas), there are adverse factors in the market such as smuggling, counterfeiting, and unfair competition, currently affecting market prices. Despite this, the sector has grown between 5% and 8% per annum, regarding employability, ASOCOLDRO issued figures show that 52% of the workforce was female and 48% was male (Dinero, 2015).

The competition in Colombia of the large areas in 2016 is comprised of, according to ASOCOLDRO, 3000 small commercial establishments, of which

98% belong to urban areas and 2% to rural areas, 25% of these establishments are single family and 75% belong to pharmaceutical chemists (Espectador, 2016).

5.2 The Process

The process of supplying is carried out in a frequent way but not for the same reference, the moment a product is about to go out of stock or has been completely out of stock, it proceeds to request the transport of the storage using a forklift from the corresponding rack to be supplied with the

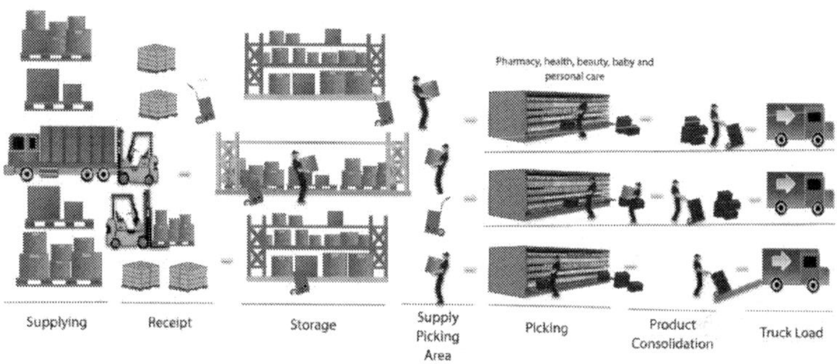

shelves of picking

Figure 1: Distribution center process

Once any of the points of a sale send the record of the sales made of any product, a request of supply of that specific SKU (Stock Keeping Unit) arrives, as well as the quantity. This data is registered in the technology platform of the pickers indicating this information to supply this product at the specific point of sale; by using a container which will have the products requested by a store, the operator proceeds to collect these items.

When the container is ready, the process of labeling is carried out with the supplied information about the products, as well as the destination, they are then taken to the loading area, where they will be waiting to be transported to the different points distributed in different cities of the country. See "Figure 1".

Table 1: Times of process

Process	**Time (s)**
Storage per box	37,5
Supply picking area per product	63,4
Picking per product	25,92
Order consolidation	37.68

The times of the processes are evidenced in Table 1, these times have direct implications associated with operation costs.

6 Methodology

6.1 Data Gathering

The products chosen for this research are OTC (Over the counter), within this group are medicines that do not require a medical prescription, personal care products, beauty, and baby care products. The data obtained was the result of the collection of information during fieldwork in January

2019. Additionally, classified products of the 3 reference types were taken (A, B and C), and because the mode of operation is cross docking, none of the references have storage. This indicates that the least amount of time a product takes to leave the distribution center is 3 days, and the most are 7 days; for this reason, all products were considered as having a high turnover (Aldana, 2014).

6.2 Forecast of Demand

To make the demand forecast, the linear regression method was used, which consists of finding the relationship between one variable with another, in the case of this study, the existing relationship is between the demand and the days, generating, as a result, the following mathematical formula:

$$y = ax + b \tag{1}$$

$$a = \frac{n\sum_1^6 xy - \sum_1^6 xy}{n\sum x^2 - (\sum x)^2} \tag{2}$$

$$b = \frac{\sum_1^6 y - a\sum_1^6 x}{n} \tag{3}$$

Where,

X= The number of days from Monday to Saturday (1 and 6).
Y= The daily demand that was presented in the company's database.
n= The number of demand history data that was used to find the linear regression. See Table 2.

Table 2: Forecast of Demand – Linear Regression

	Demand data						Forecast					
Description	Monday	Tuesday	Wednesday	Thursday	Friday	Saturday	Monday	Tuesday	Wednesday	Thursday	Friday	Saturday
Acetaminofen adultos 500mg tabletas caja	8	0	0	0	0	0	5	4	2	1	-1	-2
Atiban 2mg tabletas caja x30tab. pfi	0	6	6	0	0	0	4	4	3	2	1	1
Epítasis 3g sobres caja x6sob.	2	8	8	2	2	2	6	6	5	4	3	3
Esparadrapo leukoplast 7.5x4.57cm. x1und.	0	1	1	0	0	0	1	1	1	1	1	1
Aspan 50Mg Tabletas Caja x30Tab. Siegfried	0	1	1	0	0	0	1	1	1	1	1	1

Demand data: shows the BUO (Basic order unit) daily demand of each reference per day, in some cases, the demand is equal to zero given that at the points of sale there has been no sale of the product.
Forecast: shows the BUO (Basic order unit) daily forecast of each reference per day, taking into account the database of the demand of the company, through linear regression, in some cases the forecasts are negative numbers, this is due to the fact that the demand is zero units, which means that for that day it is not necessary to make an order.

6.3 Model formulation

The construction of the model went through several stages; these are described in detail below.
Stage 1: Gathering the necessary information for the implementation of the model correctly as historical demand, in-place inventory for each reference, storage capacity in picking and distance from the stock storage area to the corresponding picking storage area.
Stage 2: Based on the data found, we proceed to determine the costs of maintaining the units in inventory, of making an order and the waiting time to supply, and the lead time of the product, with which we proceeded to calculate the minimum security stock to have a 95% service level.
Stage 3: Through linear programming a supply planning model is built with the Solver tool in which the decision variables are established such as the optimal quantity of inventory, order size, restrictions of maximum inventory capacity, productivity and minimum level of inventory whose objective is to minimize the costs of the operation of supply and storage in picking. See Table 3.

Table 3: Solver supply planning model

Reference	**I_i**	**Q_i**	**S_i**	**L_i**	**Restriction on Inventory**	**Restriction on Capacity**	**Restriction on Productivity**
Acetaminofen adultos 500mg tabletas caja.	*	*	12	4,9	8	72	94
Atiban 2mg tabletas caja x30tab. pfi	*	*	4	0,7	51	360	686
Epítasis 3g sobres caja x6sob.	*	*	4	0,7	60	370	686
Esparadrapo leukoplast 7.5x4.57cm. x1und.	*	*	2	4,6	28	60	105
Aspan 50Mg Tabletas Caja x30Tab. Siegfried	*	*	1	0,7	23	96	686

* Take into account that, to solve the model in the solver, these spaces are left, since, in these, the optimal solution appears

Description of Solver supply planning model:

Ii: Optimal units in the inventory of each reference.

Qi: Optimal order size for each reference.

SSi: Security stock is the minimum level of inventory that allows meeting the customer's demand with a service level of 95%, a value of k and d of 1.64 for a normal distribution corresponds to this percentage.

$$SS_i = k \cdot \sigma_d \cdot \sqrt{L} \tag{4}$$

Li: Lead time is the time that elapses between the order being placed and the delivery of the order.

Restriction Inventory: This restriction guarantees that the customer's demand is 100% satisfied.

Restriction Capacity: This restriction ensures that the units in the inventory do not exceed the maximum capacity of the distribution center.

Restriction productivity: With this restriction, it is ensured that the productivity goals will be met within the available times of the working day. Later in this document, section 6.3.3 explains the formulation of each of the restrictions.

Stage 4: The model of the horizon is constructed that will show the programming of the assortment for the following week, in this the references are related to the corresponding data that allow me to find the variables that will allow the operation of the model as they are; available per hours of inventory, daily demand and its deviation, and basic unit of order, among others. Behind it, the planning model that will provide the data for the column of the size of the order is linked, and the reorder point is calculated from the inventory given by the solver. See Table 4, 5 and 6.

Table 4: Horizon Model part 1 –Product Characteristics

Product Characteristics									
		Product Location							
A B C	**Reference**	Sector	**Hall**	**Rack**	**Level**	**Position**	**BUO**	**Available. (Und.)**	**Demand (Und.)**
C	Acetaminofen adultos 500mg tabletas caja	1	13	11	4	4	9	21	1
A	Atiban 2mg tabletas caja x30tab. pfi	4	2	1	1	5	70	57	2
C	Epítasis 3g sobres caja x6sob.	4	1	1	2	7	74	68	4
C	Esparadrapo leukoplast 7.5x4.57cm. x1und	1	4	14	1	3	50	29	0
B	Aspan 50Mg Tabletas Caja x30Tab. Siegfried	1	7	16	3	10	12	24	0

* Take into account that, to solve the model in the solver, these spaces are left, since, in these, the optimal solution appears.

Table 5: Horizon model part 1 – Inventory Policy Configuration

Inventory policy configuration										
			Reorder		**Order**					
ABC	**Daily Demand+**	**Daily demand**	**Hours**	**Days**	**Days**	**Hours**	**Q_i**	**Shelving**	**Available/Hour in-**	**Demand / Hours**
C	5	1.0	*	*	7	93	*	27	72	1
A	5	1.0	*	*	3	187	*	70	152	1
C	7	1.0	*	*	7	118	*	74	109	1
C	1	1.0	*	*	7	300	*	50	174	1
B	1	1.0	*	*	5	144	*	24	144	1

* Take into account that, to solve the model in the solver, these spaces are left, since, in these, the optimal solution appears.

To develop the aggregate planning model, the planning horizon is established for a day and the demand to be supplied is identified, based on the demand histories given by the organization.

Supply Program								Horizon								inventory at the end of the week (Days)
								Friday				Saturday				
Reference	Monday	Tuesday	Wednesday	Thursday	Friday	Saturday	Cartons	8	11	14	15	8	11	14	15	
Acetaminofen adultos 500mg tabletas caja	Supply						1	174	171	168	166	166	163	160	158	16
Atiban 2mg tabletas caja x30tab. pfi							5	72	69	66	64	64	61	5	56	3
Epítasis 3g sobres caja x6sob.							1	152	149	146	144	144	141	138	136	13
Esparadrapo leukoplast 7.5x4.57cm. x1und							2	109	106	103	101	101	98	95	93	8
Aspan 50Mg Tabletas Caja x30Tab. Siegfried							1	144	141	138	136	136	133	130	128	12

Table 6 : Horizon model part 2

To make a better description of the model, each column will be described:
Horizon model part 1:

ABC: Classification of products in categories ABC according to the rotation of the said reference.

Code: «Article code» or «Reference number».

Product Location: Coordinates of the product location in the distribution center as its respective one; sector, hall, level, rack, and position.

Sector: Zone of the winery to which it belongs.

Hall: Hall in which the shelf is located.

Level: Level of the rack in which it is found numbered from 1 which is the closest to the ground.

Rack: Subdivision of the shelf where it is located

Position: Specific position of the product inside the cubicle

BUO: Basic ordering unit.

Available: Units of the product in the storage area.

Demand: Units demanded by the client; information of its historical data granted by the company.

Daily demand +desviation: It is equivalent to the daily demand summation of said reference plus the deviation of that demand.

Daily demand destivation: It is the deviation of the daily demand for that product.

Reorder: This point indicates when the order must be placed either in hours or days of optimal inventory using the tool Solver

$$ROP_i = \frac{I_i}{D_i} \tag{5}$$

Order: Q Is the optimal order size for each reference, this value is extracted from running the optimal supply model by means of the solver tool.

$$Q = Q_i \tag{6}$$

To this item belong the following four columns: Hour, Days, Q, and shelving.

Hour: Duration of my inventory in stock in hours.

$$\text{Duration inventory}=\frac{\text{Inventory in shelving(units)}}{\frac{\text{Demand+desvaition(units)}}{\text{8 (hour)}}} \tag{7}$$

Days: Duration of the inventory in stock in days

Q: Corresponds to the optimal order size for each reference. This value is extracted from the execution of the optimal supply model by means of the Solver tool.

Shelving: Round the order size Q, according to the multiple of the BUO.

$$\frac{Q}{BUO} = \text{Enterer} \tag{8}$$

$$\frac{Q}{BUO} * BOU \tag{9}$$

$$\text{Available /hour inventory} := \frac{\text{Available(units)*Hour}}{\text{Gondola(units)}} \tag{10}$$

$$\frac{\text{Demand}}{\text{hour inventory}} := \frac{\text{Demand+desviation(units)*Hour}}{\frac{\text{Shelving(units)}}{\text{8(hours)}}} \tag{11}$$

Available/Hour inventory: Equivalent to the product among the reference units on the shelf for the hours of inventory on shelving.

$$\text{Available/Hour inventory} = \frac{\text{Available*Hour}}{\text{Shelving}} \tag{12}$$

Demand/Hour inventory: It is equivalent to the product between the daily demand and the deviation from the reference for the hours of duration of the inventory on shelving, all the above about 8 hours a day of the working day.

$$\text{Demand/Hour inventory} = \frac{\frac{\text{Daily demand +desviation*Hour}}{\text{Shelving}}}{8} \tag{13}$$

Horizon model part 2:

Supply program: This section of the Table, the planner calculates the day to provisioning on the shelf. This allows the process of preparation of all references to be carried out in an automatic way with the ability to exercise greater control of the workforce. Likewise, the units to be supplied in the form of corrugated boxes are shown.

Horizon: In this, the working days of the week are divided into 4-time slots of three hours each, and the units in stock are calculated with the following conditional equation.

$$\frac{\text{AV}}{\text{Hours}} \leq \text{Reorder(Hours); Order(hours)} + \frac{\text{AV}}{\text{Hours-(Demand*3)}} \tag{14}$$

$$\frac{\text{AV}}{\text{Hours}} \geq \text{Reorder(Hours);} \frac{\text{AV}}{\text{Hours-(Demand*3)}} \tag{15}$$

The column inventory at the end of the week shows the equivalent of the units in inventory in days, for how many days this inventory reaches me to meet the demand of the customers.

Furthermore, it is possible to observe the items that were used to find the costs associated with the supply plan. See Table 7.

Table 7: Costs associated with the supply plan

Item	Price	Unit
Inventory maintenance cost	0,001	€ /unit/day
Logistic operator direct	0,02	€ /hour
Forklift cost	0,04	€ / hour
Ordering cost	0,01	€ / unit/ chance
Lead time	0,07	minutes

Inventory maintenance cost:

$$\text{Inventory maintenance} = 33{,}6\left(\frac{\text{Cm}^3}{\text{unit}}\right)*0{,}008 = 0{,}001\frac{€}{\frac{\text{unit}}{\text{day}}} \quad (16)$$

Logistic operator direct workforce cost: It has one operator.

$$\text{Logistic operator direct workforce} = 1\text{hour}*1{,}22\frac{€}{\text{hour}} = 1{,}22\left(\frac{€}{\text{hour}}\right) \quad (17)$$

Forklift cost: It has one machine.

$$\text{Forklift cost} = 1\text{hour}*2{,}49\frac{€}{\text{hour}} = 2{,}49\left(\frac{€}{\text{hour}}\right) \quad (18)$$

Ordering cost: It has one machine and one operator.

$$\text{Forklift \& logistics operator cost} = 1\text{hour}*3{,}71\frac{€}{\text{hour}} = 3{,}71\left(\frac{€}{\text{hour}}\right) \quad (19)$$

Total costs: It is equivalent to the sum of the product between units in inventory by the cost of keeping the inventory plus order size by cost make an order.

$$\sum_{i=1}^{12.000} I_i \cdot C_{si} + Q_i \cdot C_{Pi} \quad (20)$$

Initially, the company established an ABC classification for its products, the ABC analysis consists of applying the Pareto 80/20 principle to segment the products. In this case, the inventory is classified according to its importance, based on the inventory value of each reference, the inventory value is calculated by multiplying the annual demand by the unit cost. Then, the results are sorted and grouped according to the percentage they represent with respect to the total; this percentage is accumulated until completing the approximate for each category as follows:

- Category A: It is estimated that 20% of the references are equivalent to approximately 80% of the value of the inventory.
- Category B: 30% of the references are equivalent to approximately 15% of the inventory value.
- Category C: 50% of the references are equivalent to approximately 5% of the inventory value.

For this situation, a reorder point of 6 hours is applied, and the size of the order Q depends on the ABC classification. That is, the order represents 3 days of inventory for category A products, 5 days of inventory for category B, and 7 days of inventory for category C. Using the maintenance costs of the inventory and its supply in the current planning model, the associated costs were calculated, resulting in € 17,907.48 per day.

Table 8: ABC Cost Classification

Classification	**Inventory**	**Q_i**	**ROP (hour)**	**Cost keep inventory**	**Cost order**	**Total**
C	203	9	6	0,0021	0,0000082	0,0021082
A	4	73	6	0,00036	0,0000027	0,0003627
C	94	36	6	0,00078	0,0000027	0,007827
C	61	219	6	0,0056	0,0000087	0,0056087
B	94	18	6	0,00024	0,0000027	0,0002427

Applying the self-regulating planning model, optimal amounts of maintenance and supply were determined, obtaining a cost of € 14.852.17 per day, showing a savings of 17%, annually represented in € 1.07 million. See Table 8.

6.3.1 Decision variables and vectors

Variables

I_i The number of units in stock in picking area for the reference i.

Q_i The number of units to be ordered from the storage area for the reference i.

Vectors

C_i Maximum capacity in units of reference i.

Cs_i Cost of maintaining a unit in stock of reference i.

Cp_i Cost of ordering a storage unit of reference i.

L_i Lead time of the reference i.

Ii_i The initial invention in the picking area of reference i.

D_i Daily demand per reference i.

SS_i Security stock of reference i.

The next step in the construction of the model is establishing the objective functions.

6.3.2 Objective function:

Minimize the total cost of the organization during the time of the planning, in this case, it is one day.

$$\text{Min Z} = \sum_{t=1}^{8311} I_i \cdot Cs_i + \sum_{t=1}^{8311} Q_i \cdot Cp_i \qquad (21)$$

6.3.3 Subject to:

The productivity of the workforce: The productivity of the workforce is defined as the relationship between time, work time, delivery time, and the number of units ordered. The working day is 8 hours daily.

$$\frac{(8\cdot 60)}{L_i} - Q_i \geq 0 \tag{22}$$

Inventory: The inventory of the reference I must be equal to the inventory in the immediately previous period, plus the order size made in that period, minus the required amount, and minus the inventory that must remain in stock, in order to guarantee the production flow.

$$Ii_i + Q_i - D_i - I_i = 0 \tag{23}$$

Picking capacity: The number of daily units in inventory cannot exceed the capacity of the picking area.

$$I_i \leq C_i \tag{24}$$

Security stock: The daily units in the inventory of reference, cannot be less than the security stock.

$$I_i \leq SS_i \tag{25}$$

Constrain positive: Ensure that the variables take positive values.

$$I_i, Q_i \geq 0 \tag{26}$$

7 Results

The quantity obtained with the solver model of the supply plan of the inventory units and order size for the first six SKU's (Stock Keeping Unit), is detailed in Table 8 and the reorder point is detailed in Table 9, from these data the costs of maintaining and ordering were calculated. The cost for the first six SKU's (Stock Keeping Unit). with the proposed model is detailed in Table 11.

Table 9: Available inventory and quantity to order

Reference	I_i	Q_i	**Units**
Acetaminofén adultos 500mg tabletas caja.	12	4	Units
Atiban 2mg tabletas caja x30tab. pfi	51	0	Units
Epítasis 3g sobres caja x6sob.	60	0	Units
Esparadrapo leukoplast 7.5x4.57cm. x1und.	28	0	Units
Aspan 50Mg Tabletas Caja x30Tab. Siegfried	23	0	Units

Table 10: Reorder point by product

Reference	ROP	Units
Acetaminofén adultos 500mg tabletas caja.	0	Days
Atiban 2mg tabletas caja x30tab. pfi	8	Days
Epítasis 3g sobres caja x6sob.	7	Days
Esparadrapo leukoplast 7.5x4.57cm. x1und.	28	Days
Aspan 50Mg Tabletas Caja x30Tab. Siegfried	23	Days

Table 11: Inventory costs

Reference	Inventory maintenance cost	Ordering cost
Acetaminofén adultos 500mg tabletas caja.	€ 0.024	€ 0.33
Atiban 2mg tabletas caja x30tab. pfi	€ 0.018	€ *
Epítasis 3g sobres caja x6sob.	€ 0.046	€ *
Esparadrapo leukoplast 7.5x4.57cm. x1und.	€ 0.016	€ *
Aspan 50Mg Tabletas Caja x30Tab. Siegfried	€ 0.0054	€ *

*Cost equal to zero since there is no requirement for that product, this scenario is possible because there is an inventory available.

The product “Acetaminofén” in Table 8 has an inventory (Ii) of 12 units and to order (Qi) of 4 units. In Table 9 the reorder point is equal to 0 since the order is the same day. This reorder point is calculated from the units in the inventory of the solver and the associated daily demand. The cost associated with an inventory of this reference in Table 10, shows an inventory maintenance cost of € 0.024 per reference per day and a cost of the ordering of € 0.33 per reference.

The total costs of maintaining inventory and the total cost of orders was € 14.852.17 per day for all SKUs.

8 Improvements

The development of such applications is characterized by economic and standardized, such as Excel, it makes the operation of the company faster and more effective through planning and an autonomous system that automatically updates, having information current in the cloud. and building a competitive advantage over the supply chain. However, Industry 4.0 talks about cutting-edge practices with software and advanced robots, creating a paradigm that is almost impossible for medium and small companies. This publication is intended to convey the opposite message. Industry 4.0 is also knowledge, regardless of the medium or the tool, as long as a practice determined by this fourth generation is guaranteed, in the particular case, through an approach to artificial intelligence. For companies that do not have the necessary resources to invest in an ERP (enterprise resource planning), this application can guarantee the planning of a high number of SKU’s (Stock Keeping Unit), with an attractive cost-benefit ratio for the small companies.

The aggregate planning has an important role in the economic field since it allows to meet the requirements of the clients and to save resources. The optimization achieved through fields of artificial intelligence as it is the machine learning which uses multiple algorithms such as linear regression based on the principle of minimization of linear programming allows companies to:

- Determine the optimal level of inventory in real time without any intervention.
- Reduce storage costs based on demand data in real time.
- Adapt inventory policies autonomously depending on the market situation.
- Improve productivity at levels that surpass the best possible case without the use of 4.0 technology
- Assertive management of demand variability.

9 Relevance in the Sector

In Colombia, logistics costs are equivalent to 13.5% of the final price paid for the product, keeping this in mind, a reduction in the percentage directly and positively impacts the acquired profit; i.e. the lower the production costs, the higher the profit. This allows their prices to be more competitive compared to the drugstore and retailer sector, which creates a healthier competition that benefits consumers and the whole drugstore market economy. This also encourages the growth of better planned supplying systems, propelling this sector to be at the forefront of the logistics field.

Taking into account that technology in distribution centers is very expensive for small and medium enterprises, this application shows through the guild, that tools and technology 4.0 are available to organizations for free or economic. For example, Excel, cloud, QR readers, etc. With this development, the intention is based on stimulating innovation in the planning of complex operations, which have a high cost in their processes, and generating creativity with low-cost technological tools, breaking the paradigm that technology and revolution 4.0 are a high-cost fashion.

10 Conclusions

This autonomous system, with clear policies in the optimal management of inventories, especially in the picking areas, pulls the operators to be more continuous in their operation, in such a way that they can finish their processes at the end of the shift, without needing to generate over time.
The implementation of the proposed plan is expected to achieve a service level of 95%, as well as an optimal minimization of costs at 6.28 % associated with the logistics process of storage.
The indirect cross docking process seeks to reduce the inventory at the minimum or to an almost non-existent point, with the quantity of inventory and the size of the order indicated in the plan, it is reduced to the quantity of inventory that best allows to carry out the operation of supply.
The decrease in the costs associated with the logistics process will generate greater profitability in the sector due to the impulse of a much better-structured supply. By applying the proposed model, a saving of € 1 million is evidenced by year, representing a 17% decrease in costs associated with the process, generating a strong competitive advantage in the market.

References

Alonso, Martínez, Dorado,Páez, Lota, 2018. National logistics survey 2018, Bogotá: www.puntoaparte.com.co.

Aldana, P., 2014. The cross docking as an important tool in the chain, Bogotá: University of Nueva Granada.

Anon., 2017. ingenieriaindustrialonline. [Online] Available at: https://www.ingenieriaindustrialonline.com/herramientas-para-el-ingeniero-industrial/producci%C3%B3n/planeacion-agregada-mediante-programacion-lineal/ [Accessed 01 May 2019].

Borissova, D., 2008. Bibliography. Cybernetics and information technologies, 8(2), pp. 102-103.

Chopra, M., 2008. Bibliography. En: L. M. C. Castillo, ed. Supply Chain management. Naucalpan de luárez (Mexico state): Pearson Education, pp. 56-57.

Columbus, 2018. 10 Ways Machine Learning Is Revolutionizing Supply Chain Management, New York: Forbes.

Dinero, 2015. Competencia ragulacion farmacias. [Online] Available at: https://www.dinero.com/edicion-impresa/negocios/articulo/competencia-regulacion-farmacias/215331 [Accessed 01 May 2019].

Dinero, 2019.Accelerated expansion plan in Farmatodo, Bogotá: s.n.

Espectador, E., 2016. El Espectador. [Online] Available at: https://www.elespectador.com/noticias/economia/colombia-hay-menos-3000-droguerias-de-barrio-articulo-654947 [Accessed 01 May 2019].

Fernández, I. A., 2011. Production and consumption: 49(1), pp. 179-191.

Gandhi, R., 2018. towards data science. [Online] Available at: https://towardsdatascience.com/introduction-to-machine-learning-algorithms-linear-regression-14c4e325882a [Accessed 25 April 2019].

Gholamian, M.-M., 2015. Comprehensive fuzzy multi-objective multi-product multi-site. 134(42), pp. 585-607.

Granja, A.-L., 2014. An optimization-based on a simulation approach to patient admission. Journal of Biomedical Informatics, Issue 52, pp. 427-437.

Julian, D., 2016. Designing Machine Learning Systems with Python. 1 ed. Birmingham B3 2PB: Packt Publishing Ltd.

Pereira, J., 2018. BigData mazine. [Online] Available at: https://bigdatamagazine.es/utilizacion-de-big-data-y-machine-learning-en-la-industria-4-0 [Accessed 01 May 2019].

Rüssmann, L., 2015. Bibliography. En: I. 2. A. r. r. The Boston Consulting Group, ed. The Future of Productivity and Growth in Manufacturing Industries. Boston: The Boston Consulting Group, p. 5.

Souza, C., 2018. Direct stockpile scheduling: Mathematical formulation •. 85(204), pp. 296-301.

IV.
Risk and Security Management

New Concepts for Cybersecurity in Port Communication Networks

Nils Meyer-Larsen[1]*, Rainer Müller*[1] *and Katja Zedel*[1]

1 – ISL Institute of Shipping Economics and Logistics

Purpose: Seaports are to a growing extent controlled by IT systems. Smooth information exchange is crucial for business. Downtimes give rise to substantial financial losses and supply bottlenecks, due to the interconnections in a complex alliance of port stakeholders' systems. Hence, Cyber security in an important aspect in Port Communication Systems.

Methodology: The project SecProPort will develop a holistic IT security architecture for port communication networks, which will support the security requirements of the stakeholders' operating procedures, protect them against sabotage, and prevent third parties from illicitly gathering sensitive data or getting unauthorized access to the communications network.

Findings: The desired architecture is to be implemented by first analyzing typical attack scenarios targeted at the data processed in the port communication alliance. The next step entails designing the actual security architecture for the alliance and installing a prototype in collaboration with the application partners.

Originality: SecProPort aims at a holistic approach with respect to secure port communications networks, rather than addressing individual stakeholders. The architecture will also provide resilience measures for minimizing the impact on other actors in the alliance in case of an incident, and returning to normal operation in a controlled manner.

Keywords: Port Communication Network, Maritime Cybersecurity, Port Community System, Maritime Blockchain

First received: 10.May.2019 **Revised:** 22.May.2019 **Accepted:** 02.June.2019

1 Introduction

In today's maritime logistics, cybersecurity is an issue of high importance. A number of incidents such as the NotPetya attack on Maersk in summer 2017 impressively demonstrate that cyber threats impose a high risk of considerable financial and reputational damage for the maritime industry. In 2016, the first maritime cyber-security survey conducted by IHS Markit in association with BIMCO among maritime-related businesses concluded that more than 20% of respondents had been a victim of a successful cyberattack, causing damage to their IT systems (Safety at Sea, 2016). Consequently, sustained efforts are needed to be prepared for cyberattacks. This paper presents the recently started project SecProPort (SecProPort, 2018), co-funded by the German Federal Ministry of Transport and Digital Infrastructure in the IHATEC program. The aim of the project is to systematically develop a security architecture for the communications network in sea ports and inland ports, based on an in-depth process and threat analysis.

2 Current Status of Cybersecurity in Maritime Transport

Maritime transport is central to the world economy. More than 90 percent of intercontinental goods are transported by sea. In that way, ports are a key prerequisite for economic success. Consequently, significant disruptions to large ports therefore can negatively impact maritime supply chains and can cause damage to the maritime and trading industries. One possible target of attacks is information and communication technologies, as in modern ports, the entire cargo handling system today is based on IT systems and the data exchange between a large number of involved partners is centrally organized. Port community systems (PCS), centralized information and data hubs, which provide data exchange within the port communications network in many ports, play an important role here. They by definition have a large number of interfaces to many different partners: Customs, terminal operators, ship owners, ship brokers, truck operators, rail operators, port railway, inland waterway operators, forwarding agencies, port authorities and other authorities as well as other companies. The interfaces are technically heterogeneous: UN/EDIFACT (United Nations Electronic Data Interchange for Administration, Commerce and Transport), XML (Extensible Markup Language), RPC (Remote Procedure Call), SOAP (Simple Object Access Protocol), and other protocols, but also e-mail. In addition, there are bilateral communication channels that bypass the PCS but are nevertheless relevant for overarching security considerations (Figure 1). Securing the PCS or individual partners alone against cyberattacks therefore does not necessarily lead to a secure overall system. According to (DVZ, 2019), a recent survey concluded that cybercriminals are often using smaller companies which possess a limited cybersecurity standard as a gateway to get access to larger companies. Conventional IT security measures such as firewalls only provide limited security, because in the course of digitization the different systems in question are supposed to communicate with each other. Consequently, deliberately open access paths through firewalls must be created in order to facilitate the automation of port processes.

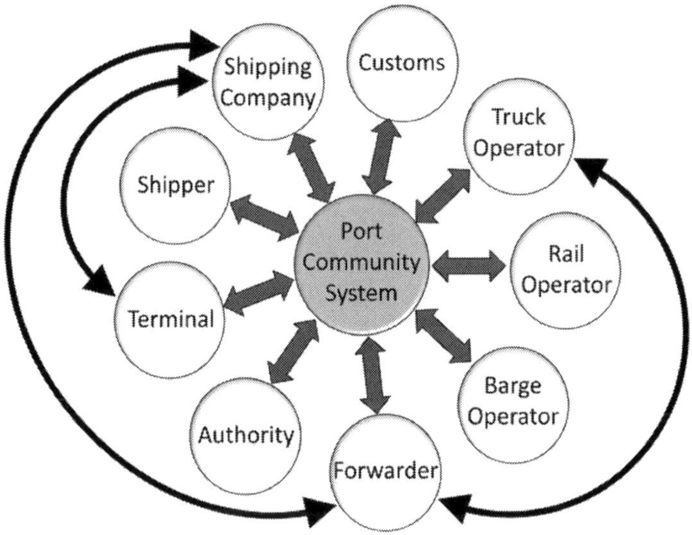

Figure 1: Multilateral communication network in port traffic with exemplary bilateral communication processes (ISL, 2019)

Cyber attacks nowadays are a high risk, as attacks can be executed from a secure distance with relatively little risk, unlike traditional physical attacks (Sensiguard, 2017). Systems can be observed over a longer time, and the discovered weaknesses can later be exploited for a large-scale cyber attack. Another problem with the continuing lack of cyber-security awareness is that attacked companies are often reluctant to report incidents "as they fear reputational damage" (Meyer-Larsen, 2018).

According to the German Bundeskriminalamt, 85,960 cases of cybercrime were perpetrated in 2017 in Germany alone, an increase of 4% compared to

2016. In the area of computer fraud alone, the damage amounted to € 71.4 million (BKA, 2017). The number of unreported cases is estimated to be very high. The case of the NotPetya attack on Maersk Shipping Company in 2017, which left several important shipping tradelanes unavailable for several days worldwide, is estimated to have resulted in a loss of approximately $ 200-300 million (Heise, 2017). During the attack, malicious software infiltrated a large number of computers from various companies and caused, among other things, loading and unloading processes of container vessels to stop. According to various experts, NotPetya was a Wiper Trojan of Russian origin, with the primary goal to disrupt operations and cause financial damage.

An important fact is that, due to the increasing interlinking between the various systems of each port operator, the attackers' modi operandi have changed. If an attacker succeeds in becoming a member of the network - be it through an external attack on the IT system of a participant or as an innate perpetrator - he can try to import manipulated messages into the communication network. Although these cannot at first sight be identified as harmful, they can potentially lead to undesirable effects, disturb operations, and support criminal activity. Consequently, even if the individual systems of the port's operators are protected according to the state of the art, this does not automatically guarantee optimal protection of the entire port communications network with its complex interactions.

Failure of the port suprastructures would not only result in financial consequences, but could also lead to supply shortages of industry and population. Confidential data may be tapped via manipulated user accounts to enable or support criminal acts, e.g. drug smuggling. Last but not least, serious safety risks can arise if dangerous goods are not handled and monitored properly as a consequence of data manipulation.

There are a number of different motivations for cyber attacks on ports and their systems, as already outlined in earlier articles. First, there are criminal groups with financial or ideological motivation. In general, criminal organizations seek to obtain information through cyber attacks, such as spyware data, to facilitate cargo theft, smuggling, or even terrorism. Another variant of attack results in the encryption of data of the port systems by utilizing ransomware, requiring the victim to pay a ransom fee to regain access to his productive data. A second group is commonly referred to as hacktivists, whose main motivation is to demonstrate their capabilities by detecting weaknesses in IT systems and conducting cyber attacks, which, in the case of ports, can cause significant disruption of port operations with the above-mentioned consequences. The third group are foreign governments and competing industrial companies. Their goals are "espionage and the identification of possible vulnerabilities of foreign port systems", which can be exploited for subsequent attacks (Meyer-Larsen, 2018).

As already explained above, the IT security of a port communications network can not be guaranteed solely by individual security measures of single actors - rather, an overarching coordinated concept of security requirements and guarantees between the actors involved is necessary. The communications network in the port is usually a structure that has evolved over decades, which has been expanded and adapted over the years according to the requirements of the involved actors. Overarching security concepts were rarely used. However, an appropriate IT security architecture is mandatory if port applications are to be protected against increasingly sophisticated cyberattacks. The system-inherent openness of the port communications network to new, potentially untrustworthy, actors requires the introduction of preferably automated verification procedures in order to be able to detect any (insider) attacks such as espionage and sabotage in due time and successfully defend them. Furthermore, the IT security architecture should implement resilience strategies in order to limit the effects of successful attacks and to maintain the working capacity of companies and port operations as much as possible. Regular operating conditions should be re-established as soon as possible after an attack. In addition, effective security mechanisms for the defense against external attacks have to be implemented, for example, in order to detect malicious software such as viruses, trojans and worms in a timely manner and to disable them.

The security of the entire communications network appears in an interplay of security requirements, measures of the involved actors, and the network infrastructure itself. The described security architecture thus places demands on the internal security mechanisms of the individual communication partners with their specific roles in the network and requires the willingness to cooperate of all involved parties. Since in many cases the involved communication partners are competitors in their businesses, a major challenge in building a security architecture is ensuring the confidentiality of information while maximizing transparency.

3 Methodology

Especially with reference to the transportation sector, an improved understanding of cybersecurity-related mechanisms is required (Chiappetta, 2017). Thus, the SecProPort project aims to investigate and further develop the IT security of relevant port processes by developing an IT security architecture for the various processes in cooperation with individual port operators. The further development of IT security in the port domain is facilitated by developing adequate migration plans and supporting software tools. A preventive approach is pursued by addressing current weaknesses in the port processes, thereby reducing the likelihood of cyber attacks on the port processes and their supporting systems in the future. Furthermore, the process-oriented approach to IT security allows a synergistic optimization of the existing work processes in ports.

As explained above, SecProPort will define an IT security architecture for the port communication network. It will be implemented in demonstrators that relate to specific scenarios. The IT security architecture will include cryptographic building blocks (encryption, cryptographic hash functions, digital signatures, public key infrastructures) as well as comprehensive role-based authorization concepts and federated identity management. Other aspects are related to information flow control, which involves intelligent monitoring of IT components within the port communications network together with attack detection in the form of intrusion detection systems and intrusion prevention systems (IDS/IPS).

The goal of SecProPort is to implement the project results into companies' internal security concepts wherever possible and feasible. The partners represented in the project will examine to what extent the developed solutions and concepts can be integrated into the companies' internal security policies. In particular, this concerns the results of the work on the security architecture and corresponding mechanisms, on incident response management, and on migration strategies. In addition, these results will also be made available to other companies in the port environment and to other industries.

The IT security architecture to be developed is based on the following four scenarios, which are subject to a requirements analysis with regard to the architecture and later will be utilized for the evaluation of the solutions developed in the project:

1. Dangerous goods registration via the National Single Window
2. Container logistics, including direct communication between ship owner and terminal, bypassing the PCS
3. XXL logistics, involving the transport and shipment of large goods such as wind turbine or aircraft parts
4. Inland port terminal, including respective communication processes in an inland port without PCS functionality.

These scenarios in particular imply the following communication framework to be investigated within the port communication network. A bilateral communication between ship owner and terminal operator, communication via the PCS from a forwarding agent or a logistics provider and a shipowner's dangerous goods declaration via the PCS are considered. The inland port scenario considers a fully distributed communication of the partners involved.

Furthermore, the identified security requirements will contribute to an industry-specific security standard that will be developed in cooperation with key stakeholders in IT security such as the German Bundesamt für Sicherheit in der Informationstechnik (BSI) or the European Union Agency for Network and Information Security (ENISA) at European level.

4 The Potential of Blockchains to Solve Cybersecurity Issues

Blockchain technology is one of the disruptive approaches that could fundamentally change the way electronic business communications operate, as the World Economic Forum writes in a publication on the potential of blockchain technology (World Economic Forum, 2017). The blockchain methodology can be understood as a distributed audit-proof database, which consists of a linked list of data blocks. In the respective next block, the predecessor block is linked in a cryptographically secured way, so that a subsequent change of a block in the chain either causes an interruption of the chain or requires amendments to all subsequent blocks. In any case, changes of the information contained in the blockchain can be determined. The blockchain network consists of a set of peer-to-peer networks that synchronize with each other according to established rules in order to verify the data contained. Each of the nodes participating in the blockchain network holds a complete copy of the entire blockchain. Compared to conventional centralized systems, the decentralized blockchain therefore has a higher reliability and an improved availability.

In addition, the integrity of a blockchain, i.e. protection against subsequent manipulation, is ensured by built-in cryptographic concatenation, which improves and simplifies transaction security in distributed systems

(Pilkington, 2016; Prinz, 2017). The important security aspects of blockchain technology include:

- manipulation and auditing security
- resilience and optimized availability
- the security of users' cryptographic identities

As explained above, the blockchain's predecessor block is linked to the next block in a cryptographically secured way. Each subsequent modification of a block alters the cryptographic key value of the block. Since this key is contained in the following block, this change is clearly recognizable and would immediately lead to discarding the changed blocks.

Blockchain mechanisms are ideal for securing transactions in distributed systems. In that way, blockchain methodology is considered suitable to protect data exchange related to supply chain operations against manipulation. For example, many research projects in the area of Internet of Things (IoT) and also in the field of transport and logistics are currently developing and evaluating blockchain mechanisms (BASTONET, 2019; Eurotransport, 2019; IBM, 2018). Nevertheless, their compatibility with organizational and legal requirements must still be investigated. Technical issues in this area include crypto procedures, audit security, reliability, availability, vulnerability and tamper resistance of blockchain-based systems.

IBM and Maersk have set up a joint venture with the goal of using blockchain technologies in the supply chain, especially in container traffic, and to secure the transactions during communication procedures accompanying the supply chain (Computerwoche, 2018). Even with medium to long-term penetration of blockchain technology, however, long transition times must be expected in which classical EDI-based port communication and

new blockchain approaches coexist, e.g. in the communication of mandatory reports with authorities. In this respect, the use of converters to communicate between "old" and "new" world is an important aspect.
Migration strategies are required. Dobrovnik (2018) suggests single-use cases as first steps with regard to the implementation of blockchain technology in logistics in order to minimize the risk of failure, because appropriate implementations can be based on existing conventional applications and thus support a smooth transition to the new technology. A respective approach will allow involved parties to gain experience and skills required for more advanced applications (Iansiti, 2017). SecProPort will follow this approach by investigating aspects of the implementation of blockchain technology in restricted scenarios.

5 Conclusion

The methods of cybercriminals are constantly evolving. In order to ensure that the maritime transport industry, with its dependence on electronic information and communication flows, is protected against respective attacks in the future, appropriate efforts must be taken to safeguard the respective communications processes in terms of confidentiality, integrity and authenticity and to make them resilient to attacks. These security mechanisms include the constant monitoring of the communication infrastructure, cryptographic encryption, as well as reliable detection and removal of malware. Certifications and regular safety audits can be used to check and document adherence to the security measures introduced. Corresponding regulations should be combined with recognized certification standards such as ISO / IEC 27001.

Furthermore, improved methods for authenticating the various involved partners, such as digital signatures, are needed in order to prove the identity of the respective communication partners beyond doubt. In this context, the implementation of blockchains can also help to ensure the authenticity or liability of transactions within the port communication network. With regard to the application of blockchains in maritime logistics, however, many legal and organisational issues remain open. The decentralized structure as well as transparency and immutability raise numerous problems and questions in legal terms, especially with respect to liability and data protection. Nevertheless, blockchains are expected to facilitate the digitization of supply chains, improve transaction visibility, and provide improved security against criminal activity. The use of blockchain technologies seems particularly useful in scenarios that are decentrally organized, such as in inland ports, which, unlike seaports, operate without a central PCS. Various current research projects are currently investigating these aspects. In the medium term, respective results can be expected, which will form a basis for further developments.
SecProPort contributes to respective efforts by investigating aspects of the implementation of blockchain technology and other IT security technologies in restricted scenarios. In the first project phase, a detailed analysis of relevant business processes in different scenarios was performed, which will form the basis for a requirements analysis with respect to the security architecture to be defined, followed by the development of respective concepts and tools. In the final project phase, the developed tools and methodologies will be validated and implemented within the SecProPort scenarios. Furthermore, the identified security requirements will contribute to an

industry-specific security standard that will be developed in cooperation with key stakeholders in IT security.

Financial Disclosure

SecProPort is co-funded by the German Federal Ministry of Transport and Digital Infrastructure in the IHATEC program.

References

BASTONET, 2019. [online] Available at: <https://bastonet.com/> [Accessed 10 May 2019]

BKA, 2017. Bundeslagebild Cybercrime 2017. [online] Available at <https://www.bka.de/DE/AktuelleInformationen/StatistikenLagebilder/Lagebilder/Cybercrime/cybercrime_node.html> [Accessed 10 May 2019]

Chiappetta, 2017. Chiappetta, A. (2017), Hybrid ports: the role of IoT and Cyber Security in the next decade. Journal of Sustainable Development of Transport and Logistics, 2(2), 47-56. doi:10.14254/jsdtl.2017.2-2.4.

Computerwoche, 2018. IBM baut mit Maersk Blockchain Plattform. [online] Available at <https://www.computerwoche.de/a/ibm-baut-mit-maersk-blockchain-plattform> [Accessed 10 May 2019]

Dobrovnik, 2018. Dobrovnik, M., Herold, D.M., Fürst, E., Kummer, S., Blockchain for and in Logistics: What to Adopt and Where to Start, Logistics 2018, 2(3), 18, 2018

DVZ, 2019. DVZ-Brief Nr. 18 vom 02.05.2019 / LOGISTIK, DVV Media Group GmbH

Eurotransport, 2019. Logistik sagt Manipulation den Kampf an. [online] Available at <https://www.eurotransport.de/artikel/ministerium-foerdert-blockchain-projekt-logistik-sagt-manipulation-den-kampf-an-10381602.html> [Accessed 10 May 2019]

Heise, 2017. NotPetya: Maersk erwartet bis zu 300 Millionen Dollar Verlust. [online] Available at <https://www.heise.de/newsticker/meldung/NotPetya-Maersk-erwartet-bis-zu-300-Millionen-Dollar-Verlust-3804688.html> [Accessed 21 March 2019]

Iansiti, 2017. Iansiti, M., Lakhani, K.R., The truth about blockchain. Harvard Business Review. 2017. [online] Available at https://enterprisersproject.com/sites/default/files/the_truth_about_blockchain.pdf> [Accessed on 1 June 2019]

IBM, 2018. Transform supply chain transparency with IBM Blockchain. [online] Available at <https://www.ibm.com/downloads/cas/1VBZEPYL> [Accessed on 10 May 2019]

ISL, 2019. Meyer-Larsen, N., own illustration, 2019

Meyer-Larsen, 2018. Meyer-Larsen, N./Müller, R.: Enhancing the Cybersecurity of Port Community Systems, in: Freitag, M., Kotzab, H., & Pannek, J. (2018). Dynamics in Logistics – Proceedings of the 6th International Conference LDIC 2018, Bremen, Germany. Springer, Cham, S. 318-323.

Pilkington, 2016. M.Pilkington: Blockchain Technology: Principles and Applications. Research Handbook on Digital Transformations, edited by F. Xavier Olleros and Majlinda Zhegu. Edward Elgar, 2016

Prinz, 2017. W.Prinz, A.Schulte, Blockchain – Technologien, Forschungsfragen und Anwendungen, Blockchain Positionspapier, https://www.aisec.fraunhofer.de/content/dam/aisec/Dokumente/Publikationen/Studien_TechReports/deutsch/FhG-Positionspapier-Blockchain.pdf, 2017

SecProPort, 2018. [online] Available at <https://www.isl.org/en/projects/secproport> [Accessed on 9 May 2019]

Safety at Sea, 2016. IHS Fairplay Maritime Cyber-security Survey – the results. [online] Available at <https://safetyatsea.net/news/2016/ihs-fairplay-maritime-cyber-security-survey-the-results/> [Accessed on 9 May 2019]

Sensiguard, 2017. SensiGuard Supply Chain Intelligence Center, Global Intelligence Note 6, October 2017

World Economic Forum, 2017. Tapscott, D. and Tapscott, A., Realizing the Potential of Blockchain, White Paper, Cologny/Geneva, 2017

Smart Risk Analytics Design for Proactive Early Warning

Katharina Diedrich[1] and Katja Klingebiel[2]

1 – TU Dortmund

2 – FH Dortmund

Purpose: Automobile manufacturers are highly dependent on supply chain performance which is endangered by risks. They are not yet able to proactively manage these risks, often requiring reactive bottleneck management. A proactive and digitalized early warning method is needed.

Methodology: The publication provides methodological-empirical contribution to proactive early warning resulting in a smart risk management approach. The methodological approach is carried out according to the design science research approach.

Findings: The developed smart risk management enables an automated, objective and real-time ex-ante-assessment of supply chain risks in to secure the supply of the automobile manufacturer. Smart risk analytics based on artificial intelligence is shown with its suitability for proactive early warning using the example of inaccurate demand planning.

Originality: The analytical approach provides insights into the flexibility of supply chains under risk and the impact over time, which is applied in the proactive early warning design. Artificial intelligence is applied to predict and assess supply chain risk events.

Keywords: Supply Chain Risk Management, Proactive Early Warning, Smart Risk Analytics, Machine Learning

First received: 19.May.2019 **Revised:** 24.May.2019 **Accepted:** 12.June.2019

1 Motivation

The profitability of automobile manufacturers (original equipment manufacturers, OEMs) highly depends on the reliability and performance of suppliers and supply chains (SCs) due to e.g. outsourcing, lean management and low inventory levels (Khan, 2018, pp. 141). The value added in the automobile industry originates by 30 percent from the OEM and by 70 percent from the SC (VDA, 2018). Therefore, the monitoring of supply chain risks (SCRs) which endanger the performance of the SC is of great relevance to OEMs.

An ordinary car consists of about 5000 parts (Kırılmaz and Erol, 2017, p.62). Most parts are provided by a large number of suppliers. This implies complex SC networks, which require considerable planning and control effort on the one hand and are prone to significant risks on the other hand.

These SCRs may be classified as internal process and control risks, supply risks, demand risks and environmental risks (Christopher and Peck, 2004, p. 10). However, OEMs are not yet able to fully manage these risks proactively. Currently, this implicates a reactive bottleneck management in the event of disruptive risk impacts. There is a need for a scientific method that supports a more proactive management to avoid or minimize the risk impacts and can be implemented easily in practice.

SCR impacts can be classified as financial, performance, operational, demand, informational, relational, hybrid and other (Oliveira, et al., 2019 p. 24). Economic impacts are connected to cost, profit and revenue and operational impacts are related to e.g. stockouts, reliability, capacity, which are features of the processes (Oliveira, et al., 2019 pp. 24). Financial impacts in the automobile industry are for example an increase in procurement costs or an increase in transportation costs (Kırılmaz and Erol, 2017, p.62). One missing part in car production can lead to delays (Lopez and McKevitt, 2017). The protection of assembly lines from delays due to missing parts, which is one aspect of SCR impacts for automobile manufacturers, is an important concern. A stopped assembly line leads to high costs in the dimension of more than $ 100 million per day of production breakdown (Kırılmaz and Erol, 2017, p. 61). In the worst case, SCRs lead to a complete production shutdown of the OEM, which is associated with immense costs (Khan, 2018, p. 141). For example, BMW had to stop the production of several models in summer 2017, as first tier supplier Bosch was unable to supply steering gears (Lopez and McKevitt, 2017). The reactivity of existing methods to manage SCRs at the OEM poses a challenge with regard to the high impact of the risks, the limited possibility to predict potential losses and the lack of opportunities to act. In result, due to the high dependency on SC performance, the variety of highly specific parts, the complexity of global SC networks and the immense damage caused by supply risks, the proactive management of fluctuations in demand planning is of crucial importance in supply chain risk management (SCRM). Digitalization is expected to further increase the availability of data in SC networks. Through the application of suitable technologies such as data mining (DM), big data (BD), machine learning (ML) and artificial intelligence (AI), risks can be detected and controlled more quickly (Schlüter, Diedrich and Güller, 2017). DM, ML and AI technologies provide new opportunities for early warning in SCRM to turn data into correlation and new knowledge.

Consequently, this contribution focusses on the supply risks resulting from inaccurate automotive demand planning and the opportunities which arise through the new technologies of ML and AI for SCRM.

The developed methodological-empirical contribution to proactive early warning is subsumed into a smart risk management approach. The methodological approach is carried out according to the design science research (Hevner, 2007). It results in a scientific concept which applies ML and AI technologies on empirical company data for demonstration and evaluation. The underlying potential for automatization and assistance supports the proactive SCRM and underlines the practical relevance.
This paper proceeds as follows: Chapter 2 reviews automotive demand planning, early warning and SCRM, which provides a basis for the paper. Section 2.3 summarizes the need for proactive early warning and design requirements. In chapter 3 the smart risk analytics design for proactive early warning is developed. Demonstration and evaluation is presented in chapter 4 based on empirical company data. Finally, a discussion and conclusion is presented within chapter 5.

2 State of the Art

2.1 Automotive Demand Planning

In build-to-order production, demand planning is a key component of the order management. The automotive demand planning is situated between the conflicting priorities of supporting maximum demand flexibility and long-term stable control of supply. Demand flexibility is needed for the realization of technical trends and customer requirements. At the same time, a long term stable control of supply is required because of the long lead times in global SCs and the high degree of product specialization. Rolling demand planning is followed by capacity planning to ensure that suppliers can deliver the demands within the scope of a certain contractually agreed flexibility (Volling and Spengler, 2011; Volling, et al., 2013). In order to secure the supply of the OEM, a long-term pre-planning of customer orders and market demands is carried out. Later on, SCs are controlled on the basis of delivery call offs to first tier suppliers. Rolling demand planning is used to achieve accurate control of supply for volatile demands in global customer markets. A more detailed description of the process can be found in Gehr and Hellingrath (2006, p. 52). A study conducted by IBM / Aberdeen suggests that a rolling forecast improves the accuracy in comparison to a less static process by roughly 14 % (Cavicchi, 2018, p. 4).

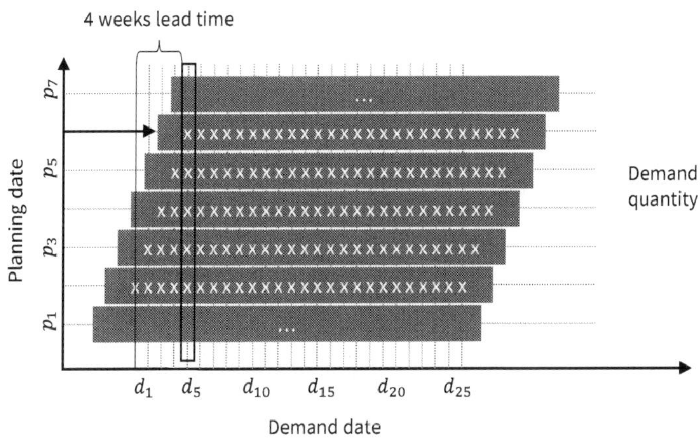

Figure 1: Data in the Rolling Process of Automotive Demand Planning

An example of the availability of demand planning data in relation to the demand date in a rolling process is depicted in Figure 1. Every horizontal line represents a planning stage of the demand quantity with regard to a specific demand date (x-axis). Until the demand date is reached, the demand quantity is planned on a weekly basis.

The pre-planned demand may differ significantly from later occurring actual customer orders, for example due to unexpected volumes of volatile, global sales markets and technological trends. The great variety of selectable car equipment results in an immense planning complexity (Thun and Hoenig, 2011, p. 244, Harland, Brenchley and Walker, 2003, Tang and Musa, 2011, p.28). According to a study, the top challenge relating to SC flexibility is the demand volatility and poor forecast accuracy with 74 % (Geissbauer and D'heur, 2010 p. 4). The risk of change in customer demands is assessed as high impact high probability in the automobile industry (Thun and Hoenig, 2011, p. 246). If the demand changes, a certain flexibility of the SC is required to ensure the ongoing supply of the OEM. Flexibility is an individual attribute of every SC. Beyond that, if flexibility is exceeded because of inaccurate demand planning, further fluctuations in demand may lead to the significant, internal risk of insecurity of supply. Demand uncertainties arise frequently in volatile automobile sales markets. This requires high flexibility of the SC to adapt the production volume or the production product mix, in order to counter demand uncertainties with volume flexibility (Thomé, et al., 2014, p. 102).

Hence, demand risks arise for example from uncertainties in demand forecasts, which in turn might result in delivery bottlenecks (Thun and Hoenig, 2011, p. 244). The customer risk of placing orders (Harland, Brenchley and Walker, 2003) is part of this risk of inaccurate demand pre-planning. The demand risk can be differentiated into volume mismatch and mix mismatch (Ceryno, Scavarda and Klingebiel, 2014, pp. 5, pp. 11)

In order to avoid the risk of inaccurate demands, Thun and Hoenig, 2011 recommend a focus on secure sales markets and on a small number of variants with as little as possible demand volatility (Thun and Hoenig, 2011, p.

245). However, this is no option for German OEMs due to their position in highly competitive sales markets and their focus on high variety as a key selling argument. Furthermore, customer preferences necessitate regionally tailored product variants(Geissbauer and D'heur, 2010 p. 2), which is increased by legal requirements.

The key figures forecast quality key figure (FQ) and the tracking signal (TS) describing the quality of automobile demand pre-planning and the respective over- or underestimation have been defined by VDA (2008). FQ describes how the forecast demand quantity, marked in Figure 1 by the black frame, has matched the later occurring real demand in the retrospective. TS describes the over- or underestimation of the demand in quantity (VDA, 2008). Both key figures may be applied usefully in a proactive SCRM approach.

2.2 Early Warning and Supply Chain Risk Management for Automotive Demand Planning

In the context of this publication, the understanding of proactive SCRM is aimed at preventing or minimizing the damage caused by risk events, while the structure of the automotive SC remains unchanged. The need for proactive SCRM, especially in the automobile industry, has already been recognized as a relevant issue in the academic field some time ago (Henke and Besl, 2008). Nonetheless, there is still a lack of suitable models and methods which support OEMs to proactively manage SCRs.

The basic idea of proactive risk management using big data technology is shown by Leveling, Edelbrock and Otto (2014), but there is still a considerable need for research on design and conception. A proactive planning procedure is proposed by Kırılmaz and Erol (2017) for shifting orders among suppliers to mitigate supply risks. Yet, this approach is unsuitable in the case of single sourcing, which often occurs in the automobile industry, because demands cannot be distributed to other suppliers. Risk monitoring has received the least attention (Ho, et al., 2015). This leads to a research need on early warning monitoring systems and their empirical validation (Ho, et al., 2015). SCRM solutions do not yet provide a sufficient time period for taking measures against the risk impact, require very high manual effort, and do not yet provide sufficient specific information on forthcoming risk events. This leads to the need to provide better information and more assistance for a proactive SCRM. An increasing amount of data is available in the company, but there is currently a lack of suitable methods which take advantage of this data in SCRM.

The main objective of early warning is the evaluation and transmission of information on already latent present risks in an early stage, leaving enough time to plan and initiate response strategies and actions. The differentiation of scientific approaches can be made with regard to the available time period to act, the reliability of the warnings and the automation of the transfer of the warnings into existing information and management systems.

An approach for early warning systems has been developed by Moder (2008): this approach is based on questionnaires and lacks technological aspects of digitalization for automatization, quantitative assessment and real-time capability. The evoluionary stages of early warning have been described by Moder (2008) in three generations (Moder, 2008 p. 22): (1) key figures, (2) indicators, (3) weak signals. This distinction supports the view of a more reporting based approach as given by descriptive analytics. Nev-

ertheless, even these approaches are usually limited to critical commodities in practice. Because of the considerable effort involved, thus, a SCR monitoring of all suppliers is currently not possible (Meierbeck, 2010).
The earlier a risk event with a relevant impact is identified, the more time is available to initiate appropriate measures in the SC to minimize its impact, and the better the chances of success (Henke, et al., 2010). Especially the aspect of providing a sufficient time period to react represents a methodological challenge.

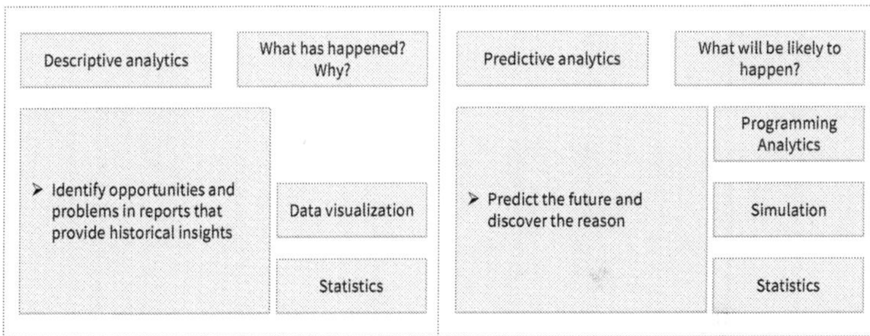

Figure 2: Descriptive and predictive analytics

Figure 2 distinguishes descriptive and predictive analytics based on Tiwari, Wee and Daryanto (2018) Fi. Business intelligence (BI), Analytics, BD, ML and AI methods are increasingly being discovered for supply chain management (SCM) to give valuable insights (Addo-Tenkorang and Helo, 2016; Arunachalam, Kumar and Kawalek, 2018; Tiwari, Wee and Daryanto, 2018), but there are currently only few research approaches for SCRM and early warning. In SCM the focus is shifting away from risk reports to assistance systems. The need for risk information has been researched by Fan, et al., 2017.

The availability and timely utilization of SCR information is extremely important and deserves more research (Fan, et al., 2017). The applicability of predictive analytics for supply risks is shown in general by He, et al. (2014). Traditional descriptive approaches for early warning apply e.g. balanced scorecards or questionnaires (see for example Czaja (2009)), having a descriptive function. Material flow simulation can be applied for similar time periods of transportation times (see for example Hotz (2007)). For longer periods of early warning, new approaches using analytics need to be developed. Predictive analytics approaches can be classified into machine learning techniques and regression techniques. Machine learning techniques are applied due to their performance in handling large-scaled datasets with uniform characteristics and noisy data (Nassif, et al., 2016 p. 2153).

2.3 Summary

Existing scientific approaches for early warning lack information on the risk cause and effect and do not provide sufficient time to react in order to minimize the risk impact. Furthermore, the prediction of the exact time period, i.e. demand date of the OEM, is not sufficiently included in early warning. So, it is difficult to identify the periods in the production program which is threatened by risk impacts regarding security of supply.

A suitable method for providing risk information at an early stage is required. This would support OEMs to follow a proactive SCRM approach in order to mitigate risk impacts. Considering industrial practice, the main target of the early warning should be the security of supply of the OEM in order to depict the connection between the risk cause and the impact of the risk. The approach should provide a sufficient time period to react before any risk impact on security of supply occurs. Furthermore, it should provide information about risk cause and effect in order to choose sustainable, cause-related actions to minimize the forthcoming risk impact.

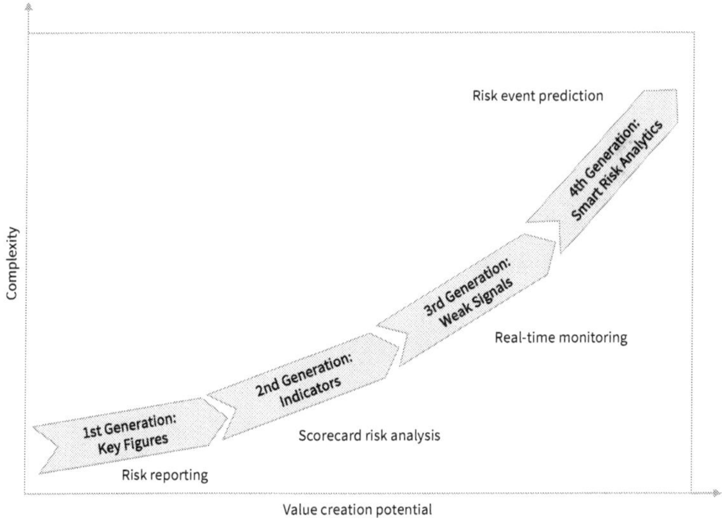

Figure 3: Stages of Evolution in Early Warning

There is a need for research in the field of the methodological-technical design in order to meet these requirements. For the development of the design approaches from digitalization and the field of ML and AI can be used. A systematic approach for DM, ML and AI application in early warning is needed to advance the quality of early warning and to reach a new level of assistance. Here, the data, which is increasingly available in the course of digitalization and currently do not add any value to SCRM, should be applied to generate new risk knowledge. The three stages of early warning described by Moder (2008) can be extended by another level for automatization, improved reaction time, reliability of information and completeness of included suppliers, visualized as smart risk analytics in Figure 3. In the following, this smart risk analytics approach has been technical-methodological designed.

3 Concept

There is a variety of SCRs that can affect the SC performance and thus the security of supply of an OEM. The aim of the concept is to forecast the security of supply as the key target, i.e. KPI, of early warning for application in an assistance system. The security of supply describes the availability of the parts required by the planned production program according to the planning and control processes of the OEM. The methodological design of proactive early warning is developed on the example of the SCR of inaccurate automotive demand planning. In order to make sustainable decisions regarding proactive SCRM actions, specific risk event information is needed at an early stage. This information will allow early action to prevent or minimize risk impacts, costs and supply bottlenecks of the OEM in the given action time (cf. Figure 4).Sufficient reaction time is an elementary factor in proactive SCRM.

The smart SCRM design aims to secure existing SC networks against risk impacts in an event-oriented manner. At the same time SC structures are not changed by measures for risk prevention or impact minimization due to economic opportunities and despite the risk liability of these structures. Executing proactive SCRM requires an information provision, which is fast, on-time, reliable and dynamic, to render assistance for management decisions.

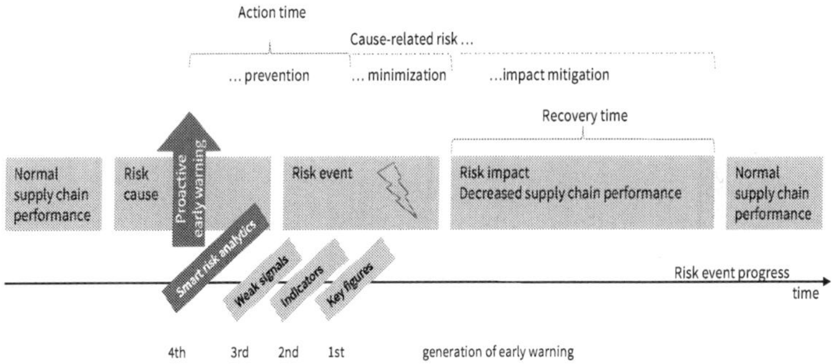

Figure 4: Proactive early warning at the risk event progress

Proactive early warning enables a reaction to forthcoming risks as visualized in Figure 4. For the smart SCRM design proactive early warning is an elementary module. Proactive early warning provides the action time period to prevent or minimize the risk impact. The design is thus geared to recurring risk events. The methodological-technical design of proactive early warning, called smart risk analytics (SRA), is introduced hereafter. The design follows the early warning requirements described in chapter 2.3.

In advance of proactive early warning, the identification of the SCRs of the OEM is required. This is followed by the prioritization of the risks that - due to recurring events - are of relevance to the OEM and should be included in the proactive early warning.

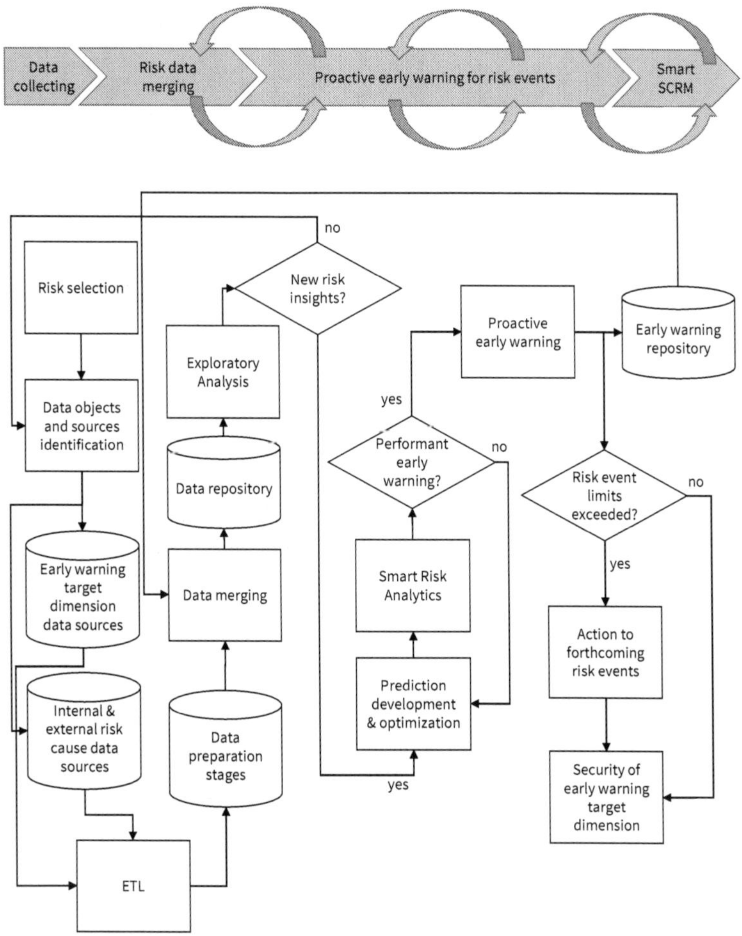

Figure 5: Processing Architecture of the Smart Risk Analytics Design

The processing architecture is visualized in Figure 5 and contains three iterative cycles. The agile approach allows to incorporate findings from one cycle into another cycle and thereby improve the quality of early warning.
The first iterative cycle (cf. Figure 5) describes the explorative analysis of risk cause and risk impact. The data basis is compiled (extract, transform, load - ETL) from internal and external data sources on risk causes and KPIs for the early warning. Data preparation is a core process for SRA and includes assessing the raw data from the original data source and converting the data into a risk data repository. Iteratively, the data basis can be expanded and the analysis repeated, so that the results may be used as a knowledge basis for the continuous improvement.
The second iterative cycle (cf. Figure 5) is applied for the SRA model development. The core of the SRA design is a forecast algorithm. Here, algorithms are trained by application on historical data of risk causes and targeted KPI. Hereby the algorithm learns the patterns and interrelations between the variables in the analyzed demand planning processes.
Applicable algorithms for SRA are decision tree, random forest, support vector machine, k-nearest neighbors, regression and artificial neural networks (cf. Backhaus, et al., 2016). A more detailed discussion of these algorithms is carried out in chapter 4. The algorithm should be selected based on performance and optimized regarding its features and parameters. The performance should be streamlined according to the early warning criteria, which are to be distinguished from the usual data mining criteria. With regard to the early warning application, incorrect warnings are to be avoided at all costs, even at the expense of missed correct alerts, as they lead to a loss of user confidence in the solution. The dynamic effects of risk causes on KPI representing the security of supply are mapped by ML or AI.

The algorithms are to be individually selected and trained for each SC as the lead time of the SRA has to reflect the SC specific dynamic risk behavior. Feature engineering is used for the development of KPI represented in risk cause data. Key figures serve as condensed input from the original process data for the algorithm. For optimization of prediction within SRA, feature engineering, feature selection and parameter optimization has been conducted for the mentioned algorithms and the performance has been compared for all algorithms. The methodological-technical design requires an iterative development and an specific implementation for each risk and for each supply chain.

The third iterative cycle (cf. Figure 5) allows the application of the trained algorithms on new data for a proactive early warning. Furthermore, all data and results are stored in a data basis in order to continuously improve the SRA. The proactive early warning for risk events can be bundled and controlled in an event-based manner in the smart SCRM. In a cloud solution, the smart risk analytics service for early warning, can be used by OEMs as well as suppliers.

4 Demonstration

For the demonstration, the concept has been implemented prototypically. By using empirical company data of an automobile manufacturer, the functionality of SRA for proactive early warning can be demonstrated. The demonstration is carried out according to the reliable and reproducible Cross-Industry Standard Process for Data Mining (CRISP DM), which asks to analytically apply, demonstrate and evaluate the concept. The CRISP DM consists of six phases, which are not run sequentially, but repeatedly in an iterative manner. The phases are (1) business understanding, (2) data understanding, (3) data preparation, (4) modeling, (5) evaluation and (6) deployment (Chapman, et al., 2000). Iterative steps are possible between all the phases, in particular between (1)-(2), (3)-(4) and (5)-(1), which allows the integration of new insights into the development. For more in-depth information about CRISP DM, please refer to suitable literature such as Chapman, et al. (2000) and Wirth and Hipp (2000).

The extent of the data basis for the demonstration is shown in Table 1. The data basis contains of a variety of part families, part numbers, demand planning dates, demand dates and demand planning values. The availability of the demand planning data for the SRA depends on the lead time of the early warning forecast in comparison to the actual time of the OEM's demand (see Figure 1). This results in the dynamic KPI calculation depending on the lead time for the SRA. The explorative analysis is conducted on the data repository for correlation analysis, and visualized in Figure 6.

Table 1: Objects of the Data Basis for Smart Risk Analytics Demonstration on the Risk of Inaccurate Demand Planning

Object Classes	**Objects**	**Data Basis**
Early warning target KPI	Security of supply	303.288 values (4 key figures for each part number and demand date)

Object Classes	Objects	Data Basis
Supply information	Forecast date Forecast period Lead time of planning Demand planning date (p_n) Demand date (d_n) Demand values Actual demand	October 9, 2017 to March 29, 2019, 76 planning dates within 1,5 years Forecast period of 26 weeks 1.971.372 demand values 371 days of production 75822 actual demand values
Demand information	Supply chain 1st tier supplier Part number Part family name	39 supply chains and 1st tier supplier 312 supply parts (differentiated by part numbers)
Features	Demand planning key figures	7.051.446 values (93 key figures calculated for each part number and demand date)

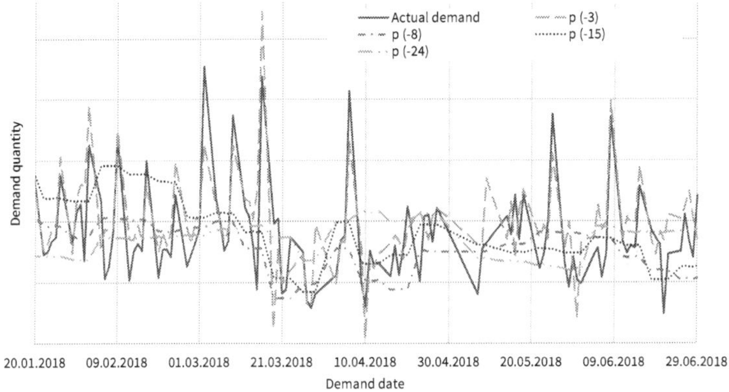

Figure 6: Explorative Analysis of Demand Planning Data

In feature engineering a variety of key figures have been defined and 7.051.446 values have been calculated. The subsequent feature selection allows the identification of relevant features depending on the algorithm. Features have been identified for each algorithm and part family for an optimization of the SRA. The demand planning forecast accuracy (FQ) was analyzed depending on the part family of the respective SC. The analysis of the forecast quality of the OEM has shown that the quality (FQ, TS) varies depending on the parts considered. A distinction can be made for e.g. part families or low-runner and high-runner.

The selected algorithms map the interrelation between automotive demand planning as the risk cause, the spread of risk and the risk impact on the KPI representing the security of supply of the OEM. The algorithm applies the trained risk behavior and can predict when and how strong a risk

impact may occur based on a certain risk cause. The behavior is to be determined for each SC and can be described as SC dynamic under risk. The dynamic behavior is depending on, among other things, safety parameters, transportation times, replenishment lead times and production times.
In the demonstration of the developed models according to Figure 5, it could be shown that the quality of the prediction of the KPI security of supply strongly depends on different demand planning dates and on the respective SC and its flexibility. This leads to different time horizons for proactive early warning. Accordingly, for the suitability of the concept, no general statement about the time horizon of proactive early warning can be made for all SCs. The dynamic behavior of the respective SC results for example from technical characteristics of the products, production and transport processes. The performance of SRA depends on the SC, the planning and procurement lead times of suppliers. Nevertheless, the resulting reaction time provided by the concept is usually up to several weeks and results from the SC's individual time horizon of proactive early warning. In addition to the function of proactive early warning, the concept also provides knowledge through insight into the flexibility of suppliers and SCs for the various products. It allows the identification of SC-specific flexibility depending on the planning cycles in demand planning.
The results of SRA are shown exemplarily in Figure 7. Similar to the classification of FQ (VDA, 2008, p. 15), limit values for the early warning risk score must be determined to distinguish between a low criticality and a high criticality which requires action. This is visualized by the dashed line. The limits of the classes should be based on the risk tolerance of the OEM. The results of SRA show the potential for dynamic risk assessment and dynamic proactive early warning by digitalization of SCRM using technological designs.

The resulting prototypical implementation has allowed a demonstration but also an assessment of applicability in practice. The demonstration has proven that the SRA concept is well suited for risk assessment and risk forecast related to the risk of inaccurate demand planning and generation of proactive early warning. Furthermore, an objective and quantitative assessment and of the expected risk impact including its timing is possible.

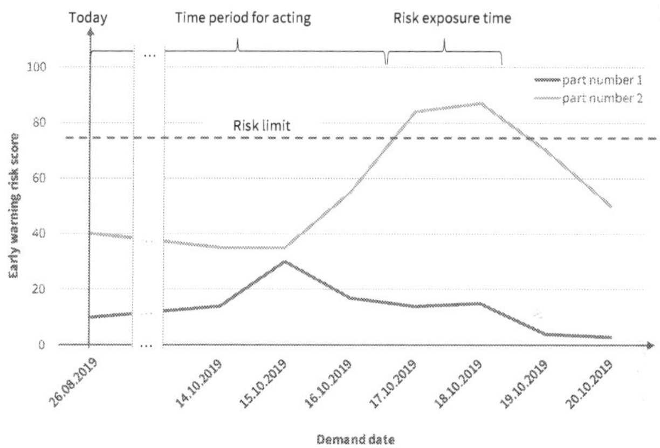

Figure 7: Smart Risk Analytics Result: Proactive Early Warning

The concept was demonstrated in its suitability for operational (short term, 0 to 2 weeks) and tactical (medium term, 3 to 8 weeks and long term 9 to X weeks) planning horizons (cf. VDA, 2008, p. 15). Accordingly, the technical concept of forecasting extends traditional approaches with regard to the available time horizon for reaction.

5 Conclusion

The SRA concept enables the objective assessment and forecasting of forthcoming risk events in an risk-cause-related proactive early warning. The concept can be technically implemented with a high degree of automatization for information and management assistance in the practice of OEMs.
The proactive early warning on the specific date of demand, which is threatened by SCRs, allows sustainable measures and collaborative management in the SC. SRA is shown with its suitability for proactive early warning using the risk example of inaccurate demand planning by application of the concept to OEM company data. Previous challenges of implementation in practice can be solved with this concept, which shows a way of digitalization in early warning and SCRM based on AI and ML to SCR data.
The analytical approach provides new insights into the flexibility of SCs under risk and the impact of risks over time. A new area of digitalized, real-time and objective proactive early warning is opened up by the concept.
For supplying automobile production with products from worldwide SCs, these supplies need to be protected against a variety of risks. Smart SCRM provides a solution to proactively secure the supply of the future production program. Depending on the risk and possible proactive measures, various company divisions of the OEM – from logistics, procurement and program planning to sales and finance - may be involved in the smart SCRM. Furthermore, the processes and organizational structures across the departments (logistics, procurement, finance and more) can be aligned to the more anticipatory reaction rather than reactive bottleneck management. The approach follows the digitalization paradigm and demands for new

competences of the employees. Due to the increasing complexity of the developed 4th early warning evolution stage, it is necessary to combine expert knowledge in SCRM and AI in the OEM's department.

The transferability of the concept to other OEMs is allowed by individual training of the algorithms on the respective SC. The further digitalization of SCs and business processes progresses, the more potential for SRA emerges and allows for the methodological concept to be expanded by taking on additional SCRs.

As it is known from other AI research projects, the generation of a suitable data basis requires a considerable amount of effort. For the sustainable application of AI in the company, the availability of data, the period of data storage and the resistance to use must be considerably optimized on the course of digitalization. Further research is needed on the assistance and automatization of optimized reactions to secure the supply against risk impacts within the smart SCRM concept.

References

Addo-Tenkorang, R. and Helo, P. T., 2016. Big data applications in operations/supply-chain management. A literature review. Computers & Industrial Engineering, 101, pp. 528–543.

Arunachalam, D., Kumar, N. and Kawalek, J. P., 2018. Understanding big data analytics capabilities in supply chain management. Unravelling the issues, challenges and implications for practice. Transportation Research Part E: Logistics and Transportation Review, 114, pp. 416–436.

Backhaus, K., Erichson, B., Plinke, W. and Weiber, R., 2016. Multivariate Analysemethoden: Springer Gabler.

Cavicchi, B., 2018. You can't afford to be static: Rolling with the punches in forecasting. ABERDEEN, 2018. <https://www.ibm.com/blogs/business-analytics/aberdeen-reasons-adopt-rolling-forecasts/; https://www.ibm.com/downloads/cas/BRE7ALNK> [Accessed 13 May 2019].

Ceryno, P. S., Scavarda, L. F. and Klingebiel, K., 2014. Supply chain risk. Empirical research in the automotive industry. Journal of Risk Research, 18(9), pp. 1145–1164.

Chapman, P., Clinton, J., Kerber, R., Khabaza, T., Reinartz, T., Shearer, C. and Wirth, R., 2000. CRISP-DM 1.0. Step-by-step data mining guide. 2nd ed. [e-book]: CRISP-DM Consortium. <https://www.the-modeling-agency.com/crisp-dm.pdf> [Accessed 2 May 2019].

Christopher, M. and Peck, H., 2004. Building the resilient supply chain. The International Journal of Logistics Management, 15(2), pp. 1–14.

Czaja, L., 2009. Qualitätsfrühwarnsysteme für die Automobilindustrie: Gabler.

Fan, H., Li, G., Sun, H. and Cheng, T., 2017. An information processing perspective on supply chain risk management. Antecedents, mechanism, and consequences. International Journal of Production Economics, 185, pp. 63–75.

Gehr, F. and Hellingrath, B., 2006. Logistik in der Automobilindustrie. Berlin: Springer Verlag.

Geissbauer, R. and D'heur, M., 2010. 2010 - 2012: Global Supply Chain Trends. Are Our Supply Chains Able to Support the Recovery? Lessons Learned from the Global Recession. [e-book]: PRTM Management Consultans. <https://www.consultancy.nl/media/PRTM%20Supply%20Chain%20Trends%202010-2012-1019.pdf> [Accessed 13 May 2019].

Harland, C., Brenchley, R. and Walker, H., 2003. Risk in supply networks. Journal of Purchasing and Supply Management, 2003(9), pp. 51–62.

He, M., Ji, H., Wang, Q. and Ren, C., 2014. Big data fueled process management of supply risks: sensing, prediction, evaluation and mitigation. Proceedings of the 2014 Winter Simulation Conference, pp. 1005–1013.

Henke, M. and Besl, R., 2008. Supplier Risk Management in der Automobilindustrie. Risiken im Vorfeld erkennen, 2008(März). <https://beschaffung-aktuell.industrie.de/allgemein/risiken-im-vorfeld-erkennen/> [Accessed 25 April 2019].

Henke, M., Blome, C., Seifert, M., Grötsch, V. and Sauerbier, M., 2010. Intensives Risikomanagement oder flexibles Krisenmanagement? Überlebensstrategien im Automobileinkauf in der Wirtschaftskrise. Wiesbaden: EBS Business School, Supply Chain Management Instutitue (SMI).

Hevner, A. R., 2007. The three cycle view of design science resarch. Scandinavian Journal of Information Systems, 19(2), pp. 87–92.

Ho, W., Zheng, T., Yildiz, H. and Talluri, S., 2015. Supply chain risk management: a literature review. International Journal of Production Research, 53(16), pp. 5031–5069.

Hotz, I., 2007. Simulationsbasierte Frühwarnsysteme zur Unterstützung der operativen Produktionsstuerung und -planung in der Automobilindustrie. phd thesis. Magdeburg.

Khan, A., 2018. Managing risks under highly dependent supplier-producer relation in modern automotive industry. In: I. Dovbischuk, G. Siestrup, and A. Tuma, eds. 2018. Nachhaltige Impulse für Produktion und Logistikmanagement. Wiesbaden: Springer Gabler, pp. 141–152.

Kırılmaz, O. and Erol, S., 2017. A proactive approach to supply chain risk management. Shifting orders among suppliers to mitigate the supply side risks. Journal of Purchasing and Supply Management, 23(1), pp. 54–65.

Leveling, J., Edelbrock, M. and Otto, B., 2014. Big Data Analytics for Supply Chain Management. IEEE International Conference on Industrial Engineering and Engineering Management, pp. 918–922.

Lopez, E. and McKevitt, J., 2017. BMW halts production as Bosch fails to meet supply needs. 01.06.2017. SUPPLYCHAINDIVE, 2017. <https://www.supplychaindive.com/news/BMW-production-halt-Bosch-supplier-shortage-parts/443949/> [Accessed 11 May 2019].

Meierbeck, R., 2010. Strategisches Risikomanagement in der Beschaffung. Lohmar-Köln: Josef Eul Verlag.

Moder, M., 2008. Supply Frühwarnsysteme.

Nassif, A. B., Azzeh, M., Banitaan, S. and Neagu, D., 2016. Guest editorial: special issue on predictive analytics using machine learning. Neural Computing and Applications, 27(8), pp. 2153–2155. <https://doi.org/10.1007/s00521-016-2327-3>.

Oliveira, J. B., Jin, M., Lima, R. S., Kobza, J. E. and Montevechi, J., 2019. The role of simulation and optimization methods in supply chain risk management. Simulation Modelling Practice and Theory, 92, pp. 17–44.

Schlüter, F., Diedrich, K. and Güller, M., 2017. Analyzing the Impact of Digitalization on Supply Chain Risk Management. IPSERA Conference, Hungary.

Tang, O. and Musa, S. N., 2011. Identifying risk issues and research advancements in supply chain risk management. International Journal of Production Economics, 2011(133), pp. 25–34.

Thomé, A., Scavarda, L. F., Pires, S., Ceryno, P. S. and Klingebiel, K., 2014. A multi-tier study on supply chain flexibility in the automotive industry. International Journal of Production Economics, 2014(158), pp. 91–105.

Thun, J.-H. and Hoenig, D., 2011. An empirical analysis of supply chain risk management in the German automotive industry. International Journal of Production Economics, 2011(131), pp. 242–249.

Tiwari, S., Wee, H. M. and Daryanto, Y., 2018. Big data analytics in supply chain management between 2010 and 2016. Computers & Industrial Engineering, 115, pp. 319–330.

VDA, 2008. VDA 5009. Forecast-Qualitätskennzahl: Definition und Anwendung. Frankfurt: Verband der Automobilindustrie.

VDA, 2018. Automobilindustrie und Märkte, 2018. <https://www.vda.de/de/themen/automobilindustrie-und-maerkte.html> [Accessed 12 May 2019].

Volling, T., Matzke, A., Grunewald, M. and Spengler, T. S., 2013. Planning of capacities and orders in build-to-order automobile production. European Journal of Operational Research, 224(2), pp. 240–260.

Volling, T. and Spengler, T. S., 2011. Modeling and simulation of order-driven planning policies in build-to-order automobile production. International Journal of Production Economics, 131(1), pp. 183–193.

Wirth, R. and Hipp, J., 2000. CRISP-DM: Towards a standard process model for data mining. Proceedings of the Fourth International Conference on the Practical Application of Knowledge Discovery and Data Mining.

ISBN 978-3-7502-4947-9
9 783750 249479 00010
www.epubli.de

Printed in Great Britain
by Amazon

50030366R00352